AA003395

Education Division 2013

Core Programming Area at the 2013 AIChE Annual Meeting: Global Challenges for Engineering a Sustainable Future

San Francisco, California, USA
3 – 8 November 2013

ISBN: 978-1-63439-037-8

Printed from e-media with permission by:

Curran Associates, Inc.
57 Morehouse Lane
Red Hook, NY 12571

Some format issues inherent in the e-media version may also appear in this print version.

Printed by Curran Associates, Inc. (2014)

For permission requests, please contact AIChE
at the address below.

AIChE
3 Park Avenue
New York, NY 10016-5991

Phone: (203) 702-7660
Fax: (203) 775-5177

www.aiche.org

Additional copies of this publication are available from:

Curran Associates, Inc.
57 Morehouse Lane
Red Hook, NY 12571 USA
Phone: 845-758-0400
Fax: 845-758-2634
Email: curran@proceedings.com
Web: www.proceedings.com

TABLE OF CONTENTS

(4am) Structure/Property Relationships in Polymer Membranes for Water Purification and Energy Applications

Geise, G. M., The University of Texas at Austin

Global demand for clean water and renewable energy sources is increasing, and both demands are inherently related because energy is required to purify water and, in most cases, purified water is required to produce energy. Polymer membranes are widely used in desalination applications, such as reverse osmosis (RO), nanofiltration (NF), forward osmosis (FO), and electrodialysis (ED), and they could be critical for many renewable power generation technologies, such as pressure-retarded osmosis (PRO), reverse electrodialysis (RED), membrane capacitive deionization (MCDI), fuel cells, and batteries. All of these applications rely, at least to some extent, on a membrane to control rates of water and/or ion transport.

Fundamental structure-property relationships are needed to guide the development of polymer membrane materials for these technologies. For desalination applications, polymers with simultaneously high water permeability and low salt permeability are desired in order to prepare membranes that can efficiently desalinate water. In potential field-driven applications, such as ED, RED, MCDI, and batteries, polymers with low ionic resistance are needed to preferentially transport certain ions while retaining others. Much remains unknown about the influence of polymer structure on even basic water and ion transport properties, and these relationships must be developed to assist the design of next generation polymer membrane materials for a variety of applications. The ability to tailor both the water and ion transport properties of a polymer will be critical for extending and optimizing technologies that can help address global needs for clean water and energy.

Selected Publications:

(12 total, 7 first author, 129 total citations ? ResearcherID, May 2013, 2 first author submitted)

G.M. Geise, H.B. Park, A.C. Sagle, B.D. Freeman, J.E. McGrath, Water permeability and water/salt selectivity tradeoff in polymers for desalination, *Journal of Membrane Science*, 369 (2011) 130-138.

G.M. Geise, L.P. Falcon, B.D. Freeman, D.R. Paul, Sodium chloride sorption in sulfonated polymers for membrane applications, *Journal of Membrane Science*, 423-424 (2012) 195-208.

G.M. Geise, H.?S. Lee, D.J. Miller, B.D. Freeman, J.E. McGrath, D.R. Paul, Water purification by membranes: The role of polymer science, *Journal of Polymer Science Part B: Polymer Physics*, 48 (2010) 1685-1718.

W. Xie, H. Ju, G. Geise, B. Freeman, J. Mardel, A. Hill, J. McGrath, Effect of free volume on water and salt transport properties in directly copolymerized disulfonated poly(arylene ether sulfone) random copolymers, *Macromolecules*, 44 (2011) 4428-4438.

G.M. Geise, B.D. Freeman, D.R. Paul, Sodium chloride diffusion in sulfonated polymers for membrane applications, *Journal of Membrane Science*, 427 (2013) 186-196.

(4dd) Micro-Scale Flow and Transport Phenomena in Chemical and Biological Systems

Rosenfeld, L., Stanford University

The effect of gravity and inertia dominate our experience of the physical world. However, as systems are reduced in size, phenomena such as diffusion, surface tension and viscosity become ever more important; at the micrometer scale they can dominate and result in a world that operates very differently from the macroscopic world we perceive and live in. There are numerous processes encountered in nature and industry where the microscale interaction of fluid-fluid interfaces is of central importance. These systems can involve fluid droplets and bubbles such as deformation of red blood cells in small capillaries, liquid membranes and phase separation processes.

My research interests are microscale transport phenomena, microstructure and physical properties of complex fluids and droplet-based microfluidics as a tool for chemical and biological systems.

I will describe here three different fundamental studies. In the first study I have addressed the interfacial flow and transport phenomena of droplets and bubbles in macro and micro gravity environments. The effort was directed to two avenues: spontaneous interaction due to interfacial transport and interaction induced by external gradients. In such drop-based systems, the presence of hybrid drops is usually of great interest. Their dynamics and deformation in viscous fluids is complex because of the existence of several interfaces. The study of these systems was done using conformal mapping techniques and numerical methods.

In the second study I have explored the transition from reversible to chaotic behavior in the oscillatory shear flow of water-in-oil emulsions. Emulsions are an important class of complex fluids with many applications in chemical, biological and industrial processes. The bulk rheological properties of these fluids are directly influenced by the dynamics at the microscopic level; hence a fundamental understanding of the microscopic dissipative rearrangements and specifically the transition from hydrodynamic reversible to irreversible behavior is of central importance. Here I have applied microfluidic techniques to study this fundamental behavior of the emulsion microstructure.

In the third study I have focused on the interfacial dynamics, rheology and stability of thin aqueous films covered with small molecule amphiphile monolayers. It was designed in order to study the stability of the tear film and dry eye disease treatment. A novel methodology to characterize dewetting of thin aqueous layers covered with insoluble surfactants was developed. Optical methods such as interferometry in order to detect the film thickness, scattering techniques in order to determine the monolayer's molecular structure and interfacial and bulk rheology measurements to determine the macroscopic, mechanical properties of the films were employed.

My goal is to establish a research program that will continue to pursue the dynamics of fluid-fluid interfaces. I am particularly interested in the effect of non-Newtonian rheology on interfacial flows as well as microhydrodynamic phenomena and its applications in microfluidic devices.

(4by) Microfluidic Transport Phenomena for Nanomedicine and Micro-Total Analysis Systems

Lim, J. M., Massachusetts Institute of Technology

Microfluidics is a powerful technology that can control micro-scale environments in an extremely precise manner. The microfluidics is widely studied for seeking answers to various fundamental questions in the field of chemical engineering and bioengineering. Based on the knowledge, the microfluidic technology is widely adopted to various applications including fine chemical synthesis, chemical and biomolecular analysis, and healthcare. My postdoctoral research with Professor Rohit Karnik at MIT, Professor Omid Farokhzad at Brigham and Women's Hospital - Harvard Medical School, and Professor Robert Langer at MIT focuses on the synthesis of advanced engineered nanoparticles for drug delivery systems using novel microfluidic systems. My postdoctoral and Ph.D. research at KAIST with Professor Seung-Man Yang and visiting scholar at MIT with Professor Todd Thorsen focused on optofluidic platform, a multidisciplinary area combining optics and microfluidics, for integrated micro-total analysis systems. I also studied on the colloidal self-assembly in confined geometry for photonic crystals and superhydrophobic surfaces. In this poster session, I will highlight aforementioned my graduate and postdoctoral researches. I will also present my vision for independent research program on the study of microfluidic transport phenomena and functional soft matters with broad range of application in healthcare, lab on a chip, and micro-total analysis systems that encompasses: 1) Microfluidic nanomedicine discovery and screening systems, 2) Development of functional microparticles and capsules, and 3) Integrated optofluidic platform for adoptive optical systems and micro-total analysis systems.

SELECTED PUBLICATIONS:

Jong-Min Lim, John Paul Urbanski, Todd Thorsen, and Seung-Man Yang, "Pneumatic Control of a Liquid-Core/Liquid-Cladding Waveguide as the Basis for an Optofluidic Switch," *Applied Physics Letters*, **98**, 044101 (2011)

Jong-Min Lim, John Paul Urbanski, Jae-Hoon Choi, Todd Thorsen, and Seung-Man Yang, "Liquid Waveguide-Based Evanescent Wave Sensor That Uses Two Light Sources with Different Wavelengths," *Analytical Chemistry*, **83**, 585–590 (2011)

Jong-Min Lim, Se-Heon Kim, and Seung-Man Yang, "Liquid-Liquid Fluorescent Waveguides Using Microfluidic-Drifting-Induced Hydrodynamic Focusing," *Microfluidics and Nanofluidics*, **10**, 211-217 (2011)

Jong-Min Lim, Se-Heon Kim, Jae-Hoon Choi, and Seung-Man Yang, "Fluorescent Liquid-Core/ Air-Cladding Waveguides towards Integrated Optofluidic Light Sources," *Lab on a Chip*, **8**, 1580-1585 (2008)

Jong-Min Lim, Gi-Ra Yi, Jun Hyuk Moon, Chul-Joon Heo, and Seung-Man Yang, "Superhydrophobic Films of Electrospun Fibers with Multiple-Scale Surface Morphology", *Langmuir*, **23**, 7981-7989 (2007)

Jong-Min Lim, Jun Hyuk Moon, Gi-Ra Yi, Chul-Joon Heo, and Seung-Man Yang, "Fabrication of One-Dimensional Colloidal Assemblies from Electrospun Nanofibers," *Langmuir*, **22**, 3445-3449 (2006)

(4ds) Design and Development of Engineered Nanomaterials Through Understanding Nano-Bio Interaction Mechanisms for Biomedical Applications

Sun, B., University of California, Los Angeles

The engineered nanomaterials (ENMs) could interact with biological entities at the nano-bio interface, and the interaction could induce cascades of signaling pathways leading to the initiation of immune responses. However, the mechanisms by which ENMs could generate immunostimulatory effects at the nano-bio interface remain elusive. NLRP3 inflammasome is drawing significant attentions since it responds to a wide range of ENMs; and it has been considered to play an important role in the induction of immune responses upon ENM-cell interactions. However, how different molecular mechanisms are involved in the ENM-induced NLRP3 inflammasome activation and how ENM physicochemical properties differentiate in their ability to activate the inflammasome remain poorly understood. Therefore, this research is mainly focusing on the elucidation of ENMs-NLRP3 inflammasome interaction mechanisms, which will help to link physicochemical properties of ENMs to immune responses at the nano-bio interface, and will provide strategies to engineer nanomaterials with customized immunostimulatory effects for biomedical applications.

(4cy) Polymer-Based 'Soft' Materials: Coacervate Assemblies, Carbon Nanotube Nanocomposites
Priftis, D., University of Chicago

The intelligent use of processes such as self-assembly, combined with the ability to manipulate the chemical structure of polymers, can lead to a wide array of materials. Such functional materials could be a solution to many of the challenges that the modern world faces, including engineering improved biomedical devices and strategies for renewable energy. During this poster session I will highlight examples from my graduate and postdoctoral research that demonstrate how these two elements can be utilized for the development of novel polymer-based soft materials. I will also describe how based on past experience I plan to create an independent research program at the interface of chemistry and materials engineering.

During my doctoral studies in the groups of Professor Nikos Hadjichristidis and Professor Jimmy Mays (at the Universities of Athens and Tennessee respectively) my attention was focused in the design of polymer-carbon nanotube (CNT) nanocomposites. These unique materials have great potentials as they combine the physicochemical properties of polymers and the extraordinary properties of the CNTs, such as mechanical strength and electrical conductivity. My research was centered on the development of a CNT polymer functionalization strategy that helps circumvent CNTs' inherent insolubility, and considerably widens the scope of nanocomposite materials that can be produced. The strategy involved covalent attachment of substituted benzocyclobutenes to CNTs. With a judicious choice of substitution, initiators for most popular polymerization techniques were attached onto the CNT surface.[1,2] Complete control over grafting percentage of initiator and surface-initiated polymerizations allowed synthesis of nanocomposite materials with desired compositions, which is essential for any application. The resulting nanocomposite materials exhibited improved mechanical and thermal properties when compared to pure polymers.[3]

In 2010 I accepted a postdoctoral scholar position from the Bioengineering Department of the University of California Berkeley (in the group of Professor Matthew Tirrell). There I initiated a project aimed at exploring a liquid-liquid phase separation phenomenon, referred to as complex coacervation. Using polypeptides as a model system I studied many aspects of complex coacervation through a number of collaborations. I identified the external parameters that affect coacervation,[4] explored the thermodynamics of coacervate formation,[5] and studied the rheological[6] and interfacial properties[7]of polypeptide coacervates. Building on this work I am currently exploring, at the Institute for Molecular Engineering of the University of Chicago, how complex coacervation can be used for the development of other polyelectrolyte self-assembly structures. More complex molecular design can be utilized, wherein polyelectrolyte domains are connected to neutral polymer blocks. These neutral domains stabilize microphase separation of the coacervate phase. Mixing of oppositely charged block-copolymers results in the formation of new self-assembly structures such as micelles or hydrogels.

My research is driven by a strong curiosity to develop and study new materials and target specific applications by bridging knowledge from various disciplines (i.e. chemistry, materials, physics, bioengineering). In the past few years I have successfully organized and collaborated on various research efforts that have resulted in a number of peer-reviewed publications (including 10 first author and 4 corresponding author articles). By working at the interface of different subjects areas, and with the use of my expertise in chemistry, polymers and material design I can make unique and lasting contributions to chemistry and material engineering.

Selected-Related References:

1. **Priftis D.,** Sakellariou G., Baskaran D., Mays J.W., Hadjichristidis N., 'Polymer Grafted Janus Multi-Walled Carbon Nanotubes.' *Soft Matter*, **2009**, *5*, 4272-4278.

2. **Priftis D.,** Petzetakis N., Sakellariou G., Pitsikalis M., Baskaran D., Mays J. M., Hadjichristidis N., 'Surface initiated titanium mediated coordination from catalyst-functionalized single and multi-walled carbon nanotubes.' *Macromolecules,* **2009,** *42,* 3340-3346.

3. **Priftis D.,** Sakellariou G., Lorenzo A. T., Müller A. J., Hadjichristidis N., 'Surface modification of multi-walled carbon nanotubes with biocompatible polymers *via* ring opening and living anionic surface initiated polymerization. Kinetics and crystallization behaviour.' *J. Polym. Sci. Part A: Polym. Chem,* **2009,** *47,* 4379-4390.

4. **Priftis D.,*** Tirrell M., 'Phase Behaviour and Complex Coacervation of Aqueous Polypeptide Solutions.' *Soft Mater* **2012**, *8,* 9396-9405.

5. **Priftis D.,** Farina R., Tirrell M., 'Interfacial Energy of Polypeptide Complex Coacervates Measured *via* Capillary Adhesion.' *Langmuir* **2012,** *28,*8721-8729.

6. **Priftis D.,*** Megley K., Laugel N., Tirrell M., 'Phase Behavior and Complex Coacervation of Polyethyleneimine /Polypeptide Solutions.' *J. of Colloid and Interface Science* **2013**, *398,* 39-50.

7. **Priftis D.,*** Laugel N., Tirrell M., 'Thermodynamic Characterization of Polypeptide Complex Coacervation.' *Langmuir* **2012,** *28*, 15947-15957.

*Corresponding author

(4cu) Surface/Interfacial Engineering for Energy Conversion (SEEC)
Peng, Q., Duke University

The surface/interfacial properties have strong effects on the chemical and physical properties of materials. With the unique background and expertise in surface chemistry, materials engineering on atomic/molecular scales, surface chemical and physical properties characterization, electrochemical/photoelectrochemical analysis, catalysis, and chemical engineering, I am well prepared for doing cutting-edge research in advancing the fundamental understanding of surface/interfacial mechanisms in materials engineering for *renewable photoelectrochemical H_2 production* and *catalytic conversion of fuels*. My particular research interests are centered in the following related fields.

I. Hydrogen Production from Photoelectrolysis of Water

Photoelectrochemical (PEC) water splitting for H_2 production is a sustainable mean to clean energy. However, current photoelectrode materials either suffer from low solar-to-H_2 (STH) efficiency or decomposition. Two topics will be studied. In *Topic I-1*, I will focus on using atomic layer deposited metal oxide coatings to improve the corrosion resistance of photoelectrode materials of high efficiency, such as GaAs and InP. The protective metal oxide layer must be pinhole free and well-bonded, allow facile charge transfer, and have low density of interfacial defects, which pose a great challenge for materials engineering. The research will gain understanding of the effect atomic layer deposition (ALD) coatings on the corrosion resistance and efficiency of the oxide coated photoelectrodes. *Topic I-2* is devoted to develop high efficient Fe_2O_3 PEC photoelectrodes. In this project, I will study heterogeneous nucleation and growth of Fe_2O_3 nanorods through a sequential pulsing vapor deposition process. The effects of supporting scaffold, interface between materials, doping, and Fe_2O_3 nanorod morphology and surface property on STH efficiency of Fe_2O_3 photoelectrodes will be investigated systematically to improve the understanding of the mechanisms underlying Fe_2O_3 photoelectrodes and to further improve the STH efficiency of Fe_2O_3photoelectrodes through rational design.

II. Advanced Catalysts Design, Synthesis and Characterization

Catalysts are important for all chemical processes. Synthesizing designer catalysts with atomic precision is the key to understand and improve the catalytic properties of materials. Two topics will be investigated. *Topic II-1* involves designing and engineering advanced colloid catalysts with ALD and molecular layer deposition (MLD). The well-defined particle size and crystal facet of colloid particles are highly desired for fundamental studies and practical catalytic reactions. However, the stabilizing ligands often existed on the surface of colloid particles are detrimental for their catalytic properties. I will investigate the synergetic effect of surface capping ligands and ALD/MLD chemistries on the catalytic properties of colloid particles. *Topic II-2* spotlights on synthesis of model complex multiple components layered metal/oxides/sulfide/nitride tandem catalysts. The study aims to explore and understand the cooperative catalytic effects from the nanoscale adjacent catalytic domains on a single catalyst body. Emphasis will be put on fundamental understanding of the surface and interfacial mechanisms during the assembly of different components and their effects on the catalytic properties of final tandem catalysts.

(4dj) Understanding Unique Properties of Zwitterionic Materials From Their Molecular Structures
Shao, Q., University of Washington

Resisting nonspecific protein adsorption in complex media is critically important for a wide range of applications such as drug/gene delivery, therapeutic protein conjugation, biosensors, and marine coatings. Polyethylene glycol has been extensively studied and widely used for these purposes, but the amphiphilic feature of polyethylene glycol brings negative effects on bioactivity and affinity. Zwitterionic polycarboxybetaine and polysulfobetaine are a new type of superhydrophilic materials that resist nonspecific protein adsorption in complex media such as whole blood. However, the nonfouling mechanisms for zwitterionic materials are not yet illustrated.

The properties of zwitterionic moiety determine the performance of zwitterionic materials. For instance, the hydration of zwitterionic moiety is considered as the main source of the nonspecific protein resistance of zwitterionic materials. Other important properties include ion and protein interactions. Zwitterionic moieties can vary their molecular structure mainly in three respects: cationic group, anionic group, and carbon spacer length between them. Varying any of these three structural aspects will change the properties of the zwitterionic moieties. Understanding how molecular structure dictates the properties of zwitterionic moieties is therefore fundamentally important not only to reveal the nonfouling mechanisms of zwitterionic materials, but also to perform a rational design of new zwitterionic materials. In this work, we studied the effects of the molecular structure of zwitterionic moieties on their molecular properties with quantum mechanical calculations and molecular dynamics simulations.

Quantum mechanical calculations were used to examine the interference between charged groups in a zwitterionic moiety, and to develop a set of force field parameters for zwitterionic moieties. With the developed force field, we studied effects of the cationic group, anionic group and carbon spacer length on the hydration of zwitterionic moieties using molecular dynamics simulations, free energy perturbation, and metadynamics simulations. Analysis of hydration numbers, residence time of water molecules, and free energy shows that the above three structure features have distinct effects on zwitterionic moieties. We also studied the ion interactions of zwitterionic moieties by analyzing the zwitterion-ion association structure, dynamics and free energy. The simulation results show an important role for the anionic group in specific cation interactions of zwitterionic moieties. Carboxybetaine and sulfobetaine showed different ion preferences and self-associations due to different combinations of charged groups. Protein interactions of zwitterionic moieties were studied in comparison with protein interactions of the (ethylene glycol)$_4$ moiety. We performed extensively long molecular dynamics simulations to investigate any specific moiety-protein interactions and their influence on protein surface area, flexibility and structure. The simulation results show that the (ethylene glycol)$_4$ moieties influence protein structure and dynamics differently from zwitterionic moieties. A series of metadynamics simulations showed that the (ethylene glycol)$_4$ and zwitterionic moieties also have different effects on the hydrophobic interactions in a biological system. These simulation studies demonstrate how the molecular structure of zwitterionic moieties determines their molecular properties, provide insight into the nonfouling mechanisms of zwitterionic materials, and offer a guideline for the rational design of zwitterionic materials.

(4ac) Particulate Process and Product Design
Emady, H. N., Purdue University

Particles are widely used in industry, with over 50% of products and over 75% of raw materials in the chemical industry in particulate form, yet less engineering and design principles exist than for their liquid and gas counterparts. In most chemical plants, the production rate of liquids and gases is 90% on average, while it is 40% on average when particulates are involved. Therefore, there is a strong need to study particulate systems, as well as to train students in this area, especially since most undergraduate chemical engineering curriculums do not include particle technology.

Many industries, including pharmaceuticals, food, agricultural chemicals, detergents, catalysts, and consumer products utilize particulate processes to produce varying particulate products. Product specifications may include size and shape of particles, or physical stability of a suspension of particles. My research interests lie in particulate process and product design, with specific applications in wet granulation, the stability of particulate products, the fundamentals of powder/liquid interactions, and the scale-up of particulate processes.

My doctoral work at Purdue University focused on the process of single drop granule formation for regime separated granulation towards the production of round particles, and I developed a regime map that predicts particle shape based on the formation mechanism. Following my Ph.D., I joined Procter & Gamble's Microstructured Fluids group as a postdoctoral researcher, where I studied particle suspension in yield stress fluids towards product stability. I am currently a postdoctoral associate at Rutgers University as part of the Catalyst Manufacturing Science and Engineering Consortium, exploring the scale-up of rotary calciners via experiments and discrete element method (DEM) simulations.

The common theme in my work is the design of particulate products and processes via the study of fundamental phenomena involved in particle-particle, particle-fluid, and particle-equipment interactions to produce required product specifications. My unique toolkit of experimental techniques and computational modeling, coupled with my first-hand industry interactions, makes me well-qualified to tackle any particulate process and product design problem. As a professor, I plan to continue my ventures in this field and to establish a world-class particle technology research program, while simultaneously inspiring and educating students in this area as well as all other areas of chemical engineering.

(4eo) A Tale of Two Studies: Dynamics and Timing
Xu, L.,

Multi-scale material simulations comprise the bridge between the microscopic length and time scale, and the macroscopic world. The first part will show how real-time trajectory calculations have provided for the first time a detailed dynamical picture of organic reactions involving cycloadditions [1-5]. For example, in 1,3-dipolar cycloadditions, the previously reported linear correlation between activation barriers and the energy required to distort the reactants to their transition-state geometries is now understandable in terms of the requirement for vibrational excitation [2,3]. In the case of carbene cycloadditions, we explored the range of geometries sampled in productive trajectories, as well as the timing of bond formation [4].

The second part will focus on large material simulations such as TiO_2 surface and polyurea/polyimide aerogels. An analytic potential energy function for dimethyl methylphosphonate (DMMP) interacting with TiO_2surface were developed, based on accurate cluster models[6]. Another project focuses on the assembly of polyurea aerogels.[7] Atomistic simulations and micropores analysis helped determining the internal structure of polyimide nanoparticles[8].

[1] D. H. Ess, S. E. Wheeler, R. G. Iafe, **L. Xu**, N. Celebi-Olcum and K. N. Houk, ***Angew. Chem. Int. Ed.*** 2008, 47,

7592

[2] **L. Xu**, C. E. Doubleday, K. N. Houk, ***Angew. Chem. Int. Ed.*** 2009, *48*, 2746 (highlighted by ***Nature Chemistry***)

[3] **L. Xu**, C. E. Doubleday, K. N. Houk, ***J. Am. Chem. Soc.*** 2010, *132*, 3029.

[4] **L. Xu**, C. E. Doubleday, K. N. Houk, ***J. Am. Chem. Soc.*** 2011, *133,* 17848.

[5] K. Black, P. Liu, **L. Xu**, C. E. Doubleday, K. N. Houk, ***Proc. Natl. Acad. Sci,*** 2012, *109*, 12860 (featured in

Computational Chemistry Highlights, ScienceDaily)

[6] **L. Xu,** L. Yang, W. A. DeJong, W. L. Hase, N. Govind, L. Hans, D. Tunega, ***J. Phys. Chem. C.*** under revision

[7] **L. Xu**, C. Sotiriou-Leventis, and N. Leventis, manuscript in preparation

[8] C. Chidambareswarapattar, **L. Xu**, C. Sotiriou-Leventis, and N. Leventis, manuscript in preparation

(4j) Colloidal Assembly and Reconfiguration Dynamics

Beltran-Villegas, D. J., University of Michigan

Understanding concentrated colloidal dynamics in the presence of different pairwise interactions and external fields provides a basis to predict the temporal evolution of colloidal microstructures in diverse phenomena including suspension rheology and colloidal assembly. However, a microscopic theory of concentrated colloidal dynamics does not yet exist that rigorously includes both statistical mechanical (configuration dependent free energy changes) and fluid mechanical (configuration dependent multi-body hydrodynamic interactions) contributions. This work shows the implementation of a novel approach to model colloidal assembly by means of analyzing the Smoluchowski Equation to describe temporal evolution of microstructures with a small number of sytem descriptors or order parameters. Colloidal assembly is simulated via Brownian dynamic and Stokesian dynamic methods to generate high-dimensional assembly trajectories. These trajectories are analyzed and converted to order parameter evolution data and used to generate a low-dimensional decription assembly dynamics. Results of this approach to a model colloidal crystallizing system show low-dimensional dynamic models that capture relevant thermodynamic and kinetic information. This methodology is used, then, to analyze the structure re-configurability of anisotropic colloidal building blocks. Knowledge of these dynamic models enable the fundamental understanding of assembly systems, as well as the optimization, design and control of assembly systems to produce low-defect colloidal crystals.

(4aj) Dynamically Reshaping Signaling Networks to Program Cell Fate Via Genetic Controllers

Galloway, K. E., California Institute of Technology
Franco, E., Univeristy of California, Riverside
Smolke, C. D., Stanford University

Engineering of cell fate through synthetic gene circuits requires methods to precisely implement control around native decision-making pathways, and offers the potential to direct developmental programs and intervene in aberrantly activated cell processes. We demonstrate a class of genetic control systems, molecular network diverters, that interface with a native signaling pathway to route cells to divergent fates in response to environmental signals without modification of native genetic material. A readily translatable method for identifying control points within natural networks is described that enables the construction of synthetic control systems that activate or attenuate native pathways to direct cell fate. We integrate opposing genetic programs by developing network architectures for reduced antagonism and demonstrate rational tuning of performance. The integrated, optimized system allowed us to selectively activate and attenuate signaling through a native MAPK pathway to conditionally redirect, or route, cells to one of three distinct fates. Additionally, we constructed a model to guide the design of similar control systems, highlighting critical parameters for balancing the opposing diverter functions. The extension of these control strategies to mammalian systems should facilitate the engineering of complex cellular signaling systems.

(4bw) Bacterial Mechanobiology: Rotary Flagellar Motors As Mechanosensors

Lele, P., Harvard University

Mechanobiology is an emerging area in biophysics, which focuses on how mechanical forces regulate complex biological processes in living systems. I am interested in determining the molecular mechanisms by which rotary flagellar motors transduce mechanical signals and trigger biochemical signaling and cell-differentiation in motile bacteria. Determining these signaling cascades will provide molecular insights into how bacteria adapt and survive in challenging environments, including how bacteria establish infections.

A majority of motile bacteria swim by rotating helical filaments with reversible, electric motors. Changes in the direction of rotation (switching) and speed of rotation are fundamental characteristics of the flagellar motor, critical for chemotaxis and motility. However, the biophysical mechanisms of relevant motor functions are unclear. In the Berg lab, I proposed a theoretical model predicting how motor protein exchange controls the switching function, and successfully tested the model through single molecule fluorescence measurements (1). Recently, I have discovered how the motor acts as a viscometer – it remodels the stator complex when triggered by mechanical stimuli (2). To our knowledge, this is the only known biological motor which varies the number of force-generating units to tailor its response to external mechanical loads. Importantly, these discoveries provide a handle on an important developmental problem: how do flagellar motors signal cell-development?

My research program will focus on mechanobiology, in a variety of model prokaryotic systems, with emphasis on the physics of motility and self-assembly in biological macromolecules. These investigations might lead to the development of strategies to control microbial growth, based on targeting of the surface-sensing mechanisms in rotary motors.

[1] P. P. Lele, R. W. Branch, V. Nathan, H. C. Berg, "Mechanism for adaptive remodeling of the bacterial flagellar switch", Proc. Natl. Acad. Sci. USA (2012), 109 (49), 20018-20022.

[2] P. P. Lele, Basarab G. Hosu, H. C. Berg, "Dynamics of mechanosensing in the bacterial flagellar motor", under revision, Proc. Natl. Acad. Sci. USA.

Collaborators:

Prof. Urs Jenal, Biozentrum, University of Basel, Switzerland

Dr. Yuhai Tu, RSM & Manager, Theory & Computational Physics, IBM, USA

(4s) In Vivo In Vitro and In Situ Nanosensor Applications

Cash, K. J., Northeastern University
Clark, H. A., Northeastern University

Recent years have shown the incredible versatility and utility of polymeric nanosensors. Sensors have been developed for a wide variety of ionic analytes (e.g. sodium, potassium, pH) as well as nonionic analytics (e.g. glucose) and have been applied for challenging research fields such as detection of sodium sparks in live cardiomyocytes. This field of research has yielded new knowledge on cellular metabolism and function. Additionally, these sensors can be applied in vivo, through implantation in or under the skin in animals; showing potential as a research tool for continuous monitoring of analyte concentrations without the need for direct samples to be taken from the animal.

To date the choice of recognition element has limited the range of targets which can be measured with these polymeric nanosensors. This stems from the need for small molecule recognition elements (e.g. ionophores) which can be incorporated into the polymer matrix. Through the combination of these nanosensors with enzymatic recognition elements we can now detect a far wider variety of biologically relevant targets through detection of metabolic byproducts of enzymatic reactions. This poster will discuss continuous monitoring of several analytical targets using these nanosensor approaches. These include lithium monitoring for bipolar disorder treatment as well as histamine concentrations in small animals as a proof of concept for eventual monitoring of a suite of physiological parameters. This will also form the foundation of my research program which will focus on the development of novel biomedical diagnostic approaches for *in vivo* and *in vitro* monitoring, and will also leverage the advances made in biomedical research applications toward improved process monitoring and analytics.

Postdoctoral Research Advisor

Heather A. Clark, Department of Pharmaceutical Sciences, Northeastern University, Boston MA

Graduate Research Advisor

Kevin W. Plaxco, Department of Chemistry and Biochemistry, University of California, Santa Barbara

Funding

KJC: NIH/NIBIB F32 EB015270-01
HAC: DARPA, NIH/NIGMS R01 GM084366, NIH/NINDS1R01NS081641

(4cw) Biomimetic and Microfluidic Approaches to Biomolecular Function and Application

Perry, S. L., University of Illinois at Urbana-Champaign

Living cells have evolved sophisticated intracellular organization strategies that are challenging to reproduce synthetically. Biomolecular function depends on both the structure of the molecule itself and the properties of the surrounding medium. The ability to correlate structure to function at biologically relevant timescales, simulate the *in vivo* environment, and isolate biological networks for study in an artificial milieu without sacrificing the crowding, structure, and compartmentalization of a cellular environment, represent engineering challenges with tremendous potential to impact both biological studies and biomedical applications.

Efforts to understand biomolecular structure-function relationships have relied heavily on static structural depictions that do not fully describe the dynamic nature of reaction processes. This limitation is the result of difficulties in applying dynamic or time-resolved crystallographic methods to a majority of biomolecular targets. These challenges are associated with radiation damage and/or the need to simultaneously and repetitively trigger the biomolecular reaction within a crystal. X-ray transparent microfluidic platforms for protein crystallography can be used to address these challenges, enabling the serial analysis of a large number of crystals so as to avoid both radiation damage and the necessity of repeatedly cycling a reaction. The fine control over compositions on-chip enables both the easy growth of a large number of isomorphous crystals for analysis and reproducible chemical triggering of reactions (i.e. ligand addition), challenges that are difficult to address using traditional, single-crystal methods. Such chips can also enable high throughput structural analysis with respect to many variables including pH and ionic strength to better understand biomolecule function.

The challenge in designing synthetic organelles and *in vivo* microenvironments is maintaining both crowding and compartmentalization while controlling the available intermolecular interactions. Emerging experience has shown that complex coacervates (liquid-liquid phase separation) utilizing biomolecules produces an effective biomimetic microenvironment. Initial efforts are focused on understanding how functionalities, introduced through sequence-specific motifs, affect the emergent properties of these materials on both the bulk and molecular scale, and thus alter biomolecule sequestration and function. Molecular design enables the use of structure and microphase separation as additional design parameters. Using these strategies, I propose the development of artificial organelles for applications in biochemistry, bioenergetics, biocatalysis, and biomedicine. The understanding gained from the development of these systems may provide insight into the function of analogous membraneless organelles such as carboxysomes, p-bodies, and p-granules as well as elucidation of potential pathways for the evolution of prebiotic life.

(4av) Carbon Capture, Utilization and Storage: From Material Screeening to Process Optimization to Network Design

Hasan, M. M. F., Princeton University

Atmospheric CO_2 levels, which surpassed 400 parts per million this year, are at their highest in human history. The continued increase in CO_2 levels is attributed to the anthropogenic CO_2 emissions from global consumption and burning of fossil fuels. The goal of my research is to design, optimize and screen processes, materials, and supply chain networks for cost-effective CO_2 capture, utilization and storage (CCUS) to reduce CO_2 emissions from stationary sources. This would enable the production of clean energy from carbonaceous fuels, and allow both conventional fuels (coal, oil and natural gas) and nonconventional fuels (shale gas, stranded natural gas, biomass, municipal solid waste, etc.) to continue to meet the global energy demands without emitting excess CO_2, thereby reducing the risks of global warming and man-made climate change.

In a CCUS chain, CO_2 is first captured and compressed at the source of emission, and then transported to a site where it can be utilized or injected for sequestration in geological formations. The costs of CO_2 capture and compression represent 60-70% of the total CCUS cost. Through process synthesis and optimization, we have explicitly shown how the investment and operating costs of absorption-, membrane- and adsorption-based post-combustion CO_2 capture processes would change from sector to sector, even from plant to plant, and how the carbon mitigation effort could benefit by early-adopting industries which have higher feed CO_2 compositions and flow rates [1–2].

More recently, we have demonstrated that no single materials-centric metric is adequate to identify the most cost-effective materials or processes for CO_2 capture, and we must combine atomistic, geometric and process understanding to screen materials and processes in tandem for CO_2 separation [3-4]. Using a hierarchical *in silico* screening method [3] that combines material screening with process optimization, we have identified many zeolites for cost-effective CO_2 capture from power plants for the first time. This has also unfolded the possibility to quickly explore the universe of microporous materials (zeolites and metal-organic frameworks) and membranes (polymeric and zeolitic) for discovery of novel applications that include, but are not limited to, carbon capture, natural gas upgrading, air separation, H_2 purification, and xylene separation.

It is also critical to integrate capture, compression, transportation, utilization and sequestration in an optimal fashion. To this end, we have designed a nationwide CCUS supply chain network [5] by considering simultaneous selection of materials and processes for CO_2 capture, selection of CO_2 sources, selection of utilization and sequestration sites, network costs, and actual geographic locations. Results suggest that it is possible to capture 50% of the current CO_2 emissions from the stationary sources in the United States at reasonable cost. Results for the CCUS in South Central United States and Texas are also favorable. Overall, it is found to be crucial to consider both the selection of materials and processes as integral parts of a large-scale CCUS network.

References:

[1.] Hasan, M. M. F.; Baliban, R. C.; Elia, J. A.; Floudas, C. A. Modeling, simulation, and optimization of post-combustion CO_2 capture for variable feed concentration and flow rate. 1. Chemical absorption and membrane processes. Ind. Eng. Chem. Res. 2012a, 51, 15642-15664.

[2.] Hasan, M. M. F.; Baliban, R. C.; Elia, J. A.; Floudas, C. A. Modeling, simulation, and optimization of post-combustion CO_2 capture for variable feed concentration and flow rate. 2. Pressure swing adsorption and vacuum swing adsorption processes. Ind. Eng. Chem. Res. 2012b, 51,15665-15682.

[3.] Floudas, C. A.; First, E. L.; Hasan, M. M. F.; Gounaris, C. E. Database screening method for cost-effective separations. U.S. Provisional Patent Application #61/761,436 and #61/765,284.

[4.] Hasan, M. M. F.; First, E. L.; Floudas, C. A. Cost-effective CO_2 capture based on *in silico* screening of zeolites and process optimization. Submitted for publication, 2013.

[5.] Hasan, M. M. F.; Boukouvala, F.; Floudas, C. A. Nationwide, regional and statewide CO_2 Capture, utilization and sequestration (CCUS) supply chain networks. Submitted for publication, 2013.

(4al) Synthetic Micro/Nanomachines and Their Biomedical Applications

Gao, W., University of California, San Diego

The 1966 movie Fantastic Voyage captured the world's imagination, portraying a tiny submarine navigating through the human bloodstream and treating life-threatening medical conditions. This poster summarizes my PhD research advances in nano/microscale machines to realize the Fantastic Voyage vision, and highlights the challenges and opportunities in translating the progresses toward practical biomedical applications. Various areas of medicine, including targeted drug delivery, precision nanosurgery, biopsy, cell sorting, or diagnostic assays, would benefit from recent developments of efficient fuel-free and fuel-driven nano/microscale machines.

Polymer-based Catalytic Microrockets. The polymer-based catalytic tubular microrocket is synthesized using a template based electrodeposition method. The oxygen bubble propelled microrocket harvests the energy from chemical fuels (such as H_2O_2) and displays a record-breaking speed of 1400 body lengths/s. It can serve as an ideal platform for diverse biomedical and environmental applications. For example, antibody modified microrockets can isolate circulating tumor cells (CTCs) from biological media, and lectin modified polyaniline based microrockets can be used for selective bacteria (*E. Coli*) isolation from food, clinical and environmental samples. Poly(3-aminophenylboronic acid)/Ni/Pt microrocket itself provides the 'built in' glucose recognition capability for 'On-the-fly' capture, transport and release of yeast cells.

Micromotors: Toward *In Situ* Fuels. Various new hydrogen bubble propelled micromotors which can be self-propelled in their natural environments without any additional chemical fuel are demonstrated. The polyaniline/zinc microrockets display effective autonomous motion in extreme acidic environments (such as the human stomach). The attractive guided cargo transport capabilities of these acid driven microrockets hold great promise for *in vivo*biomedical applications such as targeted drug delivery. In contrast, the new water-driven Al-Ga/Ti based Janus micromotor can be propelled by the hydrogen bubbles generated from the rapid aluminum and water reaction which is promoted through 'liquid metal embrittlement'. The self-assembled monolayers (SAMs) of alkanethiols modified seawater-driven Mg Janus micromotors, which utilize macrogalvanic corrosion and chloride pitting corrosion processes, can be used for environmental oil remediation.

Magnetic Nanowire Swimmer. Magnetically powered nanomotors have attracted considerable attention due to their great biocompatibility. A high-speed magnetically-propelled nanowire swimmer which mimics swimming microorganisms by exploiting the flexible nanowire as artificial flagella under rotating magnetic field is illustrated. Potential applications of these cargo-towing nanoswimmers are demonstrated by the directed delivery of drug-loaded microparticles to HeLa cancer cells in biological media.

During my PhD research, I gained multidisciplinary knowledge (nanomaterials, electrochemistry, polymers and nanomedicine) and developed many novel approaches. I would like to continue my research further towards addressing the important challenges in this area: artificial nanomotors are expected to advance from current initial proof-of-concept studies into practical *in vitro* and *in vivo* biomedical applications for further evaluation.

Advisor: Joseph Wang, distinguished professor at UC San Diego.

For more information, please visit: https://sites.google.com/site/weigao2009/

(4be) Advanced Biomanufacturing With Genome Engineering and Synthetic Biology

Hong, S. H., Northwestern University

Rapid population growth, energy security, and climate change concerns have intensified the need for sustainable, low-cost production of energy, chemicals, and advanced materials. Industrial biotechnology, or the use of bacterial microorganisms, is one of the most attractive approaches for addressing this need, particularly when large-scale chemical synthesis is untenable. Unfortunately, engineering whole-cell microorganisms remains costly, risky, and slow. Major challenges are product toxicity to the host and low production yields and volumetric productivities. My faculty research program seeks to address these challenges by using interdisciplinary approaches to analyze, understand and control cellular processes on a molecular and systems level for advanced biomanufacturing. Specifically, I am particularly interested in three areas: i) biofilm-based biofuel production using genetically engineered biofilms, ii) microbial electrosynthesis using engineered electron-accepting microorganisms to convert electricity to biochemicals, and iii) cell-free metabolic engineering using optimized cell extracts for biopolymer production. I will apply a unique and diverse set of skills in protein engineering, genome engineering, metabolic engineering, and cell-free synthetic biology for the biosynthesis of a wide variety of important materials, chemicals, and fuels. My work will allow me to build a leading research program for compelling applications in medicine, basic science and quantitative applied biotechnology.

(4ck) Control of Stochastic Interacting Systems

Mesbah, A., Massachusetts Institute of Technology

Uncertainties and disturbances are ubiquitous in real interconnected systems, where interactions among several systems increase the overall complexity of the integrated system dynamics. System stochasticity (due to uncertainties and disturbances) can have significant economic and safety implications for system operation. Advanced control strategies with the capability to remain effective in the presence of system stochasticity and component (e.g., actuator and/or sensor) failures set new horizons for high-performance and cost-effective operation of complex dynamically interacting systems. I intend to establish a research group in control of stochastic interacting systems with a particular focus on energy and healthcare applications. Addressing the system-wide control challenges of interconnected systems that are prone to severe uncertainties and systematic failures will be my primary research interest.

During my Ph.D. in the Delft Center for Systems and Control at Delft University of Technology (The Netherlands), I investigated model predictive control of nonlinear, distributed systems. I developed a control strategy based on a full population balance modeling framework for real-time control of batch crystallization systems. The effectiveness of the developed control algorithms in terms of increased system productivity and improved product quality was demonstrated experimentally. Furthermore, I developed a new methodology for closed-loop performance monitoring and diagnosis, which is a crucial step in the maintenance of model-based controllers to sustain their life-time performance.

Throughout my postdoctoral work at the Department of Chemical Engineering at Massachusetts Institute of Technology, I have been expanding the scope of my research towards model predictive control of continuous manufacturing systems. I have been developing generic formalisms to facilitate the quantification of system stochasticity for model predictive control of complex systems in the presence of uncertainties. I have also been working on optimal failure-tolerant control of multi-objective control systems.

As a faculty member, I will utilize my systems and control expertise to conduct leading research at the intersection of control theory, applied mathematics, and process systems engineering towards the development and application of control algorithms for complex systems.

(4bu) Structure and Dynamics of Block Copolymer Based Soft Materials
Lee, S., University of Minnesota

Block copolymers, made by joining two or more linear polymer blocks, have been studied over a half century, and these compounds are still being intensively explored. We recently discovered an equilibrium phase of packed spheres in poly(1,4-isoprepe-*b*-DL-lactide) (IL) diblock and poly(styrene-*b*-1,4-isoprene-*b*-styrene-*b*-ethylene oxide) (SISO) tetrablock copolymers, known as Frank-Kasper σ-phase using small angle X-ray scattering (SAXS) and transmission electron microscopy (TEM) experiments. The σ-phase contains 30 microphase-separated spheres in a large tetragonal unit cell, nucleates and grows from the body-centered cubic (BCC) phase similar to its occurrence in metal alloys. It is also known as a crystal approximant to dodecagonal quasicrystals, indirectly predicted by the Landau theory in diblock copolymer melts.

The mean field treatment on block copolymer thermodynamics near the order-disorder transition has given valuable understandings on this subject, but it assumes an infinitely large degree of polymerization (N) with an infinitesimally small Flory-Huggins interaction parameter (χ). All experimentally accessible block copolymer compounds have N and χ far from these limits, and the fluctuation effect near ODT from finite N has not been fully elucidated. We have investigated the thermodynamics of poly(1,4-isoprene-*b*-DL-lactide) diblock copolymers near the ODT with small N and relatively large χ using SAXS, dynamic mechanical spectroscopy (DMS), and differential scanning calorimetry (DSC) techniques. These studies provide fresh insights into the role of fluctuations on the state of segregation and ordering kinetics in symmetric and asymmetric diblock copolymers, prompting new questions regarding the application of mean-field and fluctuation corrected theories to block copolymer melts with finite molecular weights.

Block copolymer surfactancy using polyalkyl-based non-ionic surfactants have been investigated over several decades, these macromolecular surfactant compounds form mostly non-equilibrium states in water and/or oil due to the extremely low critical micelle concentrations (CMC's), which are at least several orders lower than the CMC's of traditional non-ionic surfactants. We investigated the non-ergodic effects by examining two ternary mixtures, each containing the same components, poly(1,2-butadiene-b-ethyelene oxide) (OB), water, and 1,5-cyclooctadiene, but the mixtures were prepared using different protocols. Both systems split into three layers, and the middle layer was evaluated using cryo-scanning electron microscopy (cryo-SEM) and SAXS. Qualitatively different microstructures were found in each system. This difference demonstrates that three-component block copolymer/water/oil mixtures are much more significantly affected by non-equilibrium effects than the analogous block copolymer based two component systems.

Along with above research conclusions, future research plans will be discussed

(4dy) Advanced Nanocomposites and Smart Solvent Processing

Urena-Benavides, E. E., Georgia Institute of Technology
Kitchens, C. L., Clemson University
Liotta, C. L., Georgia Institute of Technology
Eckert, C. A., Georgia Institute of Technology

My future research will apply smart solvents technology to the isolation of carbohydrate nanocrystals from lignocellulosic materials. A greener processing scheme is expected for production at an industrial scale. Moreover, the isolated nanomaterials will be processed into advanced nanocomposites while studying the ability to control their mechanical and transport properties. The principles and technology developed will be used for developing new smart materials. Among the vast number of applications, the fabrication of gels and membranes for specialty separations will be of significant interest.

Smart solvents systems were the focus of my postdoctoral work. This technology aims to advance greener chemical processes that can be performed at industrial scales. For a process to be "green" it must be economical as well as environmentally friendly. Even though carbohydrate nanocrystals have limited toxicity and come from renewable resources; the process for isolating them from lignocellulosic biomass requires severely harsh acidic conditions and extensive purification. Smart solvents like sulfolenes and gas expanded alcohols can be used as self-neutralizing acid/solvent systems, which are recyclable and eliminate the need of costly purification schemes.

During my doctoral research I isolated cellulose nanocrystals (CNC) from cotton and employed them as a reinforcement of calcium alginate nanocomposite fibers. This novel material is inspired on the behavior of native cellulose fibers. The fibers' mechanical properties can be tuned by controlling their microstructure through simple processing parameters like jet stretch. Complementing the smart solvents technology with controlled processing of nanocomposites will lead to improved production schemes for new advanced materials.

(4ca) Computational Investigations Into the Chemistry of Biomolecules and Functionalized Nanomaterials

Liu, H., University of Alabama

Understanding the properties of biomolecules and nanomaterials is of great interest in the scientific community. In my work, I have been using computational chemistry methods to study various biomolecules and functionalized nanomaterials. For example, the chemical mechanism of the methemoglobinemia process caused by an antimalarial drug primaquine was investigated using docking, quantum mechanics and QM/MM methods. The methemoglobinemia toxicity is a serious concern for this drug that has been unknown for more than six decades. By studying the primaquine…hemoglobin interaction and the subsequent chemistry upon such an interaction, we proposed the first mechanism in the literature to explain the methmoglobinemia caused by primaquine. In addition to biomolecules, the chemistry of various material and nanomolecules were also studied by computational methods. For example, the fundamental acid-base properties of imidazole, the precursor for a polymer membrane material to be used for CO_2 capture, were studied by quantum mechanics methods. The effects of various substituents were investigated and the results are able to aid the future design of this material. Furthermore, the mechanism of the formation of a gold nanoparticle was studied by quantum mechanics methods. Our results explain various experimental findings and provide insights into how this nanoparticle is formed. An overview of my other relevant work will also be presented.

(4et) Materials-Inspired Energy-Efficient Separations Through Membranes and Sorption Processes

Zhang, K., Georgia Institute of Technology

My research interests are in the areas of energy and sustainability with a particular focus on advanced separation systems. During my PhD with Prof. Jerry Lin, I worked on the inorganic membranes (metal and zeolite membranes) for high-temperature H_2 purification, whereby I obtained knowledge in the area of membrane science as well as adapting them to the design and operation of membrane reactors.

I am currently a postdoctoral fellow working at Georgia Institute of Technology, where I am working on gas/vapor and liquid separations for biofuel recovery and hydrocarbon separations. This position has allowed me to expand my expertise from inorganic membranes to polymer-inorganic hybrid membranes and also from membrane-based separations to both adsorption- and membrane-based separations. By identifying new materials and processes, I developed adsorbents and membranes that are designed specifically for industrially relevant applications, such as bio-ethanol purification, olefin/paraffin separation and xylene enrichment. By combining the materials research with polymer processing technologies, hybrid materials such as mixed matrix membranes and hollow fiber sorbents have been created.

My future research will address separation issues with emphases on materials innovation and scalable processes. With my expertise in membrane and adsorption, porous materials chemistry, organic-inorganic mixed matrix products and fiber spinning, I will endeavor to develop a unique program focusing on a fundamental understanding and control of product structure on the targeted separation devices. By coupling advanced materials with an innovative and scalable membrane or adsorption platform, the reductions in process energy requirements and carbon emissions from thermally-driven separation processes will be achieved.

Recent Publications (18 total, 10 first author, 164 total citations):

1. **Zhang K**, Lively RP, Zhang C, Koros WJ, Chance RR, Investigating the intrinsic ethanol/water separation capability of ZIF-8: an adsorption and diffusion study, *Journal of Physical Chemistry C* **2013**, 117, 7214-7225

2. **Zhang K**, Lively RP, Dose ME, Brown AJ, Zhang C, Chung J, Nair S, Koros WJ, Chance RR, Alcohol and water adsorption in zeolitic imidazolate frameworks, *Chemical Communications* **2013**, 49, 3245-3247

3. **Zhang K**, Lively RP, Dose ME, Li L, Koros WJ, Ruthven DM, McCool BA, Chance RR, Diffusion of water and ethanol in silicalite crystals synthesized in fluoride media, *Microporous and Mesoporous Materials* **2013**, 170, 259-265

4. **Zhang K**, Lively RP, Noel JD, Dose ME, McCool BA, Chance RR, Koros WJ, Adsorption of water and ethanol in MFI-type zeolites, *Langmuir* **2012**, 28, 8664-8673

5. **Zhang K**, Gade SK, Way JD, Effects of heat treatment in air on hydrogen sorption over Pd-Ag and Pd-Au membrane surfaces, *Journal of Membrane Science* **2012**, 403-404, 78-83

(4e) Synthesis and Design of Ionic Materials for Post-Combustion Carbon Capture

Adzima, B., University of Colorado

Ionic liquids possess low vapor pressures, high ionic conductivities, wide electrochemical stability ranges, broad thermal stabilities, and tunable solubilities enabling their use in a wide suite of devices such as lithium batteries, fuel cells, solar cells, light emitting diodes, and gas separation membranes.[1] Attempts to synthesis improved materials by polymerizing ionic liquid molecules have however produced mixed results. For example, in gas separation membranes a trade off exists between the mechanical integrity gained by polymerizing an ionic liquid and the permeability that is lost in the process.[2] Fortunately, phase separated polymer membranes formed by diblock copolymers may offer a suitable comprise between mechanical integrity and desired properties for all of the applications listed above. Already membranes formed from conventional diblock copolymers with poor CO_2 solubility have have shown improved separation results for CO_2/N_2 and CH_4/N_2 separations, with the increases in selectivity attributed to the orientation of cylindrical block orthogonal to the surface.[1]

At NETL I have developed techniques for preparing diblock copolymers incorporating monomers with improved CO_2 solubility, such as ionic liquids, using a combination controlled radical polymerization techniques and click chemistry. [3] The core synthesis centers on the polymerization of c-vinyl triazole monomers[4] using free radical and reversible addition-fragmentation chain-transfer (RAFT) polymerizations.[5] Quaternization with amines, followed by ion-exchange reactions with $LiTf_2N$ produces a poly(ionic liquid). Approaches for producing block copolymers from these pILs will be discussed, as will strategies for synthesizing, and subsequently crosslinking block copolymers to form membranes. Alternative approaches such as directly polymerizing of c-vinyl triazolium salts, and the incorporation of ionic liquid groups unto polymer backbones using the thiol–ene reaction will also be discussed.

This experience, combined with expertise using photo and click chemistry gained during my Ph. D. will support my goal as an assistant professor of starting a research group focusing on the development of materials with improved properties for a myriad of end applications.

[1] D. Mecerreyes, *Prog. Poly. Sci.* **2011**, *36*, 1629.

[2] J. E. Bara, D. E. Camper, D. L. Gin, R. D. Noble, *Acc. Chem. Res.* **2009**, *43*, 152.

[3] H. C. Kolb, M. G. Finn, K. B. Sharpless, *Angew. Chem. Int. Edit.* **2001**, *40*, 2004.

[4] H. Nulwala, K. Takizawa, A. Odukale, A. Khan, R. J. Thibault, B. R. Taft, B. H. Lipshutz, C. J. Hawker, *Macromolecules* **2009**, *42*, 6068.

[5] G. Moad, E. Rizzardo, S. H. Thang, *Aust. J. Chem.* **2005**, *58*, 379.

(4an) Multi-Scale Modeling for Energy Applications: CO2 Sequestration, Shale Gas Recovery and Lithium-Ion Batteries

Gor, G., Rutgers, The State University of New Jersey

Sustainable use of existing energy sources and the development of novel sources is one of the main current challenges for researchers. As a faculty, I'm going to tackle this challenge from the perspectives of both theory and modeling. The particular areas of my interest are:

1) Geological CO_2 sequestration
2) Shale gas recovery
3) High energy storage batteries

My research primarily focuses on modeling chemical engineering processes in porous materials ranging from nanoporous adsorbents and battery electrodes to geological formations: adsorption, transport, phase transitions, deformation, etc. My unique experience allows me to approach these phenomena at multiple scales. My toolset involves a number of computational chemistry and physics methods, including first principles (*ab initio*) simulations, molecular simulations (Monte-Carlo and Classical Density Functional Theory), finite element analysis, and computational thermodynamics methods for fluid-phase equilibrium calculations.

St. Petersburg State University, Russia: I received fundamental education in theoretical physics at St. Petersburg State University, Russia. There, besides a profound expertise in various branches of physics, I acquired comprehensive knowledge of mathematics. My research was in the field of first-order phase transitions. Particularly, I studied how a system consisting of liquid droplets and vapor evolves in time. In my PhD thesis, supervised by Prof. A. P. Grinin and Prof. F. M. Kuni, I developed a novel approach to the kinetic theory of non-isothermal nucleation using methods of thermodynamics, mathematical statistics, and transport phenomena.

Rutgers University: During the following two years I worked as a postdoctoral researcher at the Department of Chemical and Biochemical Engineering at Rutgers University (group of Prof. Alexander V. Neimark). My main focus was on modeling adsorption in nanoporous materials. I achieved the following results: I developed macroscopic and molecular theories of the deformation of nanoporous materials during adsorption; using Classical Density Functional Theory and Monte Carlo simulations I modeled nitrogen adsorption in novel types of nanoporous carbons and developed methods for characterization of these materials (which have been further successfully implemented in commercial instruments); I also modeled adsorption and transport of chemical warfare agents in polyelectrolyte membranes.

Princeton University: Currently I work at the Department of Civil and Environmental Engineering at Princeton University (group of Prof. Jean H. Prevost). My first research project at Princeton is on modeling CO_2 leaks from geological storage. It involves the development, implementation, and application of methods of computational thermodynamics and mechanics for modeling the compositional flow of CO_2-brine mixtures. Another project I'm currently working on is on developing a multi-physics model for lithium-ion batteries electrodes to study the mechanical reasons of batteries aging and degradation.

Apart from research activities, I have always enjoyed teaching. I have taught a number of courses to a variety of different audiences: from high-school students to Princeton graduate students.

At the poster session, besides giving an overview of my past and current research, I will discuss my plans in research and teaching as a prospective faculty.

Selected publications (from 23 total):

[1] **G.Y. Gor**, T.R. Elliot, J.H. Prevost, Effects of thermal stresses on caprock integrity during CO_2 storage, International Journal of Greenhouse Gas Control, 12 (2013) 300-309.
[2] **G.Y. Gor**, C.J. Rasmussen, A.V. Neimark, Capillary Condensation Hysteresis in Overlapping Spherical Pores: A Monte Carlo Simulation Study, Langmuir, 28 (2012) 12100-12107.
[3] **G.Y. Gor**, M. Thommes, K.A. Cychosz, A.V. Neimark, Quenched solid density functional theory method for characterization of mesoporous carbons by nitrogen adsorption, Carbon, 50 (2012) 1583-1590.
[4] **G.Y. Gor**, S. Tapio, A.V. Domanskaya, M. Rasanen, A.V. Nemukhin, L. Khriachtchev, Matrix-isolation study of the phenol-water complex and phenol dimer, Chemical Physics Letters, 517 (2011) 9-15.
[5] **G.Y. Gor**, A.V. Neimark, Adsorption-Induced Deformation of Mesoporous Solids, Langmuir, 26 (2010) 13021-13027.

(4ag) Systems Biomedicine and Pharmaceutics

Ford Versypt, A. N., Massachusetts Institute of Technology

I am interested in research at the intersection of chemical engineering, computational science and engineering, applied mathematics, and biology. Many applications with great societal impact lie within the intersection of these disciplines, such as advanced pharmaceutical and medical technologies. The current pharmaceutical drug discovery and approval pipeline averages 13 years, has an attrition rate of 95%, costs more than $1 billion per approved drug, and has major impacts on clinical patient care. Predictive mathematical modeling has the potential to make significant improvements on this pipeline in several domains that often involve reacting systems with mass and energy transport on multiple length and time scales making them particularly challenging to understand mechanistically from experiments alone. Specifically, I am interested in modeling downstream processes related to how candidate drug compounds are put into formulations for dosing (manufacturing), how the drugs are released from the formulations inside patients (drug delivery), how the drugs are distributed throughout the human body and tissues (pharmacokinetics), and how the drugs dynamically affect physiological functions (pharmacodynamics).

Computational models for evaluating drug delivery from pharmaceutical formulations can narrow the range of experiments needed to identify successful formulation designs by predicting performance and thus can reduce the development time and drive down costs. In my PhD thesis research with Prof. Richard D. Braatz at the University of Illinois at Urbana-Champaign, I developed and solved novel, computationally-tractable, mechanistic mathematical models for autocatalytic reaction (hydrolysis) and diffusion (erosion) to predict the drug release dynamics from biodegradable polymer microspheres for use in human patients. These biodegradable polymer microspheres are used for controlled-release drug delivery—administering medicine over extended periods with a single dosage. The project was collaborative with Prof. Daniel W. Pack's experimental drug delivery group. Also during my PhD research, I formulated a dynamic pharmacokinetic model used for controlling and optimizing the spatial distribution of growth factors delivered to stem cells for growing engineered tissues.

Models coupled with sophisticated process control strategies allow for careful monitoring of manufacturing to reduce waste of materials and energy and to adhere to strict quality standards. My postdoctoral work at the Massachusetts Institute of Technology has focused on modeling of the physical and chemical processes involved in manufacturing pharmaceutical small molecules and biotherapeutics in order to make the production of pharmaceuticals more efficient and economical, particularly through innovative processing techniques shifting to continuous manufacturing or manufacturing on demand. In my project in the Novartis-MIT Center for Continuous Manufacturing, I developed a control-relevant, first-principles dynamic mathematical model for a continuous thin-film pharmaceutical manufacturing process. The polymer thin-film is drug-loaded and is to be used as a drug dosage formulation. The project is a collaborative effort with control engineers to design process control systems for quality-by-design industrial pharmaceutical manufacturing. I am currently working on another project with the MIT Biomanufacturing Research Program to develop a dynamic model and control system for the small-scale on-demand production of biotherapeutics.

In my research program, I aim to develop mechanistic models of dynamic physiological processes, including metabolism and transport in biological tissues, that consider the effects of pharmaceutical treatments across multiple time and length scales to translate cellular- and tissue-level insights to clinical application. These models offer ways to predict the pharmacokinetic and pharmacodynamic interactions between drugs and their distribution and impact on the human body, the impact of controlled-release drug delivery dosages on tissue-level responses, and role of therapeutics and metabolites in the pathology of diseased populations. I am currently building collaborations with experimentalists at other universities for validating the modeling efforts.

(4dr) Microbial Caffeine Junkies: Growth of Pseudomonas Putida CBB5 On Caffeine and Metabolic Engineering of E. Coli for Production of High-Value Methylxanthines From Caffeine

Summers, R. M., University of Iowa

Alkylxanthines are naturally occurring compounds that are used in pharmaceutical preparations as diuretics, asthma treatments, and cardiac, pulmonary, and neurological stimulants. The most common of these compounds, caffeine (1,3,7-trimethylxanthine), is also found in many common food and beverage products such as coffee, tea, soft drinks, and chocolates. Because of their wide use, caffeine and related alkylxanthines enter the environment through human waste water and both solid and liquid wastes from coffee and tea processing facilities. Not surprisingly, bacteria capable of growing on caffeine have been isolated previously. However, the mechanism of bacterial caffeine degradation has not been elucidated until now.

Metabolism of caffeine by new enzymes in *Pseudomonas putida* CBB5.

We have discovered a novel caffeine-degrading bacterium, *Pseudomonas putida* CBB5, from soil at the University of Iowa by enrichment on caffeine as sole carbon and nitrogen source. CBB5 utilizes five novel enzymes belonging to the Rieske oxygenase (RO) family, NdmABCDE, to metabolize caffeine by sequential *N*-demethylation to xanthine, which enters normal purine catabolism. NdmA, NdmB, and NdmC are positional-specific *N*-demethylases that remove the methyl groups from caffeine as formaldehyde. All three enzymes require an unusually large redox-dense reductase protein, NdmD, which transfers electrons from NADH to the non-heme iron at the catalytic site of NdmA, NdmB, and NdmC. NdmA and NdmB, which catalyze the first two steps in the caffeine degradation pathway, contain the basic RO structure with both Rieske [2Fe-2S] and non-heme iron domains. NdmC, however, contains only a non-heme iron domain. Curiously, the Rieske [2Fe-2S] domain for NdmC is fused to the N-terminus of NdmD, which is the only characterized RO reductase that contains an extra Rieske [2Fe-2S] cluster. NdmE is a novel protein belonging to the glutathione-*S*-transferase (GST) superfamily that may serve a structural role in bridging the Rieske [2Fe-2S] from NdmD and non-heme iron from NdmC. Together, NdmCDE form a large protein complex that carries out the N_7-demethylation of 7-methylxanthine to xanthine, the last step in the caffeine *N*-demethylation pathway. This is the first report of a GST-dependent oxygenase.

Metabolic engineering of *E. coli* to produce high-value methylxanthines.

Many of the methylxanthine metabolites produced in the caffeine *N*-demethylation pathway, such as paraxanthine (1,7-dimethylxanthine), 1-methylxanthine, 3-methylxanthine, and 7-methylxanthine, are high-value compounds used in the pharmaceutical industry. We have engineered over 40 strains of *E. coli* with various combinations of *ndmABD* to create these compounds from caffeine and theophylline (1,3-

dimethylxanthine), a 200- to 5,000-fold increase in value. We are currently performing protein engineering experiments to increase production of paraxanthine from caffeine. The caffeine-degrading genes from *P. putida* CBB5 also provide many more potential applications, including (i) environmental remediation of coffee and tea waste and by-products, (ii) bio-decaffeination of caffeine, (iii) biosynthesis of pharmaceuticals, (iv) decaffeination of coffee and tea waste prior to fermentation, (v) development of caffeine biosensors, and (vi) regulation of synthetic gene networks using caffeine- or theophylline-responsive RNA switches.

(4dk) Understanding Nature's Catalysts: Chemistry of Transition Metal Containing Enzymes

Sharma, S., Massachusetts Institute of Technology

Enzymes with transition metal containing prosthetic groups are found in all forms of life and facilitate some of the most remarkable chemistry on earth. For example, iron-sulfur containing enzymes are able to reduce atmospheric N_2 to ammonia at ambient conditions, as compared to Haber-Bosch process that is carried out at 400 C and 200 bar and uses more than 1% of global energy production. Other examples include, nickel and molybdenum containing carbon monoxide dehydrogenase (CODH), iron containing Cytochrome-P450 catalyzes C-H bond functionalization. A deeper understanding of these catalysts is not only central to problems of biology but can also revolutionize the chemical industry.

The common feature of these enzymes is the presence of transition metal atoms, that give rise to multiple low lying electronic states, which is the key to the their rich chemistry. However, determining the details of these states is a highly non-trivial problem. Traditional approaches like Density Functional Theory (DFT), although cheap and usually quite accurate, fail spectacularly when applied to transition metal containing systems. With recent advances in electronic structure theory, in particular with the advent of Density Matrix Renormalizaton Group (DMRG), we now have a method that is able to correctly describe the challenging chemistry of these enzymes.

In my postdoctoral work I have made significant contributions to the development and application of the DMRG to the quantum chemical systems. To the best of my knowledge I have written the most efficient implementation of DMRG, which is furthermore capable of targeting specific spin states in molecules. I have also demonstrated the power of the DMRG to model small molecules containing Fe_xS_y groups, which will be described in greater detail in the poster.

In my phd I studied the complex reaction mechanisms of soot formation and proposed pathways for formation of soot precursors using quantum chemical methods. Here I worked in close collaboration with experimentalists to provide mechanistic insights that were tested against their data. These results were then used in chemical mechanisms to design efficient combustors with higher efficiency and lower emissions.

During my faculty appointment I intend to combine my expertise in advanced electronic structure methods and experience in solving practical problems in collaboration with the experimentalists. In particular, I will work to bridge the gap between the kinds of model problems that new electronic structure methods are developed for, and the challenges of describing kinetics in real materials and biological problems. This will require not only the applications of the current tools, but also development of new methodologies. In conjunction with this I will use our theoretical understanding of biological motifs to collaborate with engineers and experimental chemists to guide the design of synthetic bio-mimetic catalysis.

(4dz) Tissue Engineering: From Microfluidic Devices to Biopreservation

Usta, O. B., Massachusetts General Hospital/Harvard Medical School/Shriners Hospital for Children

With the ever increasing cost of medical therapies and the nearly impossible odds of developing one successful drug that can pass the regulatory screenings, the time is right to search for cheaper, more efficient and higher throughput solutions for our screening and therapeutic needs. The micro-nano fabrication revolution in engineering and the ever improving computational capabilities are enabling technologies that now make this search more practical and feasible. Tissue engineering, a field that can take advantage of these advances, is still in its infancy despite being viewed as the future of medicine. As an engineer who has been trained extensively in computational modeling and has recently become experienced in bio-preservation and microfabrication; my ultimate goal is to contribute to tissue engineering finally bearing complete practical solutions and fundamental understanding by integrating all three areas of my current expertise; tissue engineering, biopreservation and computational modeling.

My long term vision is one where we can provide live tissue engineered products (TEPs), for clinical, research and industrial use, to anyone and anywhere in the world in a reliable and reproducible fashion. Such a vision requires a multidisciplinary research program which encompasses 1) tissue engineering 2) biopreservation 3) computational and theoretical modeling to accelerate experimental work. Accordingly my long term goal, as an independent faculty member, is to create a 3 thrust research program where I a) build micro/macro fabricated TEPs as organ models and assist/therapy devices (tissue engineering) b) devise optimized preservation methods to make the "off-the-shelf" dream a reality (bio-preservation) c) conduct coarse grained and detailed modeling to understand and accelerate experimental work (modeling).

(4bv) Biomaterials and Stem Cells for Tissue Regeneration

Lei, Y., University of California, Berkeley

Research Overview: Human pluripotent stem cells (hPSCs) have the capacities for indefinite *in vitro* expansion and differentiation into presumably all cell types of the human body. They therefore represent highly promising cell sources for numerous biomedical applications, such as cell replacement therapy, tissue engineering, and pharmacology and toxicology screens. However, before achieving their potential, we have to solve many technical challenges including: a) produce large number of various progenitor/mature cells from hPSCs and b) make these cells survive, mature, integrate with the host tissue and function after implanting *in vivo*. My researches focus on designing and using biomaterials to solve these problems. In this poster, three current projects and the future research will be presented. **1) A scalable, GMP compliant 3D culture system for the production of hPSCs and their progeny:** The above mentioned applications require large numbers of cells of high quality. For instance, $\sim 10^5$ surviving dopaminergic neurons, $\sim 10^9$ cardiomyocytes, or $\sim 10^9$ β cells are required to treat a patient with Parkinson's disease (PD), myocardial infarction (MI), or type I diabetes respectively. Analogously, $\sim 10^{10}$ hepatocytes are need for an artificial human liver, and $\sim 10^{10}$ cells may be required to screen a million compound library. Considering the large patient populations with degenerative diseases or organ failure as well as the millions of chemical/peptide/nucleotide compounds that can be screened against thousands of molecular targets, massive numbers of hPSCs or their progeny are needed. It is becoming clear that the current 2D-based cell culture systems are incapable of producing sufficient cells and are becoming a bottleneck. Using biomaterials, we developed a simple, defined, scalable, GMP compliant 3D culture system for the production of hPSCs and their progeny. With this system, we have achieved long-term, serial expansion (> 60 passages) of multiple hPSC lines with a high replication rate (20-fold over 5-day per passage), yield (2.0 x10^7 cells/ml), and purity (95% Oct4+), all of which offer considerable improvements over the current approaches. After long term culture in 3D, hPSCs remained pluripotent as shown by their ability to form all the 3 germ layers during the *in vitro* EB assay and *in vivo* teratoma assay. More importantly, directed differentiation of hPSCs into their progeny (e.g. dopaminergic neuron progenitors) could be conducted efficiently within the system following expansion, enabling us to produce large numbers of cells with direct therapeutic relevance. This system will be of broad interest from the laboratory to the biotechnology scale. **2) Scalable production of dopaminergic neuron progenitors and their application in treating Parkinson's Disease:** Implanting fetal ventral mesencephalic tissue containing dopaminergic (DA) neuron progenitors into the putamen of Parkinson's (PD) patients are effective to relief the motor symptoms. However, this treatment hasn't been widely used partially due to the ethic problem and limited cell source. hPSCs are promising to overcome these problems with recent breakthrough on efficient differentiating them into DA progenitors or neurons. However, a GMP compliant and cost effective culture system is required to produce these cells at various scales before they can be reached in clinic. In this project, we used our 3D culture system along with a small molecule cocktail to produce DA progenitors. We then used these cells to treat PD with rodent model. Research on improving the treating efficiency through combining protein delivery with cell implantation is also undergoing. **3) Synthetic ECM hydrogel for stem cell therapy:** *In vivo*, stem cells reside in complex 3D niches with well-regulated biochemical & biophysical cues that control cell maintenance, self-renew and differentiation. We developed synthetic hydrogels to mimic these niches for 3D stem cell culture and delivery. Polyethylene glycols or hyaluronic acids are cross-linked with protease degradable peptides. Peptides derived from ECM proteins are introduced as adhesion points for cells. Protein factors can be incorporated through physical or chemical immobilization. With these hydrogels, we built a 3D model for studying the calcification of vascular stem cells. We also cultured mesenchymal stem cells in these gels and found the 3D environments including the mechanics, adhesion peptide identity, concentration and presentation significantly affected their proliferation, spreading and migration. Currently, we are using these hydrogels to deliver stem cells with the objective to improve the cell survival rate and integration in vivo. **Future research will focus on** combining stem cells, protein factors and biomaterials to treat degenerative diseases including: Parkinson's, stroke, spinal cord injury, Type I diabetes and myocardial infarction

Education:

- UCLA, Chemical and Biomolecular Engineering, PhD, 2010
- UCLA, School of Medicine, M.S., 2006

Research Experience:

- UC Berkeley, 2010- (Postdoc, David Schaffer Lab)
- UCLA, 2006-2010 (Tatiana Segura Lab)

(4dh) Advancing State-of-the-Art Fields With Innovative Rheological Methods

Samaniuk, J. R., Katholieke Universiteit Leuven

The field of rheology, which endeavors to understand the governing relationships behind the deformation and flow of matter, is a relatively mature one. My research has used this to advantage by developing and applying new experimental methods based on rheological principles to advance state-of-the-art fields of science. As a graduate student I invented a novel rheometer capable of making measurements in concentrated lignocellulosic biomass. Those measurements have improved our understanding of biomass rheology and have led to new ideas for processing biomass into renewable products. My postdoctoral work has focused on the behavior of fluid-fluid interfaces. The rheological properties of interfaces affect the behavior of materials like foams, emulsions and biological membranes. Areas of science that rely on interfacial structures, like synthetic biology, require a better understanding of how interfacial rheological properties influence membrane and bulk-material properties. Currently, the rheological properties of interfaces can be examined with a number of macrorheological methods. Micro-scale methods that are based on particle-tracking experiments have recently been developed, and very recently the use of optical tweezers has allowed for active microrheological measurements. I will present an overview of my recent and past research results in these fields and indicate areas with great opportunity for advancement.

(4cp) Computational Studies On Self- and Directed-Assembly of Soft Matter Nano Building Blocks

Nguyen, T. D., Oak Ridge National Laboratory

Fabricating nanostructures that exhibit a well-defined ordering and are adaptive to environment has been highly desirable for a huge array of novel applications ranging from photonics, bio-sensing to drug delivery. Using computer simulation as a design tool, I have explored numerous approaches to engineering such structures from bottom up perspectives.

In my PhD studies, I have demonstrated that various nanostructures with one-, two- and three- dimensional periodicity can be obtained from the self-assembly of soft matter building blocks such as polymer-tethered nanoparticles and colloidal particles[1,2]. Remarkably, the interplay between the energetic and entropic interactions between the building blocks can lead to unexpectedly higher order structures that are uniform in geometry even when the building blocks are highly polydisperse[2]. In many cases, the assembled structures such as helical ribbons and bilayer sheets are shown to reversibly reconfigure upon toggling building block interactions and/or morphing their geometry[3,4]. I have also proposed unconventional, yet efficient, pathways along which target assembled structures are obtained by shifting the building block geometry[5]. Alternatively, the formation of nano- and micro-structures can be induced either by applying an external field on a system of Janus particles[6], or by taking advantage of the inherent instability of nanometer-thick films on solid substrates[7]. Specifically, as a postdoctoral researcher at ORNL I have been investigating the interplay between spinodal instability and thermal nucleation in thin liquid films wetting solid substrates. Using large-scale Molecular Dynamics simulations, I have focused on understanding the mechanisms by which metallic and liquid crystal films first rupture, and how instability develops in the presence of thermal fluctuations. The results will provide useful information at the atomistic level, which is inaccessible with experiment and continuum models.

Throughout my study, I have also implemented and re-evaluated new computational methods and models to exploit the potential performance of new computer hardware such as GPUs and coprocessors[8,9]. Among my contributions to the open-source community, I have involved in implementing the rigid body integrators for LAMMPS and HOOMD-Blue[8], and various GPU-accelerated force fields for LAMMPS. These GPU-accelerated force fields allow for simulations performed at a much larger scale and at a much more affordable expense than before. Overall, my research serves not only as a framework for designing functional nanostructured materials via bottom-up approaches, but also to connect and inspire further theoretical and experimental investigations.

References

1. T. D. Nguyen, Z. Zhang, S. C. Glotzer, J. Chem. Phys., 129, 249903, 2008.

2. Y. Xia, T. D. Nguyen, M. Yang, B. Lee, P. Podsiadlo, A. Santos, Z. Tang, S. C. Glotzer, N. A. Kotov, Nature Nano, 6, 580-587, 2011.

3. T. D. Nguyen, S. C. Glotzer, Small, 5, 2092-2096, 2009.

4. T. D. Nguyen and S. C. Glotzer, ACS Nano, 4, 2585-2595, 2010.

5. T. D. Nguyen, E. Jankowski and S. C. Glotzer, ACS Nano, 5, 8892-8903, 2011.

6. S. Hwang, T. D. Nguyen, S. Bhaskar, J. Yoon, D. Chickering, H. Bernstein, S. C. Glotzer, J. Lahann, In Preparation, 2013.

7. T. D. Nguyen, M. Fuentes-Cabrera, J. D. Fowlkes, J. A. Diez, A. G. Gonzalez, L. Kondic, P. D. Rack, Langmuir, 28, 13960-13967, 2012.

8. T. D. Nguyen, C. L. Phillips, J. A. Anderson, S. C. Glotzer, Comput. Phys. Comm., 182, 2307-2313, 2011.

9. T. D. Nguyen, J.-M. Y. Carrillo, A. V. Dobrynin, W. M. Brown, J. Chem. Theory Comput., 9, 73-83, 2013.

(4dm) Cloud-Based Simulations On Google Exacycle Provide Novel Mechanistic Insights Into Transitions in Complex Chemical and Biological Systems

Shukla, D., Stanford University

Complex chemical and biological processes lie at the heart of some of the most challenging problems in science and engineering. The molecular-level description of these processes is required not only understanding the physical phenomena but also for engineering novel solutions of the associated problem.

Multi-scale approaches are required for facing the challenging task of describing such processes that span several orders of magnitude in both length and time scales. These processes are being investigated by a spectrum of computational techniques ranging from quantum mechanics, molecular dynamics to multi-scale coarse graining and bridging between relevant scales. I have used computational methodologies such as world-wide distributed computing on platforms such as Google exacycle and Folding@home, rare event simulation techniques, GPU computing, network models etc. to study long time and length scale phenomena. In this poster, I have discussed the applications of these techniques to problems related to human health (such as conformational changes in G-Protein Coupled Receptors and Kinases, which are key signaling enzymes involved in diseases such as Cancer), materials (Biomineralization; Growth of inorganic crystals in presence of proteins), and energy (transport of small molecules in NiFe Hydrogeneases, which are involved in reversible oxidation and production of hydrogen).

More details can be found on my website: http://www.stanford.edu/~shukla

(4dg) Advanced and Functional Nanoporous Materials for Novel Applications

Saha, D.,

My overall research interests incorporate engineered porous materials and their wide scientific and technological applications. I like to design the shape and direction of my research area in such a way that it can born with the knowledge and skill that I have acquired during my previous research experience and can translate in to a more advanced and wide spectrum of research scope to respond to the increasingly complex scientific and technical challenges of the 21^{st}century. It is also of utmost importance to me to carefully design the research areas in such a way that it can provide a broad understanding of science and engineering and develop the experimental skills needed by the student and researchers for their future independent work.

The major theme of my research work in the academic and national laboratory sectors has been related to the *nanoporous materials* and their applications in energy, environmental, and biological sectors. During the pursuit of my doctoral degree, I worked on the synthesis and characterization of nanoporous media for the adsorption and separation of hydrogen, carbon dioxide and several other gases followed by adsorptive water purification to remove fluorides and microorganisms from drinking water. After my short postdoctoral research in University of Quebec at Trios Rivieres, Canada, where I mostly studied the synthesis of nanoporous metal-organic frameworks (MOFs), I began my research at ORNL on the studies of adsorption by neutron and x-ray scattering along with design and synthesis and of structural carbon based adsorbents and their applications in electrode materials and controlled delivery of drugs and related materials. Based on my research, I published a large volume of peer reviewed publications (mostly as first author), presented in several conference proceedings and filed a U.S. patent.

My *overarching goal* of research as a faculty member is case-specific design and synthesis of novel adsorbents and their functional applications. Typically, I would like to employ amphiphilic surfactant templating on various natural and synthetic precursors along with post-synthesis modifications to design highly engineered nanoporous carbons with tunability and nano-scale control of their pore morphology. The unique feature of the designed nanoporous carbons is expected to be case-specific design and a universal control of pore architecture which will provide their applications in myriad scientific and engineering fields. I would be interested to seek their applications in three different sectors: (*a*) Gas separation and storage by employing both size exclusion and guest species involvement within the nanopores of carbon (*b*) Electrode materials for batteries and supercapacitors (*c*) Storage and controlled release of functional molecules from carbon nanopores, including drugs, enzymes, proteins and genetic species.

Total peer-reviewed publications accepted: *30*; Under Review: *4*; Total citations: *380*; h-index: *11*

Selected publications:

1) N.C. Gallego, L. He, D. Saha, Y. Melnichenko, C. I. Contescu, ***J. Am. Chem. Soc.*** vol. 133, no. 35, pp 13794-13797, 2011.

2) D. Saha, S. Deng, ***J. Physical Chemistry Letters*** vol.1, no. 1, pp 73-78, 2009.

3) D. Saha, Z. Bao, F. Jia, S. Deng, ***Environmental Science and Technology*** vol. 44, no. 5, pp 1820-1826, 2010.

4) D. Saha, C.I. Contescu, N.C. Gallego, ***Langmuir*** vol. 28, no. 13, pp. 5669-5677.

5) D. Saha, S. Deng, ***Langmuir*** vol. 25, no. 21, pp. 12550-12560, 2009.

6) D. Saha, S. Deng, ***Langmuir*** vol. 26, no. 11, pp 8414-8418, 2010.

7) D. Saha, S. Deng, ***Langmuir*** vol. 27, no. 11, pp 6780-6786, 2011.

8) D. Saha, S. Deng, ***Carbon*** vol. 48, no. 12, pp 3471-3476, 2010.

9) D. Saha, S. Deng, ***Carbon*** vol. 48, no. 12, pp 3471-3476, 2010.

10) C. I. Contescu, D. Saha, N. Gallego, A. Kolesnikov, E. Mamontov, ***Carbon*** vol. 50, no. 3, pp 1071-1082, 2011.

11) D. Saha, C.I. Contescu, N.C. Gallego, ***Microporous and Mesoporous Materials*** vol. 155, pp 71-74, 2012.

12) D. Saha, Z. Wei, S. Deng, ***Separation and Purification Technology*** vol. 64, no. 3. pp. 280-287, 2008.

13) D. Saha, Z. Wei, S. Deng, ***Intl. J. Hydrogen Energy*** vol. 33, no. 24, pp. 7479-7488, 2008.

14) D. Saha, S. Deng, ***Journal of Chemical and Engineering Data*** vol. 54, no. 8, pp. 2245-2250, 2009.

Patent Filed: A.K. Naskar and D. Saha, *Lignin derived porous carbon composition, method of preparation, and use thereof.* U.S. Patent Filed; Application No. 13/766,292, 2013.

(4bs) "Smart" Foams As Templates for Catalysts and Biomaterials

Lam, S., North Carolina State University

Aqueous foams can find many applications ranging from cosmetics and food formulation to various separation processes. My work to date focuses on the development of novel 'smart' foam systems, which can be manipulated using a number of external stimuli – including magnetic fields and light irradiation. Such systems can be utilized in their hydrated state or can be freeze-dried and utilized in the form of solid materials. They have potential applications in water remediation processes for efficient collection of contaminants, as well as in the controlled delivery of a substance trapped within the matrix of the foam. During my PhD study in the group of Prof. Orlin Velev, I have characterized how the variation of system parameters (i.e. stabilizer concentration, responsive material content, water content) can affect the viscoelastic properties of 'smart' Pickering foams; and, how this in turn affects the mechanisms by which the samples respond to an applied external stimulus. We used this information to develop models describing the mechanism of foam collapse by either bubble expulsion from the foam matrix or by thin film rupture, depending on the water content in the foam. I have also gained expertise in the formulation of various types of Pickering foams stabilized by naturally derived materials (such as cellulose and lignin), as well as in characterizing the physics behind foam evolution processes.

The fundamental knowledge from studying Pickering foams can be applied to the development of novel porous materials for catalysis and cell scaffolds from different types of particle-stabilized foams (stimuli responsive, nutrient releasing, degradable, etc.). The synthesis of these materials from Pickering foams requires only a few simple steps - foam generation through high shear homogenization at room temperature, and curing or freeze-drying to obtain solid foam. Novel materials can be easily fabricated via foam templating or emulsifying particles or mixtures of particles of a certain chemical nature. One can control the pore sizes and lamella thicknesses in the resulting materials by tuning the formulation of the prefoamed suspension.

(4ec) Image-Based Fluid Dynamics for Biomedicine and Beyond
Voronov, R. S., University of Pennsylvania

4th Year Postdoctoral Fellow

Research Interests:

Image-Based Modeling is an emerging multi-disciplinary field that finds numerous applications in many diverse industries ranging from Biomedicine to Enhanced Oil Recovery. By using state-of-the-art imaging techniques in conjunction with high performance computer modeling the image-based approach enables researchers to gain insights into systems that otherwise would have been inaccessible. For example, using imaging techniques, such as 3D Computed Tomography or Confocal Microscopy, allows one to obtain exact geometries of complex structures of interest, such as that of living porous media in-vivo or that of inorganic porous media (such as rock formation). Perhaps the main advantage of this modeling approach is that the implementation of the obtained structures in simple robust computer simulations gives practical results that are based on actual experimental data and do not suffer from mathematical complexity or from over-simplifications that are often necessary in other modeling techniques.

Successful Proposals: NIH F32, American Heart Association Postdoctoral Training Grant, XSEDE/Teragrid.

Postdoctoral Project: "*Intra-Thrombus Chemo-Transport and Local Stress Mechanics Under Flow.*"

Under supervision of Scott L. Diamond & Skipp Brass, Chemical and Biomolecular Engineering/School of Medicine, University of Pennsylvania

PhD Dissertation: "*Fluid Shear Stress and Nutrient Transport Effects Via Lattice Boltzmann and Lagrangian Scalar Tracking Simulations of Cell Culture Media Perfusion through Artificial Bone Tissue Engineering Constructs Imaged With MicroCT.*"

Under supervision of Dimitrios V. Papavassiliou & Vassilios I. Sikavitsas, Chemical, Biological and Material Engineering, University of Oklahoma

Research Experience:

My academic career path has been a blend of many fields of science and engineering. My formal training was in simulation, which I have learned on many different physical scales: Density Functional Theory (femto), Molecular Dynamics (nano), and Lattice Boltzmann Fluid Dynamics (meso). Yet, projects that I worked on were always close collaborations with experimentalists: Friction Reducing Surfaces for the NAVY, Bone Tissue Engineering, Enhanced Oil Recovery and Blood Coagulation/Thrombosis. As a result of these collaborations I have acquired invaluable and versatile experience in many research areas: material science, biology, medicine, biochemistry, biomedical imaging, and engineering design, just to name some. Moreover, as a result of intensive training at various supercomputing centers (OU Supercomputing Center for Education & Research - OSCER, Texas Advanced Computing Center - TACC and Lawrence Berkeley National Laboratory – LBNL), I became proficient in high performance computer modeling.

Teaching Experience:

Aside from my research career, I also have extensive teaching experience. I spent a semester teaching basic science to highschool students as a part of a NSF fellowship; I TAed and guest-lectured undergraduate lecture and laboratory courses in Chemical Engineering both the University of Oklahoma and at the University of Texas (for which I won a Campus Wide Best TA Award); I was repeatedly chosen by my department to be the "Pro/II Expert" (duties included lecture undergrads, holding office hours and developing course material); and, lastly, I mentored graduate students in my group.

Future Direction:

As faculty I would like to continue applying Image-Based Modeling in different areas of Biomedicine, as well as in other industries. In particular I would like to unite the knowledge that I obtained from the two topics that I worked on during my Ph.D. and postdoc: Bone Tissue Engineering and Blood Thrombosis, respectively. I believe that the two are closely related, because, Bone Tissue Engineering grafts must interact with the patients' blood post-implantation. Proper blood flow through the artificial tissue must be ensured in order to eliminate waste products from the construct, to mechanically stimulate the bone cells, and to deliver oxygen and nutrients to them. Image-Based Modeling presents itself as a perfect tool for exploring this system, because a) growth of the of the bone tissue can be captured by imaging at different time points and b) effects of the dynamically changing porous structure on the hemo-chemical and mechanical processes within the construct can be readily calculated via high performance computing.

Besides Biomedicine my interests include flows through inorganic porous media (in particular multi-phase flow through porous rock as it is related to enhanced oil recovery), drug delivery in the brain, modeling industrial chemical processes, financial optimization and entrepreneurship. My overall philosophy is to use simple and well-developed simulation methods in order to get practical and realistic results based on actual physical data. Therefore, in the future, I foresee myself collaborating closely with experimentalists and/or having an experimental component in my own laboratory.

Selected Publications:

R.S. Voronov, T.J. Stalker, L.F. Brass, and S.L. Diamond, **2013**. "*Simulation of Intrathrombus Fluid and Solute Transport Using In Vivo Clot Structures with Single Platelet Resolution*". *Annals of Biomedical Engineering*. 10.1007/s10439-013-0764-z. [online ahead of print]

Stalker TJ, Traxler EA, Wu J, Wannemacher KM, Cermignano SL, **Voronov RS**, Diamond SL, Brass LF. "*Hierarchical organization in the hemostatic response and its relationship to the platelet signaling network*." **2013**, Blood; doi: 10.1182/blood-2012-09-457739 [online ahead of print]

R.S. Voronov, S. VanGordon, R.L. Shambaugh, V.I. Sikavitsas and D.V. Papavassiliou, **2012**. "*3D Tissue Engineered Construct Analysis via Conventional High Resolution Micro CT without X-Ray Contrast*". Tissue Engineering – Part C, doi:10.1089/ten.tec.2011.0612. [online ahead of print]

R. S. Voronov, S. VanGordon, V.I. Sikavitsas and D.V. Papavassiliou, **2010**. "*Computational Modeling of Flow Induced Shear Stresses Within 3D Salt-Leached Porous Scaffolds Imaged via Micro-CT*". J. of Biomechanics, 43 (7), 1279-1286.

R. S. Voronov, S. VanGordon, V.I. Sikavitsas and D.V. Papavassiliou, **2010**. "*Distribution of Flow-Induced Stresses in Highly Porous Media*" Appl. Phys. Lett. 97, 024101.

R.S. Voronov, D. Papavassiliou and L. Lee, **2006**: "*Boundary Slip and Wetting Properties of Interfaces: Correlation of the Contact Angle with the Slip Length*." Journal of Chemical Physics, 124, 204701.

Reviews:

R.S. Voronov, D.V. Papavassiliou and L.L. Lee, **2008**: "*Review of Fluid Slip over Superhydrophobic Surfaces and Its Dependence on the Contact Angle*." Ind. Eng. Chem. Res., 47 (8), 2455-2477. doi: 10.1021/ie0712941

(4ed) Nano-Plasmonics for Transformative Biosensors

Wang, Y., University of Notre Dame

Accurate, fast and low-cost detection of biomarkers is essiential for effective prescreening of many chronic diseases and cancers. However, current biosensors, DNA microarray (Affymetrix) and surface plasmon resonance SPR (Bia Core) for example, are too costly and too labor/time intensive to allow frequent and individualized screening. There is hence a need for a new integrated home-use diagnostic platform if personalized healthcare is to become a reality.

Nanoplasmonics involves surface electromagnetic (EM) waves confined to a nano-scale region (hot spot) sustained by a plasmon resonator, thus allowing sensitive biomolecule sensing. Many different designs of plasmonic nano-resonators for biosensing have been proposed like fano-resonator, nano-crescent, bow-tie antena, etc. However, the fabrication of these nano-structures is complicated and expensive. This is because their design has been highly empirical, with heavy reliance on FDTD (Finite Differential Time Domain) simulation. Despite well-known facts that SERS are unique to nanospheres/flowers and nanoparticle catalysts have favorable electron-transfer and bulk mass transport properties, a holistic approach that extrapolates the unique optical, solutal and electrical physics at small scales to design low-cost macroscopic sensors has yet to appear for the desired biomarker diagnostic platform.

We embark in this direction by using a multi-scale approach that exploits the separation of length/time scales of such nanostructures. For example, our analysis of SPR near geometric singularities (tips and wedges) reveal unique and useful features that have been supported by simulation (Wang et al, Optics Exp, 21, 6609(2013); AIChE J review May 2013). In addition to high-field intensification due to plasmonic focusing, broad-band resonance and red-shifted resonant are also found because of the lack of a natural length scale—or a diminishingly small one. I have used these unique plasmonic dispersion features of a single cone to build a sensor array at larger scales (mm) with optimum bulk/waveguide-plasmonic coupling, interference pattern and nonlinear optics. I also design into this biosensor array enhanced biomarker transport (by 100 fold) to the nanosensors with sub-micron diffusion lengths.

A first prototype with plasmonically for small-panel miRNA profiling is near maturity. A conical array is fabricated by wet etching commercial imaging fiber bundles, with thousands of small fiber cores at optimum spacing from my theory. The cone tip has a radius of curvature below 10 nm and a thin gold layer (at a theoretically determined thickness of 5 nm) is thermally evaporated onto the array to sustain plasmonics. Hairpin probes with fluorophore tags quenched by the gold film are functionalized to the gold surface. Plasmonically enhanced fluorescent reporting of hybridization occurs when the target molecule breaks the hairpin. Detection of multiple targets in a panel can be achieved by addressing each fiber core optically and using photochemistry to selectively attach different probes. Detailed spectral measurements show broadbanded plasmonic excitation of 3 fluorophores per cone to minimize interference. Intensity tracking in time show a signature constant rapid absorption flux to a conic tip, followed by a slow $t^{-1/2}$ diffusive flux to the substrate. The enhanced tip fluorescence allows detection of the approximately 100 miRNA at the tip in the first region within 15 minutes. Profiling data for miRNA of oral cancer cells from a heterogeneous sample are being collected.

In my future work, I intend to design smart nonlinear nanoplasmonic sensor arrays, with compound plasmonic Kerr materials, that are fully integrated with the micro/nanofluidic, surface acoustic wave (SAW) and micro-electronic circuitries on the same biochip. The individual nonlinear plasmonic oscillator can interact with each other to further maximize sensitivity and to interact with the other circuitries for a fully automated platform. I will explore the mutual resonance of SAW with low-frequency Bloch plasmon waves intensified by a nonlinear cone grid, thus allowing SAW amplification of plasmon intensity or vice-versa. I shall employ the high dielectrophoretic and plasmonic fields at the tips for molecular separation, ionization, transport and concentration at the nanoscale. An ambitious goal is to develop patterned plasmonic travelling waves in the array that will serve as molecular-specific nano-conveyor belt that will sequentially carry different species to different chemically active sites where single-electron tunneling events can be registered in a well-controlled nanoscale chemical environment. I have been exposed to all these technologies in my PhD career and will use my multi-scale analysis to facilitate their optimum integration. The goal is to design a broad-bandwidth and multi-functional screening chip that can detect a massively large number of protein and nucleic biomarkers, whose concentrations span 6 decades, by local separation/concentration and automated introduction of specific buffer and nanofluorescent reporters for enzymatic and optical amplification, preferably delivered with single-molecule/nanoparticle resolution.

Reference:

[1]Wang Y, Cheng X, Chang H-C, Celebrating Singularities: Mathematics and Chemical Engineering. AIChE J, Accepted, (Feature Cover) (2013) DOI: 10.1002/aic.14123

[2]Wang Y, Plouraboue F and Chang H-C, Broadband converging plasmon resonance at a conical nanotip. Opt Express, 21 6609-6617 (2013)

[3]Wang Y, Tan M.K, Go D.B and Chang H-C, Electrospray Cone-Jet Breakup and Droplet Production for Electrolyte Solutions. Europhys. Lett. 99 64003 (2012) (Editor's highlight).

[4] Xie F, Wang Y, Wang W, Li Z, Yossifon G, and Chang H-C, An Experimental Study on the Side-Opening Filling Process at the Interface between Microchannels with Different Widths. J. Nanosci. Nanotechnol. 10 7277 (2011).

[5] Chen Z, Wang Y, Wang W, and Li Z, Nanofluidic Electrokinetics in Nanoparticle Crystal. Appl. Phys. Lett. 95 102105 (2009).

(4p) Computational Modeling and Tomographic Imaging of Fluidized Beds
Boyce, C. M., University of Cambridge

Fluidized beds are of great industrial importance because of their critical role in applications ranging from fluidized catalytic cracking (FCC), to drying of particulates; fluidized beds also show promise in enabling chemical looping combustion (CLC), an emerging process for burning coal and natural gas with efficient, streamlined carbon capture. In addition to being of interest to chemical engineers for practical applications, the two-phase granular flow within fluidized beds appeals to physicists because the particulate phase behaves with unique properties, to the extent that it can be considered a new state of matter. Two-phase granular flow systems, such as fluidized beds, are difficult to measure because they are three-dimensional and opaque, and thus require the use of complex tomographic techniques. These systems also present many mathematical and computational modeling challenges because they simultaneously exhibit properties of multiple states of matter and involve up to millions of particles.

My doctoral research focuses on the development of computational methods that reveal the underlying physics of fluidized beds and the validation of these methods through direct comparison with tomographic experiments. In developing my own, first-of-its-kind 3D cylindrical discrete element model with computational fluid dynamics (DEM-CFD) to simulate fluid and particle flow in fluidized beds, I have acquired skills in implementing and synthesizing novel numerical techniques to ensure accurate and stable simulation of dynamic systems. My modeling developments provide the ability to directly simulate the flow conditions of laboratory-scale fluidized beds and corresponding tomographic measurements. By simulating the acquisition process of magnetic resonance imaging (MRI) in analyzing model results, I have gained experience in comparing model and experimental results with a new level of accuracy for cross-validation. This methodology has also shed light on tomographic measurements, providing a better understanding of the complex Fourier-domain averaging and Gaussian slice excitations, necessary for MRI velocity measurements, as compared to more conventional time-averaging and slice selectivity methods. Additionally, the use of detailed computational model results validated by experiment has allowed me to find novel approaches to solve age-old questions in the field, such as the physics behind slug rise velocity and the origin of pressure fluctuations in fluidized beds.

As a research faculty member, I plan to expand upon the foundation of my doctoral work to explore the critical interplay between physics and chemistry in fluidized beds over multiple length scales. My research program will involve combining insights from multi-scale modeling and multi-scale measurement efforts to provide insights on the varying physical-chemical phenomena that occur at different length scales. This research path will also entail coupling the chemical reactions with the modeling and measurement techniques in a way such that the relationship between the physical and chemical phenomena can be better understood. Finally, my program will involve direct collaboration with researchers who build fluidized beds for applications such as combustion and catalysis, in order to aid in the design and measurement of their experiments. My ultimate aim is to enable the next generation of two-phase granular flow systems, which can offer real world solutions to the energy challenge, such as chemical looping combustion for efficient carbon capture.

I also intend to use my unique educational background, coming from a double major in physics and chemical engineering as an undergraduate at MIT to a PhD in chemical engineering as a Gates Scholar at the University of Cambridge, to enhance chemical engineering education at an American university. From my experiences, I have seen a number of ways in which the on-campus undergraduate and graduate education experience in the U.S. can be further enhanced. Firstly, as a supervisor in the education system unique to Oxford and Cambridge, I have seen the benefits of students discussing material and problems in 2-to-4-person student groups with a graduate-level supervisor in an organized yet open-ended discussion setting. Secondly, as a Gates Scholar, I have had the privilege of interacting with leading graduate students in diverse disciplines, ranging from international relations to public health to engineering; this experience has shown the exceptional and expansive nature of the graduate education that comes from discussing important research problems that cut across academic disciplines. As a faculty member, I plan to take an active role in education to ensure more chemical engineering students at my university are provided with these valuable learning opportunities.

(4dv) Energy Conversion in Biological Molecular Machines

Torabi, K., Northwestern University
Schatz, G. C., Northwestern University

In order to carry out numerous biological functions, various molecular machines are at work within a cell. These molecular machines are responsible for conversion of energy, stored in Adenosine triphosphate (ATP), into various other forms of energy such as electrochemical potential across the cell membrane, work required for active transportation and etc. Although the basic duties of molecular machines are not very different form their macroscopic counterparts, these machines function in a totally different environment where electrostatic and solvent interactions replace the forces of gravity and inertia. More importantly, due to their small size, molecular machines are persistently affected by significant thermal fluctuations. Here we propose a theory that explains how various molecular machines take advantage of the thermal fluctuation to perform their tasks via a biased Brownian ratcheting mechanism. Our theory explains how the tight coupling of the ATP reaction to the molecular machines production cycle would lead to high energy efficiency.

(4az) Heath Himstedt: Meet the Candidate

Himstedt, H., University of Minnesota

I am seeking positions in chemical, environmental, and/or civil engineering.

My primary research interest is designing enhanced separations for sustainable engineering. One research focus will be enhanced filtration and chromatography using responsive polymers. Stimuli-responsive polymers can be incorporated into or onto supports—such as membranes or zeolites. This technique could be used for anti-fouling membranes of contaminated water or enhanced chromatographic purification of pharmaceuticals. A second research focus will be engineered surfaces for rapid, flow-through reactors. By incorporating reactive species into polymer chains grafted on a membrane support, a high surface area to volume ratio reactor could be achieved. Since mass transfer through this reactor would be via convection, diffusion limitations could be eliminated. A third research focus will be enhanced separation on a small scale for distributed chemical manufacture. By engineering combined production/separation systems on a small scale, production of certain chemicals could be performed at the point of use. An exciting application of this idea is a wind or solar powered combined reactor/separator for production of ammonia fertilizers in developing nations, which lack infrastructure for rapid chemical distribution.

I am passionate about engineering education. As a Ph.D. student, I took the initiative to serve as the instructor of record for four university courses. I am prepared to teach numerous undergraduate courses, particularly separations and reactor/process design. I also envision teaching a number of specialty graduate courses such as membrane separations, polymerization, and sustainable technologies, among others. I excel at multi-disciplinary collaboration, and I will seek out collaborations with associated departments and research centers, such as centers for sustainability, water research, or environmental stewardship. I also plan to collaborate with the department of education to develop courses on more effective engineering education.

(4q) Protein Functional Dynamics

Bowman, G., University of California, Berkeley

It is commonplace to assume a protein exists in a single structure when studying protein function or designing novel proteins. However, this single structure approximation often leaves us unable to rationalize the effects of mutations on function, much less to design proteins. Overcoming these limitations will require a rigorous understanding of the structural fluctuations proteins undergo at equilibrium and their role in a protein's function. Combining computation and experiment to understand protein functional dynamics and incorporate this structural flexibility into design protocols will be the core of my research program.

In this poster, I present some of my initial results on identifying new means of manipulating a protein's function, which I have generated as a Miller Research Fellow at UC Berkeley. This work builds off of methods I developed for building quantitative maps of a proteins conformational space. These models—called Markov state models (MSMs)—are discrete-time master equation models built from extensive molecular dynamics simulations using a combination of physics, Bayesian statistics, information theory, and network theory. In the past, I have used these models to advance molecular simulations from microsecond timescales to millisecond timescales. Now I am using Markov models to capture a representative sample of the ensemble of structures a protein visits at equilibrium and identify allosteric control points—regions of the protein that are distant from the active site yet can have profound effects on activity in response to perturbations like mutations or small molecules. I show that my methods identify known allosteric sites and identify new sites that I am currently testing via biophysical experimentation.

(4dn) Neutron Reflectivity in Biological Applications

Singh, S., Los Alamos National Laboratory

Neutron Reflectometry (NR) has long been used as a tool to study soft-condensed materials, including bio-related structures. For example, it has been applied to study model biomembranes to facilitate physical understanding of cellular interfaces and their interactions with proteins (Krueger, S., Curr. Opin. Colloid Interface Sci., 2001). Typically, this technique is applied to investigate samples that are homogeneous over large areas. Such systems include phospholipid monolayers at the air-liquid interface, pure and hybrid phospholipid bilayers on silicon and quartz substrates, and phospholipid bilayers on novel support systems designed to closely mimic biological membranes. The time-of-flight Surface Profile Analysis Reflectometer (SPEAR), at the Lujan Center utilizes a neutron wavelength range of 4.5 Å to 16 Å to study a broad range of materials including polymers, nano-engineered films, and bio-related systems (Dubey *et al.,* Eur. Phys. J., 2011). Here, we report on some of our recent contributions towards the investigation of biological systems using SPEAR and extending NR to off-specular scattering (OSS) and complex systems, such as living fibroblast cells.

Krueger, S., 2001. Neutron reflection from interfaces with biological and biomimetic materials. Curr. Opin. Colloid Interface Sci., 6(2), 111-117.

Dubey, M., Jablin, M., Wang, P., Mocko, M., Majewski, J., 2011. SPEAR — ToF neutron reflectometer at the Los Alamos Neutron Science Center. Eur. Phys. J. 126(11), 1-11.

(4l) Colloidal Linkers With Soft-to-Hard Interactions: Understanding and Application in the Assembly of Novel Functional Materials

Bharti, B., North Carolina State University

The interaction of charged colloids with soft matter covers a wide range of phenomena which play an important role in the natural environment as well as in various industrial and biomedical applications. My research has been focused on understanding soft matter interactions at nano and micro level and further applying this understanding to fabricate responsive functional materials. My doctoral research at Technische Universität, Berlin in the group of Prof. Gerhard Findenegg was aimed towards investigating the interaction of soft matter (surfactants and globular proteins) with hard nanomaterials (silica nanoparticles, nanoporous solids). The influence of surface curvature and surface modification on the self-assembly of the nonionic surfactant at silica nanoparticles[1] and in the nanopores of SBA-15 silica material[2] was studied using adsorption measurements and Small-Angle Scattering techniques (SANS and SAXS).We also studied the interaction of the globular protein lysozyme with silica nanoparticles over a wide range of pH. We demonstrated that the adsorbed protein induces pH-dependent bridging aggregation of the silica particles.[3] My present post-doctoral research in the group of Prof. Orlin Velev at North Carolina State University and Triangle Materials Research and Engineering Center (ΔMRSEC) deals with the programmed assembly of particles into stiff or flexible permanent chains. We have used heteroaggregating particle pairs as the chain precursor and dielectrophoresis (DEP) as a structure directing tool to directionally assemble permanent chains of oppositely charged particles. We demonstrated that the assembly process depends on the size ratio as well as the number ratio of the two particles, and based on the structures resulting from the process a set of assembly rules was derived .[4] The permanent chains formed by the pair of oppositely charged particles have a rigid structure. We have established an alternative novel route for the fabrication of *ultra-flexible* composite magnetic chains which can be assembled into a variety of hierarchal structures. A fatty acid dispersion of 12-hydroxystearic acid containing FeO nanoparticles was subjected to magnetic field. The applied field pulls the FeO particles to near surface contact point where the fatty acid irreversibly binds the particles into chains. The chains were shown to wind/unwind reversibly into rings, hooks or infinity type structures. This approach forms a basis for the fabrication of highly functional magneto-responsive flexible materials. This research in the future will be extended into the design of new functional materials with unusual optical, electrical and heat transport properties.

[1]B. Bharti; J. Meissner; U. Gasser; G. H. Findenegg *Soft Matter*, **2012**, *8*, 6573

[2]B. Bharti; M. Xue; J. Meissner; V. Cristiglio; G. H. Findenegg*J. Am. Chem. Soc.*, **2012**, *134*, 14756

[3]B. Bharti; J. Meissner; G. H. Findenegg *Langmuir*, **2011**, *27*, 9823

[4]B. Bharti; G. H. Findenegg; O. D. Velev *Sci. Rep.* (*Nature*), **2012**, *2*,1004.

(4w) Controlled Nucleation and Crystal Growth for Pharmaceutical, Electronic and Energy Applications

Diao, Y., Massachusetts Institute of Technology
Myerson, A. S., Massachusetts Institute of Technology
Hatton, T. A., Massachusetts Institute of Technology
Trout, B. L., Massachusetts Institute of Technology
Mannsfeld, S., SLAC National Accelerator Laboratory
Bao, Z., Stanford University

Molecular self-assembly processes, such as nucleation, crystal growth, aggregation, nano-/micro- phase separation, have a profound impact on the solid-state properties of materials. For instances, difference in crystal packing (polymorphism) influences bioavailability, stability and processibility of pharmaceutical compounds; morphology, molecular packing and orientation of organic semiconductors are critical in determining their charge transport characteristics; controlling the nanoscopic phase separation is key to achieving high-efficiency organic solar cells.

Solution printing offers great potential for achieving high-throughput, low-cost, large-area manufacturing of electronic and energy materials. However, rapid coating speed needed for industrial-scale production poses challenges to the control of molecular assembly during printing. Fluid-enhanced crystal engineering (FLUENCE)— a novel approach developed in my work, allows for a high degree of morphological control of solution-printed thin films, enabling record-setting material performances.

Interfaces play a pivotal role in directing the molecular assembly events. By engineering the nano-topology, microstructure and chemical patterns of surfaces and interfaces, my research has established new methodologies and demonstrated unprecedented control over nucleation kinetics, polymorphism and crystal morphology.

Understanding the crystallization dynamics is key to devising design principles for controlling molecular packing and crystal morphology. Towards this end, my research utilizes powerful *in situ* synchrotron X-ray scattering techniques for investigating phase transformations and crystal growth kinetics during solution coating of organic semiconductor thin films.

With the new physical insight obtained and the novel approaches developed, I aspire to carry this research further towards 1) developing new technology platforms for fabricating next-generation pharmaceutical and energy materials, 2) deciphering the mechanism of molecular assembly processes at interfaces.

(4bn) Engineering Nanobioreactors: Encapsulating Heterologous Proteins Into Bacterial Microcompartments

Kim, E. Y., University of California

Biological systems often use compartmentalization as a strategy for optimization. In bacteria, naturally occurring protein microcompartments (MCPs) of approximately 125nm in diameter are used to compartmentalize metabolic pathways, thereby concentrating enzymes and their substrates, sequestering potential toxic intermediates, and insulating the reactions within from competing enzymes or metabolites. For the propanediol utilization (Pdu) MCP of *Salmonella enterica*, it has been shown that an 18 amino acid N-terminal signal peptide is sufficient for localizing heterologous proteins into the Pdu MCP. We aim to use the Pdu MCP as a template to create customizable nanobioreactors in bacteria.

A key step towards encapsulating biochemical pathways is to gain the ability to tune the enzyme stoichiometry within the MCP. Towards this goal, we have constructed a library of single amino acid substitution mutants in the signal peptide and characterized its strength for encapsulation by using a flow cytometry based assay. We also use this flow cytometry based assay to investigate whether two different naturally occurring N-terminal signal peptides compete for encapsulation, or have orthogonal mechanisms for encapsulation. These results help us gain an understanding of the requirements for localizing enzymes to the interior of the Pdu MCP and gain tunability in enzyme stoichiometry within the MCP by manipulating the signal sequence.

(4k) Optical and Electric Forces At Nano-Structures and Interfaces: Novel Spectroscopic Sensors From Single Molecule Dynamics to Diagnostic Platforms

Basuray, S., University of Missouri

Polarization mechanisms and charge dynamics at interfaces and near geometric singularities drive a multitude of spatiotemporal phenomena from slow ionic charge accumulation to rapid photonic coupling. This leads to highly specific potential gradients that can drive a host of physical phenomena such as dielectrophoresis, electro-wetting, plasmonics and photonics. This spatial characteristic of the electric and optical field can be dynamically controlled by precise modulation of the geometric singularity of the structures in real-time.

In my future work, I intend to develop a new class of adaptive, stimulus responsive sensors constructed out of immiscible fluids, polymers, nano-particles to create sensitive, cheap, robust, nano-ordered and non-linear optical components. These self-sustaining sensors will be used for real time monitoring of in-situ bio-molecular dynamics, water quality and nutrient sensors. For example, dynamically discerning subtle changes in cell secretions as pathological indicators. However, to achieve this objective, it will require novel and creative application of optical (surface plasmon resonance (SPR), both short and long range) and electrical (electrochemical impedance spectroscopy (EIS), micro and nano-fluidics) forces to create dynamic interfaces. These interfaces can trap, focus and concentrate molecules, recognizing the benefits of the singular electric field arising out of optical coupling at geometric singularities. The designing of such sensors will require developing fundamental knowledge and research into material characterization, fluidics, optics, electro-kinetics, nanotechnology specifically their interaction through simulations and theoretical insights. Label-free identification methodologies from Surface Enhanced Raman Spectroscopy (SERS) to EIS, will allow the sensors to be wearable and adaptable for low-resource settings. Further, sophisticated optics will allow for sufficient spatial and temporal resolution to simultaneously scrutinize and observe (detect) complex biological kinetics over broad spatial and temporal scales with single molecule resolution.

My inspiration for the novel sensor methodology arises from my doctoral research into interfacial electro-kinetics and its application to bio-diagnostic platforms through geometric singular surfaces. I identified the origin of Stern Layer by critically revisiting classical theories and was the first person to show the double layer effects and dynamics on size-dependent dielectrophoretic behavior of nano-scale colloids. I used these fundamentals to develop label-free, sensitive, ultra-selective EIS based diagnostic microfluidic platforms and a novel qualitative and quantitative EIS nanofluidic nucleic assay as an alternative to qPCR, - that have since led to entrepreneur startups.

In my post-doctoral work, I extended my foray to study optics specifically the interaction of biomolecules with localized SPR and photonic waveguides through Fluorescence correlation spectroscopy (FCS), molecule-metal interactions, Förster resonance energy transfer (FRET) and SERS. A novel stamping process coupled with an intelligent metal deposition process was developed to fabricate nano-ordered surface morphology on a plasmonic grating and nano-defects in a 1-D photonic crystal, without using expensive nano-lithography. Both show extra-ordinary broad spectrum electromagnetic enhancement due to line and point defects. Enhancements to multiple spectroscopic techniques, from SERS to FCS using only an epi-fluorescence microscopy, were achieved from a single platform.

Teaching and Outreach:

- I manage the biological lab and material characterization lab at the Center for Nano-Micro Systems and Nanotechnology at the University of Missouri, Columbia, where I also train students on different characterization instruments like AFM, UV-VIS, Raman, Ellipsometer and Dynamic Light Scattering.
- I was a research scientist in the new technologies team for the development of third, fourth generation genetic sequencers at Roche, a pharmaceutical company.

Developed and taught a number of courses and labs, like undergraduate chemical engineering senior year lab, advanced applied mathematics, developing the lab portion of a microfluidics course.

(4bz) From Molecular Simulations to Gas Separation/Storage Applications

Lin, L. C., University of California-Berkeley

Recently, nano-porous materials such as zeolites and metal organic frameworks (MOFs) have been shown to potentially provide a more energy-efficient way for gas separation and storage applications. A huge number of material candidates can be possibly achieved by tuning both structural topologies and chemical compositions. Fully characterizing all those materials with experimental methods is therefore prohibitive. To identify promising materials, computational study plays a crucial role by utilizing various techniques in molecular simulations.

Accordingly, we have performed several large-scale computational screenings of hundred thousand structures on the basis of adsorption or membrane processes for several applications, including carbon dioxide capture, methane capture/storage, and olefin/paraffin separation. From the outcomes of these large-scale screenings, a number of promising materials toward each different application have been identified. Moreover, insights about the optimal structures were found to provide directions for future structure design and synthesis. For instance, one of our most recent works indicates that, for carbon dioxide capture, some zeolite structures that possess particular topologies resulted in constituting maximized density of ideal pockets for adsorbing CO_2 molecules, which can reduce the energy penalty imposed on a coal power plant by as much as 35% compared to the MEA technology.

A systematic and transferable methodology has also been developed to generate accurate force fields using high-level quantum chemical calculations. The method was first applied for flue gases inside Mg-MOF-74, an open-metal site MOF, and the obtained force fields can be used to well reproduce experimentally measured adsorption properties. In this original study, the Møller-Plesset second-order perturbation theory (MP2) with representative small framework fragments was used. To facilitate the automation of this methodology as well as resolve the intrinsic discrepancies between cluster and periodic calculations, a methodology based on the periodic density functional theory (DFT) calculations is therefore further introduced.

In this poster, advanced developments as well as insightful outcomes from the aforementioned screening studies and the methodology of generating accurate force fields will be highlighted. More importantly, the poster will demonstrate the potential of integrating all techniques, including those that are under development, to accelerate the search for promising materials toward various energy-related gas separation and storage applications.

Advisor: Prof. Berend Smit, Dept. of Chemical and Biomolecular Engineering and Chemistry, University of California – Berkeley

(4bi) Block Copolymer Patterns Replicated By Photochemical Transfer Printing

Janes, D. W., Columbia University
McCoy, B. D., The University of Texas at Austin
Inoue, T., University of Chicago
Madan, I., The University of Texas at Austin
Thode, C. J., University of Wisconsin-Madison
Willson, C. G., The University of Texas at Austin
Nealey, P. F., University of Chicago
Ellison, C. J., The University of Texas at Austin

Thin films of block copolymers are promising patterning templates for next generation microprocessors, hard drives, bio-arrays, and separation membranes because they rapidly self-assemble into periodic structures with characteristic feature sizes of 1-100 nm. However, guiding patterns must be employed in order to direct the self-assembly of BCP thin films into prescribed, device-oriented patterns or regular arrays with long-range order. Fabrication of these guiding patterns requires performing lithographic patterning with resolution near that of the BCP periodicity, which can be a relatively lengthy and expensive process step. Therefore, strategies to replicate the patterns formed with BCP thin films onto other substrates while recovering the original for more replication cycles could be practically useful because the guiding pattern would only have to be fabricated once. Here a procedure is described that uses light as an externally applied stimuli to facilitate the transfer printing of BCP top surface patterns from one substrate to another. Because photo-polymerizable liquid conformal layers are placed between the samples, the pattern is replicated continuously and over large areas. The strategy has been used to replicate patterns with periodicities as small as 20 nm, as well as films in which the self-assembled structures have been directed by an underlying guiding pattern. Additionally, efforts to improve process throughput and conformal layer uniformity will be described.

(4i) Engineering Nanoparticles for Drug Delivery and Imaging

Barua, S., University of California, Santa Barbara

Research Overview

My primary research interest is in the development of synthetic nanoparticles for the treatment and diagnosis of a disease associated with complex cellular microenvironment such as cancer. My goal is to investigate how geometric modifications of a nanoparticle can significantly impact targeted drug delivery to cancer microenvironment, and non-invasive molecular scale imaging. In Professor Samir Mitragotri's laboratory at University of California, Santa Barbara, I worked on the design of nanoparticle geometry, particularly nanoparticle shape to enhance biological interactions of macromolecular drugs with target cells. My Ph.D. thesis research under the guidance of Professor Kaushal Rege at Arizona State University focused on designing cationic polymers for gene delivery to cancer cells in combination with chemotheraputic drug molecules. I will utilize my experience and knowledge gained from these research projects to propel my future research endeavors in the fields of cancer therapy and diagnostics.

Education:

Ph.D., Chemical Engineering, Arizona State University, Tempe, AZ, 2011

M.S., Chemical Engineering, Montana State University, Bozeman, MT, 2007

B.Sc., Chemical Engineering, Bangladesh University of Engineering & Technology, Dhaka, Bangladesh, 2003

Experience:

Post-doctoral Researcher, Department of Chemical Engineering, University of California Santa Barbara (UCSB), Santa Barbara, CA, 06/2011-present

Teaching (Lecturer), Department of Chemical Engineering, Bangladesh University of Engineering & Technology, Dhaka, Bangladesh, 02/2003- 07/2004

Selected Publications:

Barua, S.; Yoo, J.W.; Kolhar, P.; Wakankar, A.; Gokarn, Y.; Mitragotri, S., *Particle Shape Enhances Specificity of Antibody-displaying Nanoparticles*. PNAS, 2013, 110 (9), 3270-3275

Barua, S.; Rege, K., *The influence of Mediators of Intracellular Trafficking on Transgene Expression Efficacy of Polymer-plasmid DNA Complexes*. Biomaterials, 2010, 31 (22), 5894-5902

Barua, S.; Joshi, A.; Banerjee, A.; Matthews, D.; Sharfstein, S. T.; Cramer, S. M.; Kane, R. S.; Rege, K., *Parallel Synthesis and Screening of Polymers for Nonviral Gene Delivery*. Molecular Pharmaceutics, 2009, 6 (1), 86-97

Selected Awards/Honors:

Daryl and Marguerite Errett Discovery Award in Biomedical Research, UCSB, 2012

Outstanding Research Assistant Award, Arizona State University, 2009

Participant, 57th Lindau meeting with 18 honorable Nobel Laureates, Germany, 2007

Inland Northwest Research Alliance fellowship, Montana State University, 2005-2007

(4dx) Biomodulatory Materials

Ulery, B. D., University of Chicago

As the field of biomedical engineering moves forward and tackles grand challenges like targeted intracellular delivery and organ regeneration, there exists a need for novel, multi-dimensional strategies to be developed. Current biomedical engineering solutions comprised of one biomaterial with basic structure performing a single task are suboptimal for carrying out high-order biologically-relevant functions. Instead new systems must be composed of multiple biomaterials with complex molecular-, nano-, and micro-architecture capable of performing a variety of tasks with desirable spatiotemporal control. This increase in complexity means that next generation technologies require research-based approaches from a variety of scientific fields integrated together in order to be successful.

My research to date has focused on utilizing biomaterials to induce desired biological responses through drug delivery, tissue engineering, and immunoengineering strategies. I completed my PhD in 2010 in Chemical Engineering with a minor in Immunobiology at Iowa State University under the guidance of Professor Balaji Narasimhan. In collaboration with Professor Michael Wannemuehler in the department of Veterinary Microbiology and Preventative Medicine, my research focused on the development of polyanhydride nanoparticle-based delivery vehicles for immunomodulatory applications. In specific, I investigated how polymer chemistry affected dendritic cell activation *in vitro* as well as enhanced and sustained vaccine-based immune responses *in vivo*. From 2010 to 2012, I worked with Professor Cato Laurencin as a postdoctoral fellow in the Institute for Regenerative Engineering at the University of Connecticut Health Center. My postdoctoral research focused on designing novel biomaterials-based strategies capable of providing the necessary chemical, biological, and mechanical cues necessary to regenerate musculoskeletal tissues. Specifically, I showed that ions and simple molecules can differentiate stem cells down desired lineages while simultaneously inducing these cells to produce their own growth and differentiation factors. Since 2012, I have been a postdoctoral scholar under the advisement of Professor Matthew Tirrell in the Institute for Molecular Engineering at the University of Chicago. My current research focuses on creating and optimizing self-assembled peptide amphiphile micelles for immunological applications. Specifically, I am working on projects in a variety of areas including prophylactic vaccines, cancer immunotherapies, and immunomodulatory tissue engineering. An overarching theme of my research has been and will continue to be the application of basic science and engineering principles to the development of new biomedical technologies which has so far yielded 19 peer-reviewed publications, manuscript submissions, book chapters, and technology disclosures.

During this poster session I will highlight examples from my graduate work and postdoctoral research that demonstrate the capacity of biomaterials to initiate and sustain immune responses and direct tissue development. Building on my previous work, I will outline my plans to build a distinguished, independent research program at the interface of chemical engineering, biomedical engineering, materials science, and immunology. I will discuss how immunoengineering and tissue engineering provide opportunities to apply novel biomaterials strategies. The overall goal of my independent research will be to exploit biomaterials chemistry to generate desirable bioengineering outcomes.

Related References:

1) **B.D. Ulery**, H. M. Kan, B.C. Williams, B. Narasimhan, L.S. Nair, C.T. Laurencin. "Facile fabrication of polymer/anesthetic nanoparticles with tunable release kinetics." *Biomacromolecules*. (*Submitted*)

2) E.K. Cushnie*, **B.D. Ulery***, S.J. Nelson, M. Deng, S. Sethuraman, K.W.H. Lo, Y.M. Khan, and C.T. Laurencin. "Inducerons: Simple signaling molecules for inductive tissue engineering." *Nature Communications*. (*Submitted*)

3) Q. Lv, M. Deng, **B.D. Ulery**, L.S. Nair, and C.T. Laurencin. "Nano-ceramic effects on stem cells in bioreactor bone engineering systems." *Clinical Orthopaedics and Related Research*. (***In Press***)

4) K.W.H. Lo*, **B.D. Ulery***, K.M. Ashe, H.M. Kan, and C.T. Laurencin. "Evaluating the feasibility of the small molecule phenamil as a novel osteoinductive factor for bone regenerative engineering." *Journal of Tissue Engineering and Regenerative Medicine*. (***In Press***)

5) P. L. Foley, **B.D. Ulery**, H. M. Kan, M. V. Burks, Z. Cui, Q. Wu, L.S. Nair, and C.T. Laurencin. "A chitosan thermogel for delivery of ropivacaine in regional musculoskeletal anaesthesia." *Biomaterials*. 34(10): 2539-2546 (**Mar 2013**)

6) K.W.H. Lo*, **B.D. Ulery***, K.M. Ashe, and C.T. Laurencin. "Studies of bone morphogenetic protein based surgical repair." *Advanced Drug Delivery Reviews*. 64(12): 1277-1291. (**Sept 2012**)

7) M.S. Peach, S.G. Kumbar, R. James, U.S. Toti, D. Balasubramaniam, M. Deng, **B.D. Ulery**, A.D. Mazzocca, M.B. McCarthy, N.L. Morozowich, H.R. Allcock, and C.T. Laurencin. "Design and optimization of polyphosphazene functionalized fiber matrices for soft tissue regeneration." *Journal of Biomedical Nanotechnology*. 8(1): 107-124. (**Feb 2012**)

8) **B.D. Ulery***, L.K. Petersen*, Y. Phanse, C.S. Kong, S.R. Broderick, D. Kumar, A. E. Ramer-Tait, B. Carillo-Conde, K. Rajan, B.H. Bellaire, M.J. Wannemuehler, D.W. Metzger, and B. Narasimhan. "Rational design of pathogen-mimicking amphiphilic materials as nanoadjuvants." *Scientific Reports*. 1(198) (**Dec 2011**)

9) L.K. Petersen, A.E. Ramer-Tait, S.R. Broderick, C.S. Kong, **B.D. Ulery**, K. Rajan, M.J. Wannemuehler, and B. Narasimhan. "Amphiphilic polyanhydride nanoparticle adjuvants activate innate immune responses in a pathogen-mimicking manner." *Biomaterials*. 32(28): 6815 - 6822. (**Oct 2011**)

10) **B.D. Ulery**, L.S. Nair, and C.T. Laurencin. "Biomedical applications of biodegradable polymers." *Journal of Polymer Physics Part B: Polymer Physics*. 49(12):832 - 864. (**Jun 2011**)

11) **B.D. Ulery*,** D. Kumar*, A.E. Ramer-Tait, D.W. Metzger, M.J. Wannemuehler, and B. Narasimhan. "Design of a protective single-dose intranasal nanoparticle-based vaccine platform for respiratory infectious diseases." *PLoS One*. 6(3): e17642. (**Mar 2011**)

12) **B.D. Ulery**, Y. Phanse, A. Sinha, M.J. Wannemuehler, B. Narasimhan, and B.H. Bellaire. "Polymer chemistry influences monocytic uptake of polyanhydride nanospheres." *Pharmaceutical Research*. 26(3):

683 - 690. (**Mar 2009**)

13) **B.D. Ulery**, K. Pustulka, Y. Phanse, B. Bellaire, and B. Narasimhan. "Amphiphilic polyanhydride chemistry affects monocytic association of nanospheres." *Proceedings of the 37th Annual Biochemical Engineering Symposium*. 37: 52 - 59. (**Jan 2009**)

* *These Authors Contributed Equally*

(4r) Computational Studies of Polymeric Systems

Carrillo, J. M., Oak Ridge National Laboratory

Computer experiments have become routine because of the availability of faster and larger computer clusters. Together with the development of parallel algorithms and the availability of accelerators such as GPU's, we are poised to handle larger simulations that approach sizes comparable to real-life experiments or perform orders of magnitude more smaller experiments. To illustrate I am presenting the results of my latest works. (1) I performed coarse-grained molecular dynamics of the thermal annealing of the active layer of an organic photovoltaic device. The simulations of the OPV were performed on the Titan supercomputer for up to 400 ns and the simulation sizes are comparable to the thickness of the active layer of a bulk heterojunction device (~100 nm). (2) I performed coarse-grained molecular dynamics simulations of polymer networks with varying stiffness of the networks' polymer chain to explain the highly nonlinear stress–strain behavior manifested in these networks that stiffen with increasing deformation. Although the sizes of the second set of simulations are small, we performed many samples that covered the whole range of polymer chain flexibility from Gaussian coils to rigid rods. In both cases, we have compared the results with theories and experiments in order to provide insights and get a better understanding of the physics of these processes. In the future, I hope to perform simulations that would tackle areas in polymer physics that are not yet fully understood such as charged polymers, liquid crystalline polymers and polymer crystallization. In these future works, I will actively search for areas of collaboration with theorists and experimentalists in the hope of contributing in presenting a more complete picture of what is happening in these polymeric systems.

(4da) Development of New Generation of Separation Technologies Through Energy-Efficient Processes and Advanced Materials

Rezaei, F., Georgia Institute of Technology

My research focuses on various aspects of chemistry and engineering of chemical adsorption, separation and reaction. During my PhD at Monash University and Lulea University of Technology with Prof. Paul Webley, I introduced a new methodology for optimizing the structure of adsorbents being used in cyclic gas separation units, in particular in post-combustion CO_2 capture process, whereby I adapted recent developments in adsorptive separation science to advance the development of new structured adsorbents which offer attractive characteristics compared to conventional packed bed systems. This work was successfully resulted in several collaborations with several companies to make/develop structured adsorbents and more recently was selected as the recipient of the International Adsorption Society Award for Excellence in the PhD Dissertation.

Currently, I am a post-doctoral fellow at Georgia Tech, where I am expanding my expertise in gas separation by designing next generation materials and processes for gas management in cyclic units. Based on my background and my expertise, I have been involved in both theoretical and experimental sides of this project. This position has allowed me to work as a cross-functional team member between theoretical and experimental research groups where I develop and implement computational models for rapid temperature swing adsorption (RTSA) process and in the meantime, I validate the simulation results by obtaining experimental data via synthesis, characterization and evaluation of adsorbent materials in the laboratory.

Integrating my background of material synthesis, process design and simulation, and modeling and optimization of chemical processes, my future research will address the fundamental issues concerning clean energy and global warming through development of novel materials and cost-effective technologies for gas and liquid separation applications. Given the context of the global energy security and greenhouse needs, I will strongly focus on performing cutting-edge research in the field of separation science by designing efficient and low-energy systems such as hybrid adsorbents/membranes or catalyst/membranes materials/processes with the emphasis on energy conservation and environmental footprint reduction.

Selected Publications (22 total, 11 first author, 200 total citations, h-index: 8):

1. F. Rezaei, R.P. Lively, Y. Labreche, G.Chen, Y.Fan, W.J. Koros, C.W. Jones, "Aminosilane-Grafted Polymer/Silica Hollow Fiber Adsorbents for CO_2 Capture from Flue Gas." *ACS Appl. Mater. Interfaces*, 5 (**2013**), p. 3921.
2. F. Rezaei, S. Subramanian, R.P. Lively, J. Kalyanaraman, Y. Kawajiri, M.J. Realff, "Modeling of

Rapid tTemperature Swing Adsorption Using Polymer-Supported Amine Hollow Fibers." *Chemical Engineering Science*, (**2013**), submitted.

3. Y. Labreche, R.P. Lively, F. Rezaei, G. Chen, C.W. Jones, W.J. Koros, "Post-Ppinning Infusion of Poly(ethyleneimine) into Polymer/Silica Hollow Fiber Sorbents for Carbon Dioxide Capture", *Chemical Engineering Journal*, 221 (**2013**), p. 166.
4. F. Rezaei, P. Webley, "Optimal Design of Engineered Gas Adsorbents: Pore-Scale Level."*Chemical Engineering Science*, 69 (**2012**), 270-278.
5. F. Rezaei, M. Grahn, "Thermal Management of Structured Adsorbents in CO_2 Capture Process." *Industrial and Engineering Chemistry Research*, 51 (**2012**), 4025-4034.
6. F. Rezaei, et al. "The effect of Wall Porosity and Zeolite Film Thickness on the Dynamic Behavior of Adsorbents in the Form of Coated Monoliths." *Separation and Purification Technology*, 81 (**2011**), 191-199.

(4aw) Novel Electrochemical Sensor Array for Biofluids

Halpern, J. M., Case Western Reserve University

Significance and Research:

Point-of-care diagnostics is one of the hottest growing areas in biotechnology and medicine. As of today, routine medical diagnostic techniques are still limited in many ways; they are time-consuming, costly, sometimes inaccurate, or even invasive. New tools to quickly and accurately diagnose a wide variety of diseases are currently attracting research interest and funding. Novel electrochemical biosensors detect specific enzyme and protein changes that can be indicative of disease; however, these sensors are limited to testing only one molecular interaction and are unstable over time. In addition, enzymatic sensors ignore small chemical markers within biofluids that are capable of aiding in disease analysis, such as: aldehydes, ketones, and other volatile organic compounds for cancer; hormones for stress-related disorders; and homovanillic acid, vanillylmandelic acid, and other catecholamines for neurological disorders. New sensors that can accurately and quickly measure multiple small chemical markers within biofluids are necessary.

My research will focus on the development of a novel electrochemical sensor for quick, accurate, and point-of-care disease discrimination via biofluid (*e.g. urine)* analysis of these small chemical markers. Specifically, non-enzyme electrochemical contacts that use molecular interactions (such as the affinity binding between cyclodextrin and small hydrophobic chemicals) will be used to accurately distinguish a number of different chemical markers. Initially, a single developed sensor will be tested to determine key affinity interactions for enhanced sensing capabilities towards these small chemical markers. Eventually, the new electrochemical sensor will be integrated into an array, with multiple different molecular interactions to probe, and to identify, a range of these small chemical markers accurately. Through statistical analysis (*e.g.* principle component analysis) we could then successfully determine the ?chemical fingerprint? of specific diseases. The electrochemical sensor array will be capable of a long shelf-life and use life-time, and the array will provide data to the patient within minutes instead of days or weeks. After the electrode array is developed, I will seek clinical partners to test various diseases, such as various types of cancer at different stages, neurological diseases (*e.g.* Alzheimer's, Parkinson's), and stress-related disorders.

Grants and Honors:

Fulbright Core Postdoctoral Scholar, January 2013

Lady Davis Trust Fellow, January 2013

NIH Musculoskeletal Postdoctoral Trainee, NRSA, T32, April 2011

Advanced Platform Technologies (APT), Veterans Affairs Innovative Initiative Program Role: Co-PI, August 2011

Mentors/Advisors:

Graduate Advisor: Heidi B. Martin, Department of Chemical Engineering, Case Western Reserve University

Post-Doctorate Advisor: Horst A. von Recum, Department of Biomedical Engineering, Case Western Reserve University

Fulbright Post-Doctorate Sponsor: Hossam Haick, Department of Chemical Engineering, Technion-Israel Institute of Technology

Teaching and Mentoring Experience:

I have extensive mentoring experience (over 8 students in the past 6 yrs). In addition, I have designed and co-instructed a course titled, ?Research and Innovation?, and I have been invited to give guest lectures at several universities.

My general approach to teaching is to give students the opportunity for interaction and discussion through project-based coursework. Undergraduate students typically benefit from applying the material through discussion and interaction. In large classrooms, I intend to break up the students into smaller groups (4-5 students), in order to provide class time for this approach.

Other Presentations at AIChE 2013:

JM Halpern and H Haick. Molecular-Terminated, Oxide-Free Silicon Nanowire Field Effect Transistors: Effect of the C-C bond Nearest to the Surface?

JM Halpern, M Keech, A Weinstock, R Mathers, and HA von Recum. Affinity-Based Thermoset Biodegradable Polymer?

JM Halpern, CA Gormley, M Keech, and HA von Recum. Sterilization of Anti-Microbial Surface Coatings?

Full CV: http://bit.ly/HalpernCV

(4y) Polymer Membranes for Energy Storage and Delivery

Davis, E. M., National Institute of Standards and Technology

Developing a fundamental understanding of ion and water transport mechanisms through ion exchange membranes (IEMs) is vitally important for the design and development of next-generation membrane materials for energy storage and delivery applications. However, there are very few *in situ* structural and chemical characterization methods of IEMs during device operation, when molecular-level interactions and/or structural changes may occur, ultimately leading to fundamental changes in the transport mechanisms of ions and water through the IEM. Therefore, a key part of my future research as a faculty would be developing and employing powerful *in situ* characterization techniques to help investigate the connection between material chemistry and structure to crucial performance properties, in an effort to elucidate the fundamental mechanisms by which transport occurs in these IEMs. The information obtained from these *in situ* investigations would then be used to design and synthesize next-generation materials, which provide more facile ion transport than their predecessors. I believe that the energy community will be well-served by the development of measurement tools and systematic studies aimed at gaining a fundamental understanding of the materials properties and transport phenomena which give rise to performance enhancements.

Molecular-level structural and chemical probing techniques, including time-resolved neutron scattering and infrared spectroscopy, will play a critical role in developing an understanding of changes in material chemistry and structure to changes in device performance. Without such techniques, the community will have no direct measurement tools by which to perform systematic studies to understand structure-property-performance relationships of these systems. During my PhD work, I gained extensive experience on the use of infrared spectroscopy techniques to accurately capture and measure water transport in glassy polymers with molecular-level resolution. This expertise in transport phenomena and infrared spectroscopy will greatly aid and facilitate my future research endeavors, where I propose to develop/employ such power spectroscopic techniques to measure the transport of ions and water in these membranes, providing polymer chemists a useful set of design parameters for the synthesis of new ion exchange membranes for energy applications.

Furthermore, my work as a post doctoral fellow will focus on the design and development of the aforementioned *in situ* structural and spectroscopic techniques in order to investigate the effect of structure and molecular interactions on water and ions transport in IEMs for redox flow batteries. My work will provide me with access to sophisticated neutron scattering techniques, which will allow me to probe membrane structure and polymer dynamics during device operation. Once developed, these *in situ* techniques can be utilized in my future research, providing me with access to powerful and unique experimental techniques that may otherwise be unavailable. I believe that my PhD work on water transport in glassy polymers, as well as my post doctoral work on developing *in situ* structural and chemical characterization techniques for IEMs during device operation, will provide my future work with a solid foundation in an area of research that I believe will be receiving an increasing amount of funding in the future.

(4ea) Engineering the Next Generation of Therapeutic Proteins

Van Deventer, J. A., Massachusetts Institute of Technology

Recombinant protein-based reagents, diagnostics, and therapeutics are beginning to come of age. However, the development of proteins tailored for specific applications, especially therapeutic, remains an ongoing challenge. Cell surface display technologies provide an especially powerful set of tools for protein engineers to utilize in the development of new therapeutic molecules. The work presented here introduces and exploits improvements in cell surface display technologies in order to develop proteins with therapeutic properties and novel functionalities.

In many cancers that present as solid tumors, an expanding body of work suggests that the tumor stroma plays an integral role in the initiation and development of epithelial cancers. Within the stroma, cancer-associated fibroblasts (CAFs) are known to contribute to immune suppression, metabolism, invasion, and metastasis. Despite the demonstrated importance of CAFs in the tumor microenvironment, targeting CAFs within the tumor microenvironment remains a challenge. To facilitate interference with these cells, we have developed a suite of antibodies targeting multiple epitopes of fibroblast activation protein-alpha (FAP), a membrane-bound protease selectively overexpressed on CAFs in numerous epithelial cancers. Screening a billion-member synthetic antibody library displayed on the surface of yeast yielded more than 50 unique antibodies recognizing FAP, of which 28 were crossreactive for both the human and murine forms of the antigen. We developed secondary screens to verify that candidate binding proteins could recognize the membrane-bound form of the antigen and identify candidates recognizing distinct epitopes of the antigen. These secondary screens utilized soluble, yeast-produced antibody-like structures in the efficient identification of a handful of candidate proteins with promising properties. These candidates were subjected to affinity maturation, converted to full-length IgGs, and produced with reasonable yields in mammalian cells. The full-length IgGs exhibit excellent binding properties, recognize multiple distinct epitopes of FAP, and modulate the surface levels of FAP on activated fibroblasts in culture. Further in vitro characterization and in vivo studies are underway. These antibodies are expected to provide biological insights into the roles of CAFs and therapeutic approaches to cancer treatment. Furthermore, the general antibody isolation and characterization approaches used in this work are accessible to many single-investigator laboratories and should be applicable to the development of antibodies against numerous therapeutic targets.

Protein engineering using proteins containing the conventional 20 amino acids has yielded a startling array of proteins with druglike properties. However, the chemical diversity present in the conventional set of amino acid side chains is quite limited and has motivated efforts to introduce new functionalities into proteins using noncanonical amino acids (ncAAs) and other approaches. Although therapeutic applications of proteins containing functionalities beyond those encoded in the common 20 amino acids appear promising, proteins with these added functionalities will likely require further improvement using library-based screening methods. To address this issue, we have developed an E. coli cell surface display platform for evolving antibody fragments containing analogs of the amino acid methionine. In our initial efforts, screening proteins containing various methionine analogs, including "clickable" analogs, led to the emergence of highly functional variants that exhibit unique properties modulated in part by ncAA side chains. These properties include azide-containing fragments with binding kinetics improved 2-fold relative to their methionine-containing counterparts and fragments exhibiting chemical reactivity in strain-promoted azide-alkyne cycloadditions. Characterization of the azide-containing fragments indicates that these fragments are quite functional, with both soluble and cell surface-displayed fragments retaining their ability to bind antigen after chemical modification. The development of a ncAA-compatible cell surface display platform and the isolation of clickable, functional fragments using the platform raises intriguing possibilities for future therapeutic screening efforts, including the modulation of molecular recognition events with ncAA side chains and screening libraries of chemically modified proteins for functional variants.

(4bq) Dynamics of Cells, Multicellular Organisms, and Flowing Sand Grains

Khan, Z. S., Texas Tech University

What do single cells, multicellular organisms, and flowing sand grains have in common? They are all highly dynamic materials that can display a variety of complex behaviors including collective motion with patterns of extended spatiotemporal coherence, anomalous diffusion, active migration, and morphogenesis. Developing a deeper understanding of dynamics in these complex systems has the potential to lead to considerable engineering breakthroughs. For example, investigating tumor cell transport in the microcirculation can lead to improvements in disease diagnosis and prognosis; quantifying the stereotypical motions of important multicellular biological organisms can provide information on how genes and drugs regulate locomotion; and research into bulk grain transport under flow can have considerable implications for the pharmaceutical, agricultural, and energy production industries. My contributions to date in these areas are described below. In the future, I plan to integrate my expertise on biological soft matter, microfluidics, and novel imaging techniques to advance both fundamental science and engineering applications in these areas.

(i) Cancer cell mechanics and metastasis

Alterations in the mechanical properties of cells have been implicated in several pathologies and diseases, such as cancer metastasis, where studies have indicated that cancer cells have distinct mechanical properties compared to healthy cells. I am currently investigating the potential of cell mechanics as a biophysical marker for cancer diagnosis using a microfluidic device that contains a narrow micrometer-scale pore. Fluid flow is used to drive cells into the pore mimicking the flow-induced passage of circulating tumor cells through the vasculature at a hemodynamic range of shear stresses. By integrating high speed imaging, the device allows for the simultaneous characterization of five different parameters including the blockage pressure, cell velocity, cell size, elongation and the entry time into the pore.

I have tested a variety of in vitro cell lines, including brain and prostate cancer cell lines, and have found that the entry time and blockage pressure are the most sensitive measurements capable of differentiating between cell lines with differing invasiveness[1]. By characterizing this device with simple model systems including viscous drops and soft elastic particles, I have found that the blockage pressure and entry times show no apparent dependence on elastic modulus or drop interfacial tension, but depend significantly on drop internal viscosity. Thus, our findings suggest that in the hemodynamic range of shear stresses the metastatic potential of circulating cells can be characterized by an internal viscosity[2]. This result has important implications for not only high throughput mechanical phenotying of tumor cells, but also on how circulating tumor cells are transported in microcapillaries.

I intend to continue investigating cancer cell mechanics using novel microfluidic devices. My future focus will be directed towards determining how the nucleus, nuclear structural proteins, and chemotherapy drugs affect cancer cell invasiveness and deformability. My proposed research will advance scientific understanding of cancer cell migration, metastasis, and drug efficacy, and will lead to new engineering tools for clinical diagnostics of cancer.

(ii) Behavior and locomotion of C. elegans

I am currently involved in a multi-group collaboration investigating the relationships between genetics, environmental cues and locomotion in the soil-dwelling nematode *C. elegans. C. elegans* is an important model animal in biology and medicine due to its simple nervous system, short life span and availability of tools to manipulate its genome. Not surprisingly, these attributes have made *C. elegans* the center of three Nobel prize winning investigations.

The main phenotype of this animal is locomotion, that is often used in genetic and drug screens. While the crawling motion of this simple animal has been widely studied, the details of the animal's swimming motion has not been quantitatively explored in the context of increasing the mechanical load or stimulation. Currently I am investigating the effects of varying the mechanical load of the environment on the animal's locomotion using a principal component analysis of its swimming body shape[3]. I intend to utilize the techniques that we are developing[4] to further investigate the effects of chemical and physical stimulation on this model creature. Thus, my efforts will lead to fundamental understanding of how the mechanical environment and physiochemical stimuli affect the locomotion of this important model organism. It may also provide the biomedical community new tools to better quantify the effects of gene expression and chemicals on its locomotion.

(iii) Dynamics of granular flows

Granular systems display a variety of complex behaviors including spontaneous pattern formation and anomalous diffusion. Granular materials are also the second most handled material in industry, the first being water. Indeed, granular materials are notoriously difficult to mix efficiently, and it has been estimated that nearly 10 percent of the world's energy each year is spent processing granular materials. Much of my previous research focused on a system frequently used to investigate granular flows: the rotating drum apparatus.

An interesting phenomenon is often observed when one attempts to mix grains of different sizes, densities, or surface roughness in a rotating drum; instead of mixing, the grains separate partially or completely. After hundreds of drum rotations an initially mixed binary distribution of different-sized grains sorts itself into almost periodic bands along the axis of the drum. These bands are threaded by a radial core of the smaller grains that develops prior to axial band formation[5]. Using surface imaging techniques I have discovered that the mixing of grains in a rotating drum can be subdiffusive[6]. I have also investigated single-particle dynamics of tracer particles imaged using bulk synchrotron radioscopy imaging, and have found that while the standard measure of diffusive behavior, the slope of the mean-squared displacement, is close to that expected from diffusive transport, more detailed analyses indicate anomalous transport[7]. My future research interests include investigating the connection between the occurrence of pattern formation and the presence of dynamical heterogeneities that appear under flow, and investigating how the presence of an attractive force introduced by wetting the grains[8] alters the dynamical features, when compared with the dry non-cohesive case. The understanding gained from these studies may provide insight into industrial grain-mixing procedures.

1. Z. S. Khan & S. A. Vanapalli, *Biomicrofluidics* **7**, 011806 (2013).

2. Z. S. Khan, N. Kamyabi & S. A. Vanapalli, manuscript in preparation.

3. Z. S. Khan, F. Van Bussel, M. Rahman, J. Blawzdziewicz & S. A. Vanapalli, manuscript in preparation.

4. V. Padmanabhan, Z. S. Khan, D. E. Solomon, A. Armstrong, K. P. Rumbaugh, S. A. Vanapalli & J. Blawzdziewicz, *PLoS ONE* **7**, e40121 (2012).

5. Z. S. Khan, W. A. Tokaruk and S. W. Morris, *Europhysics Letters* **66,** 212 (2004).

6. Z. S. Khan and S. W. Morris, *Physical Review Letters* **94**, 048002 (2005).

7. Z. S. Khan, F. Van Bussel, M. Schaber, R. Seemann, M. Scheel and M. DiMichiel, *New Journal of Physics* **13**, 105005 (2011).

8. Z. S. Khan, A. Steinberger, R. Seemann, and S. Herminghaus, *New Journal of Physics* **13**, 053041 (2011).

(4ax) Advanced Biomaterials for Next Generation Biomedical Devices and Biological Investigation

Heath, D., Singapore-MIT Alliance for Research and Technology

Biomaterials science is a multidisciplinary field with implications in global health, disease detection and prevention, drug delivery, regenerative medicine, and the exploration of basic biological science. Advances in this area require strength not only in materials development but also fundamental biology, disease pathology and treatment, cellular and molecular assays, and an intimate knowledge of how these two fields interface.

I feel that my graduate school experience and postdoctoral training have positioned me to make significant contributions to this field. I completed my Ph.D. research with Professor Stuart Cooper at The Ohio State University where we focused on developing new polymeric materials which specifically adhere endothelial progenitor cells, a type of adult stem cell which has been implicated in cardiovascular maintenance and repair and which holds great potential in regenerative medicine. This work culminated in the publication of six manuscripts to peer reviewed journals (two second author papers and four first author papers). Also, I wrote two book sections for the textbook *Biomaterials Science* 3rd Edition.

I continued my time in the Cooper Lab as a postdoctoral researcher for one year to work on improving the isolation and culture of rare adult stem cells from peripheral blood by separating the cells with magnetic and bioactive nanoparticles and culturing them on cell-specific electrospun scaffolds. My contributions to this research focused on synthesizing and characterizing bioactive polymers, fabricating three dimensional scaffolds via electrospinning, and assessing the response of primary human cells. These contributions resulted in two peer reviewed manuscripts (one first author and one second author).

Currently I am completing a three year postdoc with Linda Griffith and Paula Hammond of MIT and Mary Chan-Park of Nanyang Technological University through the Singapore-MIT Alliance for Research and Technology in the BioSystems and Micromechanics Interdisciplinary Research Group (SMART BioSyM). My current work focuses on the development of a novel microwell array for the high throughput *in situ* cloning of rare cells from patient samples, research which promises to provide unprecedented insight into the complex behavior of single cells from patient samples during *ex vivo* culture. As of May 2013 two first author papers from this research are near completion and two other manuscripts are under review which focus on additional work I have done in the Chan-Park lab.

In addition to developing as a researcher, I have also taken initiative to grow as an educator. During my time at OSU I served as a teaching assistant for one term of *Chemical Reaction Kinetics and Reactor Design* and two terms of *Introduction to High Polymer Engineering*. During the autumn 2008 term I received the Outstanding TA Award. In addition to my required TA activities I earned a *Graduate Interdisciplinary*

Specialization in College and University Teaching (requiring the completion of 20 credit hours of education-related classes) and participated in Ohio State's *Preparing Future Faculty* program where I learned about the life of faculty. Furthermore, I developed the *Chemical Engineering Mentored Teaching Experience*, a course where interested graduate students are exposed to educational theory and allowed to teach three courses in a chemical engineering classroom (the development of this course was published in *Chemical Engineering Education*). I subsequently participated in the *Chemical Engineering Mentored Teaching Experience.* I was also the independent instructor for the course *Advanced Separation Processes* where I organized and presented course material; developed homework sets, exams, and a distillation project which required the students to use *Aspen* software; oversaw the activities of five teaching assistants; and assigned final grades. I am currently in discussion with the School of Chemical and Biomedical Engineering at NTU to co-teach a course on polymer science during the 2013 school year.

I am excited to begin the next phase of my career as an assistant professor, and I look forward to mentoring students and building and sustaining a research program which focuses on developing the next generation of polymeric biomaterials. From a materials perspective my interests include substrate-independent and scalable biofunctionalization techniques which can be applied to a wide variety of biomaterials (an objective which builds on my existing strength in chemical surface modifications, plasma treatment techniques, and LbL thin film deposition, and micro/nano-fabrication technology). Also, my existing research has focused on isolation and expansion of rare adult stem cells. In addition to further exploring the utility of these cells in regenerative medicine applications, I am also interested in further understanding the heterogeneity of these cell populations which may provide valuable insight into how to best apply them clinically.

Due to the size limitations of this abstract I cannot go into detail about the specific research projects I have in mind; however, I hope to speak with you about them at the AIChE meeting. Though I currently live and work in Singapore, I will be in the United States for at least one month during February/March 2014 to facilitate interviewing for positions in the US.

(4d) Biomaterials and Nanostructures for Tissue Engineering

Almquist, B. D., Massachusetts Institute of Technology

Interfaces in biological systems are highly dynamic regions that act as concentrators of function and information. Understanding how to manipulate the structure and behavior of the biotic-abiotic interface is key to developing innovative devices and materials with enhanced function. My Ph.D. research focused on developing tunable interfaces that allow the seamless integration of artificial materials into cell membranes. By recreating the hydrophobic banding of transmembrane proteins on inorganic structures, I produced biocompatible membrane-penetrating interfaces while also yielding novel insights into the interfacial phenomena that govern how proteins and lipid membranes interact. These insights have subsequently been used to enable stable, nondestructive intracellular electrical access. During my postdoctoral training, I have been working to develop methods for programming materials with dynamic biological function, essentially defining the temporal interfaces between biomaterials and tissues. These materials are being used to understand how coordinated cytokine therapies affect the healing of chronic wounds, along with more general questions regarding tissue regeneration. Using this background, I plan to develop new therapeutic devices and materials to observe and influence cell and tissue behavior.

(4ef) Application and Mechanism Understanding Of NANO-Structured Catalysts In Biofuel Production

Wen, C., University of South Carolina

APPLICATION AND MECHANISM UNDERSTANDING OF NANO-STRUCTURED CATALYSTS IN BIOFUEL PRODUCTION

Cun Wen, University of South Carolina, Columbia, SC

Department of Energy has set the goal of biofuel production to 36 billion gallons by 2022, while the present production rated at 1.1 billion gallons at 2011. The capacity needs more than 3200% increase in total or 40% increase in annual for the next 10 years. One of the limiting factors in biofuel production is the low throughput of the batch reaction mode. Though it has been long known that heterogeneous processes can accelerate the biofuel production, the actual progress is hindered by poor understanding of reaction mechanisms and requires sophisticated material design and synthesis. The reaction mechanism research is intricate with numerous potential reaction pathways, similar to the possible routes to overcome Himalayas. Our strategy to tackle this issue is to combine the mechanism research with kinetics studies, which serves as loyal contour-mapper and tells us the most feasible way. This strategy has been proven to be effective.[1-10] Herein, I bring in nano-structured heterogeneous catalysts with our strength in advanced nanomaterial synthesis and fundamental reaction-mechanism studies to expedite the biofuel production.

To upgrade biomass to biofuels, the aromatic rings and the C-O bonds need to be broken down and C-C/C-H bonds need to be build up, on which stage heterogeneous catalysts can show off. Firstly, 2-methyfuran acid catalysis reaction is chosen as a model reaction to exemplify our idea of heterogenerizing biofuel production. The 2-methylfuran is a mode molecular representing biomass feed stocks, which usually need acid catalysts for deoxygenating because of oxygen as heteroatom and need chain growth for generating biofuel because of low carbon numbers (<C6). Usually concentrated sulfuric acid is used as acid catalyst to serve the purpose, but it is unfavorable for large scale industrial production. Our results show that the zeolite catalyst can replace the concentrated sulfuric acid with even better production selectivity in terms of cetane number. Furthermore, the cetane number of the product can be increased by decrease the acid strength of the zeolites. To understand the relationship between acid strength and product cetane number, the reaction mechanism on the zeolite catalyst is probed by *in situ*techniques, such as FTIR. The reaction mechanism study reveals that high acid strength increases the adsorption energy, and thus prolongs the resident time of the intermediates on the zeolite. Because longer resident time facilitates the C-C chain growth, the branched products with poor cetane number are favored on zeolites with high acid strength.

Secondly, the products from 2-methfuran acid catalysis reaction need further hydrogenation to upgrade to final biodiesel, and Co is reported to be active for the hydrogenation reaction, whose activity can be further promoted by Cu doping. Thus, bimetallic core-shell nanoparticle Cu@Co with alloy of CoCu on the shell (for high activity) and Cu as the core (for low catalyst cost) would be optimum for this reaction. Traditionally, the core-shell bimetallic nanoparticles are synthesized in two steps. In the first step, the core is formed with one metal, and then in the second step the core is coated with second metal as the shell. Thus, the two-step method not only is poor in economic, but also yields mono-metal instead of alloy on the surface, while alloy is needed for high activity in this hydrogenation reaction. Herein, I demonstrated a one-pot methodology for mono-dispersed Cu@Co Core-shell nanoparticle synthesis with alloy of CoCu on the shell as catalyst for hydrogenation reaction. Furthermore, I can not only synthesis the nanoparticle in one-step, but also keep a tight control over the size distribution (relative standard deviation 7.9%).[11]

Thirdly, biomass upgrading to biofuel is not limited to the acid and hydrogenation reaction, and other process can also serve the purpose, such as partial oxidation. The partial oxidation has been proved to be an effective way to produce biofuels from biomass aligned with Fischer-Tropsch synthesis. However, the long-standing problem in partial oxidation is the hot spot formation in the catalyst bed, which can lead to the catalyst deactivation or even explosion. It has been known that the hot spot is formed by the side reaction that is the total oxidation taking place beside the partial oxidation. The key factors guiding the reaction path through total or partial oxidation is unknown, which hinders the catalyst development. To address this issue, I looked into the mechanism of partial oxidation production of syngas ($CO+H_2$). Our research has revealed that the lattice oxygen reducibility of the catalyst is the key factors governing the reaction mechanism. With high lattice oxygen reducibility, total oxidation rather than the partial oxidation is favorite. By tuning the lattice oxygen reducibility of $Er_2O_3$with Mn doping, I successfully shift the reaction mechanism from total oxidation to partial oxidation, and eliminated the hot-spot formation.

Read Me More:

1. Wen C, Liu Y, Guo Y, Wang YQ, Lu GZ. Strategy to eliminate catalyst hot-spots in the partial oxidation of methane: enhancing its activity for direct hydrogen production by reducing the reactivity of lattice oxygen. *Chem Commun.* 2010;46(6):880-882.

2. Wen C, Zhu, Y.; Ye, Y.; Zhang, S.; Cheng, F.; Liu, Y.; Wang, P.; Tao, F., Water–Gas Shift Reaction on Metal Nanoclusters Encapsulated in Mesoporous Ceria Studied with Ambient-Pressure X-ray Photoelectron Spectroscopy. **ACS Nano**. 2012, 6, 9305-9313.

3. Wen C, Liu Y, Guo Y, Wang Y, Lu G. Synthesis of the rare earth compound nanosheets induced by lamellar liquid crystal. *Solid State Sciences.* 2009;11(11):1985-1991.

4. Wen C, Liu Y, Tao F. Integration of surface science, nanoscience, and catalysis. *Pure Appl Chem.* Jan 2011;83(1):243-252.

5. Liu Y, Wen C, Guo Y, Lu GZ, Wang YQ. Modulated CO Oxidation Activity of M-Doped Ceria (M = Cu, Ti, Zr, and Tb): Role of the Pauling Electronegativity of M. *J Phys Chem C.* 2010;114(21):9889-9897.

6. Liu Y, Wen C, Guo Y, Lu GZ, Wang YQ. Effects of surface area and oxygen vacancies on ceria in CO oxidation: Differences and relationships. *J Mol Catal a-Chem.* Feb 2010;316(1-2):59-64.

7. Liu Y, Wen C, Guo Y, et al. Mechanism of CO Disproportionation on Reduced Ceria. *ChemCatChem.* 2010;2(3):336-341.

8. Media highlights: Pichon A. Catalysis: Lower reactivity means more hydrogen. 2010; http://www.nature.com/nchina/2010/100120/full/nchina.2010.7.html.

9. Media highlights: Saxton C. Enhancing catalytic activity. 2010; http://www.rsc.org/Publishing/ChemTech/Volume/2010/02/enhancing_catalytic.asp.

10. Media highlights: Chinese New Year. 2010; http://www.rsc.org/Publishing/Journals/cc/News/2010/Chinese_New_Year.asp.

11. Wen C, Hattrick-Simpers J, Lauterbach J, Inventors. One-step synthesis of monodisperse transition metal core-shell nanoparticles with solid solution shells, US provisional patent: 61/686,288, 2012.

(4cr) From Carbon Supercapacitors for Energy Storage to Membranes for Water Desalination: Improving Materials Design With Advanced Computational Methods

Palmer, J. C., Princeton University

The electronic and surface properties of nanostructured materials can be engineered to influence the elementary sub-atomic and molecular-level processes that govern energy conversion, fluid separations, catalysis and biological function. As a result, they now find widespread use in technologies for energy storage and production from renewable sources, CO_2capture and sequestration, water purification, and drug delivery and formulation. To realize the full potential of nanomaterials, however, rational design principles are needed to guide their synthesis and application.

My research focuses on applying physics-based computational techniques to develop such design principles, helping to guide experimental efforts to produce next-generation technologies for applications ranging from renewable energy to biological therapeutic formulation. During my doctoral studies with Professor Keith E. Gubbins at North Carolina State University, I used advanced molecular simulation methods to model disordered nanoporous carbons (DNCs), which are structurally amorphous materials used in applications including gas and liquid separations, electrochemical energy storage and catalysis. Using such methods, I developed the most realistic structural models of DNCs to date, which were able, for the first time, to quantitatively predict attributes of real DNCs, including X-ray diffraction measurements and adsorptive properties. I also used the models to investigate the influence of synthesis conditions on the nanostructure of DNCs, providing insights that have assisted experimentalists in developing new materials with superior surface properties for electrochemical applications. As a postdoctoral researcher with Professor Pablo G. Debenedetti at Princeton University, I developed a new computational method for simulating physisorption processes on compliant materials (*e.g.*, polymers, proteins, and metal-organic hybrids) that undergo significant structural transformations during fluid uptake. I am currently using this method to simulate water sorption on protein systems in order to explore how biological matrices are affected by dehydration and subsequent rehydration. Understanding this process is crucial to formulating stable biological therapeutics, which are often stored as freeze-dried solids to prevent the aggregation and degradation processes that can occur in an aqueous solution. In the next stage of this project, I will use my new technique to understand mechanisms that allow excipients, such as polymers and carbohydrates, to stabilize proteins against dehydration, providing insights that will assist in the development of rational formulation strategies for therapeutics.

My future research will build upon the experience I have gained during my doctoral and postdoctoral studies in applying state-of-the-art computational techniques to understand the structural and functional properties of materials. I will use my expertise in such areas, along with the training I have received in authoring successful grant proposals and mentoring junior students, to establish an ambitious computational research program aimed at addressing emerging issues related to energy production/storage and sustainability through improved materials design. My initial research projects will focus on (1) designing disordered nanoporous carbon electrodes for use in supercapacitor devices, (2) optimizing amine-functionalized materials for carbon capture, and (3) engineering membrane technologies for water desalination and energy production. Computational techniques will be used to improve understanding of the structure-function relationship of the materials used in each application and screen candidate materials to identify those with superior performance attributes.

(4ek) Directed Assembly of Colloidal Particles

Wirth, C. L., Katholieke Universiteit Leuven

The directed assembly of colloidal particles is important to not only the fabrication of advanced materials, such as colloidal crystals, but also to microelectronics and next generation electrochemical systems, particularly when the particle is a carbon nanotube (CNT) or a sheet of graphene oxide (GO). Both my PhD and postdoctoral research has focused on the rich science underlying the assembly and structure of colloidal particle ensembles. I will present a series of vignettes summarizing my recent work in this area.

My PhD work focused on the directed assembly of particles proximate to an electrode stimulated by an electric field. The process is reversible and depends on the strength and frequency of the applied electric field. Surprisingly, the structure of the assembly also depends on the dispersing electrolyte and, until recently, the origin of this electrolyte dependence was a mystery. Negatively charged polystyrene spheres 5.7 μm in diameter dispersed in 0.15 mM KCl or $NaHCO_3$ assemble into a tightly packed hexagonal array upon application of a ~1 kV/m 100 Hz electric field, but an ensemble dispersed in 0.15 mM NH_4OH or KOH assembles into an array with a larger inter-particle spacing despite all other experimental parameters being identical to the previous case. I discovered that electrolyte dependence arises in both the polarization of the particle's diffuse layer and the conductivity of the dispersing electrolyte. My thesis included the first *a priori* model for electrolyte dependent particle motion normal to a polarized electrode.

My current work is focused on the assembly and structure of colloidal particles pinned at a fluid/fluid interface. Although it has been known for over a century that solid particles are effective at stabilizing a fluid/fluid interface, there remain several unanswered questions that are crucial to fully understanding this phenomenon. Two important research questions concern the correct expression for the interaction potential of particle pairs and how to predict surface pressure from this pair potential. In addition, I am interested in designing processes that direct the assembly of colloidal particles with bulk and interfacial flow. I am helping to develop a "capillary focusing" technique that uses the differences in capillarity experienced by fluid evaporating from a substrate with a heterogeneous surface to control the deposition and assembly of CNTs. I plan to use this technique to fabricate CNT based microelectronics in environmentally friendly conditions.

In the future, I will build on my training in electrokinetics and complex fluids to conduct research on responsive complex interfaces. In addition, I am interested in developing new techniques for the measurement of anisotropic colloidal interactions.

(4ei) Characterization of Human Diseases At Single Cell Level for Therapeutic and Diagnostic Applications

Wu, P. H., Johns Hopkins University

My research interests focus on the application of engineering and physics principles to decipher mechanisms of disease and normal cell functions at the molecular and cellular levels. I received my Ph.D. in Chemical Engineering from the University of Florida. My doctoral research focused on the quantitative characterization of cellular and subcellular dynamics. I improved resolution in particle tracking nanorheology, which allowed me to investigate the association between cell mechanics and cancer progression. During my postdoctoral work with Prof. Denis Wirtz at the Johns Hopkins University I extended this research direction to develop a systematical approach to identifying biophysical signatures cell subpopulations, specifically of metastatic cancer cells. In 2012, I received a postdoctoral fellowship award from the American Heart Association.

As an independent investigator, I plan to apply this new approach to continue the development of novel single cell analysis tools that can be effectively applied toward deciphering complex human diseases in a high-throughput context. I will integrate the extracted single-cell information into the flow of information in biological systems: genetics, epigenetics, environmental inputs, and, finally, appearance and behavior. Together, this data will complete the roadmap to understand disease mechanism. My ultimate research goals are to translate in-depth understanding of disease mechanisms into an effective and novel platform for drug discovery and a predictive model to improve diagnosis and prognosis.

The proposed research described below combines my previous expertise in cell biology and cell mechanics/motility with state-of-the-art analytical techniques, such as high throughput cell phenotyping, to identify and characterize disease mechanisms. I plan to pursue projects in the following three areas: 1) ***Dynamic process of cell functions and cell phenotypes***. 2) ***Analysis of cancer metastasis using quantitative cell mechano-phenotyping***. 3) ***Molecular mechanisms of cancer cell mechanics in vivo***. These three topics integrate biology, machine vision, and multivariate system analysis algorithms. Cumulatively, these researches will provide innovative approaches for understanding health and disease problems.

Selected Publications (14 total, first/co-first¶ author in 6 papers)

1. W.C. Chen¶, P.H. Wu¶, J.M. Phillip, S.B. Khatau, J.M. Choi, M.R. Dallas, K. Konstantopoulos, S.X. Sun, J.S.H. Lee, D. Hodzic, D. Wirtz, "Functional interplay between cell cycle and cell phenotypes", ***Integrative biology*** (2013) DOI: 10.1039/C2IB20246H.
2. P.H. Wu, C.M. Hale, W.C. Chen, J.S.H. Lee, Y. Tseng, and D. Wirtz, "High-throughput ballistic injection nanorheology (htBIN) to measure cell mechanics", ***Nature Protocols*** 7 (2012) 155-170.

3. P.H. Wu, S.H. Hung, T. Ren, I. M. Shih, Y. Tseng, "Cell cycle dependent alteration in nucleus accumbens associated protein-1 (NAC-1) nuclear body dynamics and morphology" ***Physical Biology*** 8 (2011) 015005.
4. P.H. Wu, A. Agarwal, H. Hess, P.P. Khargonekar, Y. Tseng, "Analysis of video-based microscopic particle trajectories using Kalman filtering", ***Biophysical Journal*** 98 (2010) 2822-2830.
5. P.H. Wu, N. Nelson, Y. Tseng, "A general method for improving spatial resolution by optimization of electron multiplication in CCD imaging", ***Optics express*** 18 (2010) 5199-5212.

(4ab) Benchtop Disease Models Using Tissue Engineering
Dumas, J., Georgia Institute of Technology

Benchtop disease models using tissue engineering

Jerald E. Dumas

Biomedical Engineering, Georgia Institute of Technology, Atlanta, GA

Advances in biomaterial research have greatly impacted medicine through new innovations in areas such as detection and treatment. However, many grand challenges still exist as the boundaries between biology, chemistry, and engineering continue to merge. I have used tools from each area for both tissue repair and the study of diseases.

As a doctorate student in the chemical and biomolecular engineering department at Vanderbilt University, I developed a two-component allograft mineralized bone particle (AMBP)/polyurethane (PUR) system for the treatment of bone defects. The biodegradable PUR phase, which consisted of lysine triisocyanate and polyester polyol, had tunable degradation rates and mechanical properties. The AMBP phase, which is bound to the PUR phase, was used as a filler to enhance the mechanical properties and provide a resorbable pathway for cellular infiltration. The adaptability of the AMBP/PUR composite system made it suitable for multiple applications including an implant or injectable platform. Further, the AMBP/PUR composites served as a delivery system for biologics such as recombinant human morphogenic protein (rhBMP-2), accelerating new bone formation in critical size defects in New Zealand White (NZW) rabbits.

As an Institutional Research and Academic Career Development Award (IRACDA) postdoctoral fellow in the biomedical engineering department at Georgia Institute of Technology/Emory University, I am studying the role of proteases in cancer. I am developing a detection device based from multiplex cathepsin zymography, which uses a gelatin substrate in a SDS-PAGE gel to quantify enzyme activity. Cathepsin activity has shown to be upregulated in cancer tissue, and this difference is used to distinguish normal from tumor tissue. Concurrently, I am utilizing hydrogel and polyurethane technology to create an in vitro bone metastasis model that will study key parameters such as protease (i.e., cathepsin) activity and cell invasion rate.

There have been great advances in understanding pathology using *in vivo* models as in which host animals provide a complex pathophysiological environment. However, *in vitro* systems can provide more defined information as they provide a more controlled environment that allows for the study of more isolated factors in pathological states as compared to the unpredictable variability of *in vivo* models. In my lab, I will develop benchtop tissue surrogates for co-culturing cells to: 1) model disease states, 2) identify new biomarkers of disease, 3) model drug uptake and efficacy, and 4) develop predictive computational models of disease progression and patient treatment response. This work will be accomplished with the development of responsive biomaterials that mimic the extracellular matrix (ECM), which is a key component in pathological states. The engineered benchtop systems developed in my lab will accurately predict *in vivo* results with a fraction of the time and cost investments required of traditional animal studies.

(4ap) Systems Biology Approaches for Cancer Diagnostics and Therapeutics

Graham, N. A., California Institute of Technology

New measurement technologies have transformed biology from a qualitative, descriptive field into a quantitative, multiparameter science. This transition has enabled the development of integrated, systems models of disease that bridge traditional biology and quantitative sciences including chemical engineering. I am passionate about harnessing the power of quantitative systems approaches to uncover novel biology in order to design new diagnostics and therapeutics, with a particular focus on cancer.

My graduate research at Caltech focused on crosstalk within biomolecular signaling networks, particularly between soluble and insoluble microenvironmental cues such as peptide ligands (eg, EGF) and intercellular contact (eg, cadherins). This research provided a foundation in quantitative cellular biology and shaped my interest in how cancer cells deregulate physiological signals to achieve pathological behavior.

For my postdoctoral research, I joined the Department of Molecular and Medical Pharmacology at UCLA to study translational and systems approaches to cancer diagnostics and therapeutics. Here, I will present three studies from my postdoctoral research: i) a systems analysis of positive feedback in signal transduction networks that underlies cancer cell death following metabolic perturbations; ii) development and application of a novel bioinformatic approach (self-organizing maps) to analyze single-cell, multiparameter data from microfluidic measurements of brain tumor biopsies; and iii) phospho-proteomic analysis of mouse and human prostate cancer tissues to predict therapeutic treatment modalities.

My future research program will synthesize my training in both engineering and biology. I will build a highly collaborative research program that integrates genomic, proteomic and metabolomic data to build predictive, systems models of disease. In collaboration with physician scientists and clinicians, I will generate and integrate quantitative biological data to uncover novel biological mechanisms with translational relevance. My training in chemical engineering, biological sciences and bioinformatics has positioned me at the forefront of the interdisciplinary field of cancer systems biology, where I will play a central role in discovering novel cancer biology that can be exploited for both diagnostics and therapeutics.

PhD: California Institute of Technology (2007), Dept of Chemical Engineering, Advisor Anand R. Asthagiri

Postdoc: University of California, Los Angeles, David Geffen School of Medicine, Dept of Molecular and Medical Pharmacology, Advisor Thomas G. Graeber

(4cx) Discovering New Self-Assembling Materials Through High-Throughput High-Performance Computation

Phillips, C. L., University of Michigan

The self-assembly of nanoscale components is one of the most promising routes for the controlled microfabrication of smaller and smaller devices. An expanding array of new techniques to nano-engineer materials is generating a rich design space for creating nanoscale components, with designed electronic, optical, mechanical, or transport properties. These materials can have applications in catalysis, drug-delivery, new optical materials, metamaterials, and microelectronics.

However, there are many open questions in self-assembly. How can we self assemble materials with more complexity? What forms of complexity are accessible and what are the pathways by which they assemble? And also, if a complex structure forms somewhere in the design space of a material, how do we find it?

In my research, I use and develop theoretical and computational tools to explore the design space of self-assembled materials. In my PhD dissertation under Prof. Sharon Glotzer of the University of Michigan, I studied the self-assembly of polymer-tethered nanoparticles in exotic phases and the self-assembly of terminal colloidal clusters with controllable structure. I also developed GPU-accelerated simulation tools, and a mathematical method for modeling arbitrarily shaped particles.

For my postdoctoral research, I am investigating how high-throughput computation can be used to rapidly discover new types of complex structures. Increasing computational power in the form of massively parallel supercomputers and GPU computing has enabled the ability to run massive parameter searches using simple models for interacting nano-scale components. At the Computation Institute at Argonne National Laboratory I have been researching how these searches should be organized and how the output of these searches can data-mined for new structures never identified before. By integrating techniques such as shape matching and machine learning, I have found previously unidentified complex structures in relatively simple systems that may well be realizable in experiment.

In my research program, I aim to develop new methods for coupling machine learning and pattern recognition techniques with high-throughput computation to accelerate the discovery of complex soft materials. By rapidly amassing data and forming phase diagrams, we can focus experimental effort towards the parts of design space where novel structures will assemble and advance understanding as to what drives self-assembly of the simple versus complex, and ordered versus disordered systems.

C.L. Phillips, G.A. Voth, "Discovering crystals using shape matching and machine learning," Preprint, 2013

C.L. Phillips, E. Jankowski, B.J. Krishnatreya, K. Edmond, D. Pine, S.C. Glotzer, In Preparation, 2013

R. Marson, C.L.Phillips, J.A. Anderson, S.C. Glotzer, In Preparation, 2013.

C.L. Phillips, J.A. Anderson, G. Huber and S.C. Glotzer, "*Optimal Filling of Shapes,"* Phys. Rev. Lett, **108**, 198304 (2012)

C.L. Phillips, E. Jankowski, M. Marval and S.C. Glotzer, "*Self-assembled clusters of spheres related to spherical codes,"* Phys. Rev. E, **86** (4) 1124 (2012).

C. L. Phillips and S.C. Glotzer, "*Effect of nanoparticle polydispersity on the self-assembly of polymer tethered nanospheres,"* J. Chem. Phys., **137** (10) 104901 (2012).

TD Nguyen, CL Phillips, JA Anderson, SC Glotzer, "*Rigid body constraints realized in massively-parallel molecular dynamics on graphics processing units,"* Computer Physics Communications**182**(11), 2307-2313 (2011).

CL Phillips, JA Anderson, SC Glotzer, "*Pseudo-random number generation for Brownian Dynamics and Dissipative Particle Dynamics simulations on GPU devices,"* J. Comp. Physics **230**(19), 7191-7201 (2011).

C.L. Phillips, C.R. Iacovella and S.C. Glotzer, "*Stability of the double gyroid phase to nanoparticle polydispersity in polymer-tethered nanosphere systems*," Soft Matter **6**, 1693 – 1703, 2010.

C.L. Phillips, R. Magyar, P.S. Crozier, *"A two-temperature model of radiation damage in a-quartz,"* J. Chem. Phys., **133** (14) 144711 (2010)

C.L. Phillips, P.S. Crozier, "*An energy-conserving two-temperature model of radiation damage in single-component and binary Lennard-Jones crystals,*" J. Chem. Phys., **131** (7) 074701, (2009)

(4dt) Multifunctional Polymeric Nanomaterials for Environmental and Biomedical Applications

Tang, C., North Carolina State University

Simple and accurate detection of heavy metals and biomolecules including DNA, protein, and microorganisms are of interest in environmental and biomedical applications. Use of nanotechnology is especially promising for advances in this area. My current and past work has combined polymer science, nanotechnology and biotechnology in order to functionalize nanomaterials with biomolecules. Building on this experience, *the goal of my proposed research is to develop multifunctional polymer-based nanomaterials for detection of heavy metals and biomolecules* based on fundamental understanding of material interactions. These advanced functional materials may have a variety of potential applications ranging from water purification to clinical lab diagnostics i.e. in biomarker research, cancer diagnostics, and detection of infectious diseases.

In my graduate work at North Carolina State University working under the direction of Dr. Saad Khan, the goal of my research was to capitalize on the high specific surface area of nanofibers as well as the advantages of biocatalysts by immobilizing proteins/enzymes within chemically crosslinked polymer nanofibers. Surprisingly, despite the short length scale, the catalytic activity of the enzyme within the nanofiber was significantly affected by diffusion limitations. Thus, surface modification of nanofibers should provide a route to improved performance of the biomolecule. Based on the understanding gained, nanofibers with various surface chemistries can be produced by electrospinning and functionalization with biomolecules such as oligionucleotides or antibodies for highly specific capture of molecules of interest e.g. bacterial pathogens. For example, nanofibers can be produced from polymers with an amine containing backbone such as chitosan which can used to attach antibodies of interest through bifunctional reagents such as glutaraldehyde. Chelators can also be employed for capture of heavy metals, e.g. deferoxamine for detection of iron can be attached using the same chemistry. The surface chemistry of the fibers can be modified to accommodate other molecules of interest. Further, by combining the biomolecules of interest or chelators with fluorophores, bifunctional materials capable of capture and detection can be achieved.

Multifunctional polymer nanoparticles are also of potential use in sensing applications. As a Postdoctoral Associate working with Dr. Prud'homme at Princeton University, I have been developing nanoparticles for use as contrast agents in diagnostic medical imaging applications using Flash Nanoprecipitation (FNP). In FNP, amphiphilic block copolymers direct self-assembly of biocompatible nanoparticles with high concentrations of encapsulated components. The precipitation process is controlled by carefully tuning the time scales of micromixing, self-assembly, and nucleation and growth. Further, the surface of the nanoparticles can be functionalized with desired surface chemistries or ligands of interest. Building on this experience, FNP may prove a promising route to multifunctional materials that combine the unique attributes of inorganic nanoparticles. For example, gold nanoparticles functionalized with recognition molecules such as antibodies, antigens, or oligonucleotides can be used for colorimetric detection of biomolecules as the optical properties of the nanoparticles which will change upon interactions with the target biomolecules. Magnetic nanoparticles can be used to capture and manipulate a target biomolecule. By designing a nanoparticle with a magnetic core, and then functionalizing the surface with gold nanoparticles, sensing capabilities may be enhanced as the magnetic core should facilitate separation and/or concentration while the gold nanoparticles provide optical detection. By attaching ligands of interest to the polymer nanoparticle or the gold nanoparticle, specific detection of molecules may be achieved.

The methodologies developed using nanofibers and nanoparticles are versatile and can be adapted for detection of a number of analytes. Detection and removal of heavy metals for water purification as well as detection of bacterial pathogens for biodefense applications are of particular interest.

(4cd) A Molecular Engineering Approach Against Bacterial Biofilms and Spores

Mi, L., University of Washington
Jiang, S., University of Washington

A Molecular Engineering Approach against Bacterial Biofilms and Spores

Biofilms and spores are the ultimate survival mechanisms of bacterial cells against harsh environments. Both of them defy conventional antibiotic treatments, and both of them can lead to serious clinical consequences, e.g. cystic fibrosis (CF) and *B.anthrax* infection. Here we present novel molecular engineering approaches to combat these pressing biological threats.

For bacterial biofilms, the critical step lies in the prevention of initial bacterial adhesion and subsequent surface proliferation. To this end, zwitterionic polymer-based nonfouling materials were designed and synthesized that not only can resist bacterial adhesion via polymer hydration but also carry additional biological functions, particularly the controlled release of antimicrobial agent as counter ions or leaving groups to inhibit bacterial growth close to the surface. In addition to the integration of nonfouling and antimicrobial properties, well-defined zwitterionic copolymer structures were also synthesized by means of living polymerization that are thermo- or pH- responsive, making them suitable for specific biomedical applications including wound dressing and biosensing.

For bacterial spores, the existing industrial approach relies on concentrated hydrogen peroxide solution for sterilization, which is dangerous to human as well as toxic to the environments. To remedy this obvious shortcoming, we resort to the fundamental study of dodecylamine (DDA) lethal germination mechanism. DDA was arguably the strongest antispore agent known to date that can trigger spore germination followed by inactivation at low concentration and under near ambient temperature. However, the human toxicity of DDA limited its application outside the realm of academic study. By unveiling the DDA mechanism of action, we were able to grasp the key structural features central to its bioactivity and formulate molecular principles to guide the design of novel antispore molecules. We have so far experimentally demonstrated that the by applying our molecular understanding we can successful design and discover molecules that possess strong antispore activities.

(4ai) In Situ Microscopic Characterization of Energy Materials for Grid-Scale Storage
Gallaway, J. W., Energy Institute, City College of New York

Batteries are designed to maximize interfacial area by employing thick, porous electrodes, which can display nonhomogenous reaction conditions. Non-electrochemical Information (microscopy, diffraction, etc) is needed to characterize localized, microscopic phenomena within cycling batteries. This can involve either model experimental systems or powerful cell-penetrating techniques such as energy dispersive X-ray diffraction (EDXRD).

In this presentation I report in situ EDXRD data collected on cycling LR20 Zn-MnO2 batteries using white beam X-rays at the National Synchrotron Light Source (NSLS) located at Brookhaven National Lab. This data reveals transient zinc oxide layers that form and dissolve at the anode-separator interface during cycling, as a function of battery discharge rate. This is found to be a cause of battery failure in some cases when the battery has reached hundreds of cycles.

This is important because it is generally recognized that electrochemical storage on a large scale is desirable for a number of societal reasons including the smart grid and solar cell integration. For applications with critical cost and environmental restrictions the Zn-MnO2 alkaline chemistry is well-suited. My work is broadly focused on inexpensive ways to accomplish grid-scale storage.

Additionally, the statistical aspect of batteries is often overlooked in academia. Batteries are deterministic systems, but during long cycling lifetimes battery fade and failure are subject to contingency, nucleating events, and other phenomena that must be treated statistically. Said more plainly, batteries often fail due to their materials, but even a simple battery can fail more than one way. EDXRD results confirm this and provide a way to quantify failure modes.

(4bd) Semiconductor Nanostructures: Diffusion, Impurities, and Nanowire Fabric

Holmberg, V. C., University of Texas at Austin

Developments in materials synthesis over the past decade have enabled the large-scale production of a wide variety of semiconductor nanostructures that have historically been available only in small quantities. These advances have led to the creation of a new class of materials and devices that utilize macroscopic quantities of nanomaterials. For example, solution-liquid-solid (SLS) and supercritical fluid-liquid-solid (SFLS) growth allow Si and Ge nanowires to be produced on the gram-scale. Recently, these processes were used to develop the first Si and Ge nanowire fabrics – free-standing, flexible ceramics (up to 100 cm^2 in area) made entirely of single-crystalline semiconductor nanowires. These new semiconductor fabrics are promising for future battery materials, as well as for applications requiring light harvesting and conversion.

In many cases, the physical properties of nanoscale semiconductors are drastically different from their bulk counterparts. Reducing the dimensions of a semiconductor to the nanoscale can lead to dramatic effects in optical, electrical, and mechanical properties. For example, semiconductor nanocrystals exhibit size-tunable absorption and fluorescence, with applications in bioimaging, solar energy, and display technology. Likewise, many semiconductor nanowires are incredibly flexible, exhibiting extremely high bending strengths and very large strength-to-weight ratios, unlike their brittle bulk counterparts. Phase transitions and impurity diffusion can also be quite different in nanoscale semiconductors.

Atomic impurities are particularly important in semiconductor technology. They introduce charge carriers that are used to tune electrical properties in devices such as transistors, light-emitting diodes, and lasers, and they are used to store energy in electrochemical devices like Li-ion batteries. On the nanoscale, even single atomic impurities can drastically change the optical, electrical, and magnetic properties of the material. *In situ* transmission electron microscopy (TEM) experiments can be used to address the dynamics of impurity diffusion and solid-state reactions in semiconductor nanostructures, but many challenges exist in the controlled incorporation and characterization of impurities in these systems.

Future research plans will be discussed in addition to the above topics.

SELECTED PUBLICATIONS:

Vincent C. Holmberg, Matthew G. Panthani, and Brian A. Korgel, "Phase Transitions, Melting Dynamics, and Solid-State Diffusion in a Nano Test Tube," *Science*, **326**, 405–407 (2009).

Vincent C. Holmberg, Katharine A. Collier, and Brian A. Korgel, "Real-Time Observation of Impurity

Diffusion in Silicon Nanowires," *Nano Letters*, **11**, 3803–3808 (2011).

Vincent C. Holmberg, Timothy D. Bogart, Aaron M. Chockla, Colin M. Hessel, and Brian A. Korgel, "Optical Properties of Silicon and Germanium Nanowire Fabric." *Journal of Physical Chemistry C*, **116**, 22486–22491 (2012).

Vincent C. Holmberg, Justin R. Helps, K. Andre Mkhoyan, and David J. Norris, "Imaging Impurities in Semiconductor Nanostructures," *Chemistry of Materials*, **25**, 1332–1350 (2013).

Vincent C. Holmberg and Brian A. Korgel, "Corrosion Resistance of Thiol- and Alkene-Passivated Germanium Nanowires," *Chemistry of Materials*, **22**, 3698–3703 (2010).

Vincent C. Holmberg, Michael R. Rasch, and Brian A. Korgel, "PEGylation of Carboxylic Acid-Functionalized Germanium Nanowires," *Langmuir*, **26**, 14241–14246 (2010).

Vincent C. Holmberg, Reken N. Patel, and Brian A. Korgel, "Electrostatic Charging and Manipulation of Semiconductor Nanowires," *Journal of Materials Research*, **26**, 2305–2310 (2011).

Damon A. Smith, **Vincent C. Holmberg**, and Brian A. Korgel, "Flexible Germanium Nanowires: Ideal Strength, Room Temperature Plasticity, and Bendable Semiconductor Fabric," *ACS Nano*, **4**, 2356–2362 (2010).

Damon A. Smith, **Vincent C. Holmberg**, Doh C. Lee, and Brian A. Korgel, "Young's Modulus and Size-Dependent Mechanical Quality Factor of Nanoelectromechanical Germanium Nanowire Resonators," *Journal of Physical Chemistry C*, **112**, 10725–10729 (2008).

(4cc) New Approaches for Mesoscale Modeling

Markutsya, S., Iowa State University

Computer modeling and simulation is a powerful tool for investigating the fundamental principles of physics, chemistry, and biology at different time and length scales. Recently, more attention is brought to the intermediate-scale architectures and phenomena (so called a mesoscale level description) that connects atomic scale systems and bulk materials. Understanding and control of systems at mesoscale level promotes discovery of next-generation materials and technology. For my research I have used coarse-graining (CG) of the reference all-atom system as one of the mesoscale approaches. CG models, when implemented in molecular dynamics simulation (CGMD), have successfully modeled the structure and thermodynamics of systems at mesoscale level. I have successfully applied this approach to obtain an accurate CG model of crystal cellulose microfiber to enhance biofuel production.

The major drawback of CGMD approaches developed from all-atom reference data is faster CG dynamics as a result of the reduction in degrees of freedom that eliminates much of the friction between CG beads. To resolve this issue I have developed a new systematic mesoscale approach to accurately capture dynamics in CG models in addition to accurate statistical properties. With this approach, coarse-grain computer simulations can be used not only for qualitative but also for accurate quantitative prediction of systems' properties when broadly applied to biological and engineering fields. In my research program I will be working on further refinement of the method. In meanwhile, I will apply this novel theoretical approach to different biological and engineering systems at the mesoscale level to generate new bioengineering solutions. This opens many interdisciplinary collaborative opportunities for working closely with other mechanical and biological engineers, material scientists, chemists, chemical engineers, physicists, molecular biologists, and researchers from other fields. A successful proceeding toward these goals is supported by my extensive academic training as well as experimental experience in physics and engineering as well as my solid knowledge of computer modeling and simulations. My broad background of different fields allowed me to successfully resolve many challenging problems during my research career.

During this poster session I will present details of my PhD and postdoctoral research, including the development of the mesoscale approach to reproduce dynamic properties of CGMD. A brief overview of future applications for this newly developed mesoscale approach is also presented.

(4ce) Discovery and Characterization of the Astexin Family of Lasso Peptides

Maksimov, M. O., Princeton University

Genome mining has empowered the rational discovery of novel natural products in the post-genomic era. Motivated by the unique chemical properties of lasso peptides such as their extraordinary stability and natural antimicrobial activity, we set out to develop a novel genome-mining algorithm tailored to their unique characteristics. Our results identified 79 putative lasso peptide biosynthesis clusters and were validated by the expression and characterization of a novel member of the class called astexin-1 (1). We have continued our work in this direction by characterizing two peptides encoded by the same organism called astexin-2 and astexin-3. As a result of our work and efforts in other laboratories, the number of confirmed lasso peptides has nearly doubled in the past two years. The expansion in information about the lasso peptide family led us to reexamine the gene architectures of this class of natural products. Through this line of inquiry, we were able to characterize a new enzyme involved in lasso peptide biosynthesis and glean valuable insights about the evolution of this peptide class.

1. Maksimov MO, Pelczer I, & Link AJ (2012) Precursor-centric genome-mining approach for lasso peptide discovery. *Proceedings of the National Academy of Sciences* 109(38):15223-15228.

(4bl) Interfacial Heterogeneity At the Molecular Level

Kastantin, M. J., University of Colorado
Langdon, B. B., University of Colorado at Boulder
Schwartz, D. K., University of Colorado Boulder

Interfaces are encountered in many experimental systems that alter the behavior of molecular adsorbates relative to bulk solution. For example, stable proteins in solution commonly aggregate and form films at interfaces that can inhibit industrial separations or long-term storage of drug products, alter the wear resistance of artificial joint surfaces, or mediate cellular responses to regenerative tissue scaffolds. Thus, there is intense interest in understanding exactly how interfacial properties influence molecular behavior. Common techniques for probing interfacial phenomena measure the average surface coverage, mobility (i.e. diffusion), molecular conformation, and macroscopic behavior (e.g. anti-adhesive qualities, catalytic activity, etc.). For lack of better information, it is widely assumed that all molecules behave as an "average" molecule, with a single net adsorption rate, diffusion coefficient, and so on. However, in many systems, important behavior stems from molecules that deviate significantly from the norm.

This work addresses interfacial heterogeneity at the molecular level in a variety of different proteins that dynamically explore chemically modified glass surfaces. Single-molecule resolution, provided by total internal reflection fluorescence microscopy, allows independent, simultaneously occurring behaviors to be characterized separately. We have previously observed that, while direct attractions between proteins and surfaces are relatively weak, clusters of proteins exhibit surface residence times that increase exponentially with the number of constituents. This work will present several new observations in this area, including: direct measurements of cluster formation dynamics, surfaces that promote cluster formation, and surface-induced conformational changes that may affect the tendency of a protein to aggregate. In total, these observations lead to a deeper understanding of interfacial phenomena than can be discerned from the average behavior.

(4ch) Towards Targeted Molecular Design of Functional Materials

Marshall, B. D. II, Rice University
Chapman, W. G., Rice University

Molecular design and controlled synthesis of molecules with specific properties will create new classes of smart materials in the upcoming century. By manipulating the forces between molecules, researchers can control both the microscopic and macroscopic behavior of materials in a way that would have seemed like science fiction just decades ago. The interactions between molecules can be tuned by varying hydrogen bonding, charge, rigidity, architecture etc… How do we sort through this plethora of possibilities to attain a molecule that achieves a desired goal? The three general approaches are theory, simulation and experiment. Synthetic chemistry is advancing at a blistering pace, giving researchers the ability to design ever more complex molecules to achieve specific goals; however, these design processes are very time consuming and expensive. With the infinite parameter space available for manipulation, experiment needs to be complimented by other cheaper and faster methods. These methods are theory and molecular simulation (MS), which allow for a rapid exploration of parameter space to target specific molecular structures as candidates for more expensive and time consuming experimental design. With the rapid increase in computational ability MS has become an important and widely used tool; however, the use of MS introduces another set of problems. To obtain reliable answers using MS, one must obtain good statistical sampling which can prove very difficult in many important situations. For instance, MS has difficulty with dilute components (such as the bulk surfactant concentration in an oil / water interface) and strongly associating systems (such as patchy colloids at low temperatures). Theoretical methods have no such handicap. Another advantage theory has over molecular simulation is speed of calculation. A significant challenge in developing molecular level theory for complex fluids is the anisotropy of interactions between molecules. However, advances over recent years have opened the door for model development models which account for various shape and interaction anisotropies. In this paper I review my recent contributions to the field of complex fluid theory. Specifically, I demonstrate how theory can be developed and applied to predict the interfacial properties and structure of the oil / water interface, how rod – coil copolymers align at liquid interface, and the structure and thermodynamics of patchy colloid fluids.

(4g) Hydrogel Biomaterials As Model Systems for Cancer and Stem Cell Engineering

Ananthanarayanan, B., University of California, Berkeley

Research Overview:

I am a broadly trained bioengineer interested in developing new synthetic biomaterials for fundamental studies in cancer and stem cell biology and applications in regenerative medicine.

My graduate training in Chemical Engineering in the laboratory of Prof. Matthew Tirrell was focused on engineering the self-assembly of peptide amphiphiles into multi-functional nanostructures. I developed a facile method to display peptide ligands on surfaces by incorporation of peptide amphiphiles in solid-supported lipid bilayers. In collaboration with Prof. David Schaffer and Prof. Kevin Healy at UC Berkeley, this platform was used to screen peptides for their ability to support neural progenitor cell adhesion, proliferation, and differentiation into neurons and astrocytes. I also worked on tailoring the structure and stability of micellar nanoparticles composed of peptide amphiphiles for drug delivery applications.

My post-doctoral training with Prof. Sanjay Kumar at UC Berkeley aims to delineate the role of cell-matrix mechanobiology in brain physiology and pathology. To this end, I developed a synthetic hydrogel based on hyaluronic acid (HA) as a brain matrix-mimetic model system and showed that glioma cell motility in HA hydrogels displays patterns strikingly similar to those seen *in vivo*. My current work aims to dissect the cell-matrix interactions and intracellular signaling pathways that underlie this distinct motility phenotype, with the aim of discovering new drug targets. I am also developing novel HA hydrogels based on click chemistry to enable rapid gelation and cell encapsulation. In collaboration with Prof. David Schaffer's group, I am employing these fully defined matrices for differentiation and transplantation of human embryonic stem cell-derived dopaminergic neurons as a potential therapy for Parkinson's disease.

(4at) Materials for Energy: From Catalytic to Nuclear Reactions

Hammond, K. D., University of Massachusetts

Addressing the world's future energy needs in the coming years will require both more efficient utilization of existing energy sources and development and/or expansion of current and new energy sources. This poster discusses two extremely interesting classes of materials relevant to energy applications. The first is zeolites: materials with inherent porosity on the length scale of small molecules. Zeolites revolutionized the petrochemicals industry when they were first introduced over fifty years ago, and we hope that recent attempts to react them with nitrogen to produce strongly basic catalysts for carbon-carbon bond formation will do the same for oxygenated feedstocks such as those derived from biomass. The second class of materials is plasma-facing materials, specifically those intended for the components of ITER and other nuclear fusion reactors. Nuclear fusion is, indirectly, the source of all energy we utilize every day, and harnessing it without the gravitational confinement present in stars is one of the greatest challenges humanity has ever striven to meet. Doing so will require materials, many of them perhaps yet to be discovered, that will survive the harsh environment of a nuclear fusion reaction.

(4es) Simple Physical Models of Complex Fluids Analyzed Using Classical Density-Functional Theory and Generalized Langevin Dynamics

Yu, H. Y., University of Pennsylvania

Theoretical modeling provides insights into fundamental understanding of many complex systems ranging from functionalized nanoparticle suspensions to biological fluids. Although simple, a suitably chosen coarse-grained description that captures the essential physics of the system can guide the material design for a variety of applications before conducting exhaustive experimental efforts or computational tests of the full complex problem. Here I demonstrate the application of theoretical modeling in two nanoparticle systems.

Nanoparticle-organic hybrid materials (NOHMs) are a new type of complex fluid consisting of 10 nm diameter spherical inorganic core particles surface-functionalized with oligomeric organic molecules with no other solvent. The fluidity is provided by the attached oligomers that mediate the interparticle forces and affect the equilibrium properties and dynamic behavior of the bulk system. The absence of a solvent, the small size of the nanocores and oligomers, and the incompressibility of the tethered oligomeric fluid make the oligomer-mediated interactions non-pairwise-additive. For the first time, we have developed a classical density-functional approach for model hard spheres with tethered bead-spring oligomers that allows us to have a direct description of the system free energy as a functional of the probability densities of cores and oligomers without assuming pairwise additivity. The equilibrium distributions and the static structure factor of NOHMs that can be utilized to characterize experimental systems are obtained from the minimization of the free energy. The theories also predict that the solvent-free homogeneous nanoscale mixture of organic oligomers and inorganic cores can lead to a unique thermodynamic driving force for solute uptake as the solute releases the entropic penalty of the space-filling oligomers, which makes NOHMs a candidate for carbon capture applications. Furthermore, analyzing the rheological behavior of NOHMs through solving the Smoluchowski equation for non-equilibrium probability density suggests that the information regarding the appropriate closure for non-equilibrium many-body interactions can be gained from experimental measurements of the high frequency shear modulus. The significance of this work is that the theories provide a first prediction of how the entropic forces of the tethered fluid in solvent-free conditions influence the experimentally measurable properties of the material. The theoretical approach can be widely applied to design directed self-assembly of colloids and to understand the regulation of cell membranes driven by inter- or intra-cellular forces.

The use of functionalized nanocarriers is gaining translational prominence in targeted vascular drug delivery. Owing to the complexity of the blood flow environment and a wide spectrum of time and length scales that are relevant to nanocarrier-endothelium binding dynamics, a simplified theoretical description that allows us to perform a sensitivity analysis of the parameter space is valuable. To this end, we formulate non-Markovian, generalized Langevin equations for ligand-grafted nanocarrier motion and ligand-receptor (on the endothelium) pair relaxation in the presence of red blood cell-driven marginating potential and particle-wall hydrodynamic forces. While the detailed hydrodynamic interactions of the system can be resolved using direct numerical simulations (DNS) of fluctuating fluids, for a given memory kernel (e.g. Ornstein-Uhlenbeck process) of the viscous drag on the particle the Brownian force can be obtained via continuous-time random-walk simulations satisfying the fluctuation-dissipation theorem. Analyzing the nanocarrier velocity autocorrelation functions and the potential of mean force along a specified reaction coordinate helps quantify the binding affinity in the presence of hydrodynamic interactions. In the future, we plan to integrate DNS and Brownian dynamics simulations to track the nanocarrier motion and binding in a vascular network tree. The broader impact of this approach is that it can be potentially generalized as a versatile multiscale modeling method for hydrodynamics of complex flows.

(4af) Single-Walled Carbon Nanotube Dynamics in Simple and Complex Media
Fakhri, N., University of Gottingen

The discovery of fullerenes provided exciting insights into how highly symmetrical structures of pure carbon can have remarkable physical properties. Single-walled carbon nanotubes (SWNTs) are the vanguard of such architectures. The organization of the hexagonal honeycomb carbon lattice into high-aspect-ratio cylinders with a variety of helical symmetries creates very unusual macromolecular structures representing an emerging research area in condensed matter physics and materials science: traditionally hard materials appearing in new soft matter applications and environments.

I will first show that the dynamics of SWNTs in liquids are essentially polymer-like. By exploiting the intrinsic near-infrared fluorescence of semiconducting SWNTs, we have imaged the Brownian motion of individual nanotubes in water and have measured directly the bending stiffness of SWNTs. The semiflexible chain model represents accurately the configurational dynamics of SWNTs suspended in water. Interestingly, the persistence length of SWNTs is comparable to that of biopolymers.

This finding paves the way for using SWNTs as a model system for semiflexible polymers to answer long-standing fundamental questions in polymer physics. I will discuss the confined dynamics of stiff macromolecules in crowded environments, a common feature of polymer composites and the cell cytoskeleton. In fixed porous networks, we find that even a small bending flexibility strongly enhances SWNTs motion. This ends a 30-year-old debate in polymer physics: the rotational diffusion constant is proportional to the filament bending compliance and counter-intuitively, independent of the network porosity. The dynamics of SWNTs in equilibrium and non-equilibrium biopolymer networks is more complex. At long times, SWNTs reptate in the networks. At short times SWNTs can sample the spectrum of local stresses in equilibrium networks. In the non-equilibrium networks we observe strong local shape fluctuations driven by force generating molecular motors. I will discuss a newly developed microrheology technique in which we use nanotubes as "stealth probes" to measure viscoelastic properties of the host media.

Finally, I will introduce a new single-molecule technique based on ultra-stable near-infrared fluorescence of short SWNTs, to study intracellular transport dynamics in living cells and in whole organisms. The combination of long-time stability and high signal-to-noise ratio enables the accurate long-term tracking of single motor proteins tagged with SWNTs traversing the entire cell. Remarkably, we can distinguish the motor protein's motion along its microtubule track from the track's underlying random non-thermal fluctuations.

(4dw) Understanding Solutes, Surfactants, and Nano-Particles in Bulk and At Interfaces Using Molecular Simulations

Tummala, N. R., The University of Oklahoma

Many enabling technologies haven been realized with the new class of materials (graphene, nanotubes, quantum dots, fullerenes, fullerene derivatives) within past couple of decades. However, the true potential of such materials is not fully exploited owing to the lack of understanding of how such materials aggregate and interact in bulk and at different interfaces. To understand such interactions at molecular level (from electronic, materials, and biological perspective) is becoming increasingly important to furthermore the increase in applications of these new class of materials.

Towards this end, we employed molecular dynamics simulations within my graduate studies to understand surfactant aggregates on nanotubes with different curvature,[1-5] graphene,[6] and compared them to other regular surfaces like graphite and silica with varying levels of hydroxylation.[7-11] Our results from the simulations of sodium dodecyl sulfate, sodium dodecyl benzene sulfonate, and flavin mononucleotide surfactants on carbon nanotubes (CNTs) provided the understanding of the surfactants aggregate structure on CNTs and how those aggregate structures are affected by changing curvature, ions, and concentration. SDS aggregates on graphene substrate illustrated the effect of lateral confinement and edge effects of the substrate on the admicellar characteristics.

Understanding nano-particle interactions with the surface are important when such particles flow within pores, especially in hydraulic fracturing. We evaluated such flows in structured and unstructured porous media using Lattice-Boltzmann methodology and nano-particle flow was modeled by employing Lagrangian tracking methodology with different adsorption kinetics. Finally, with Prof. BrÃ©das group at the Georgia Institute of Technology, we are trying to understand how molecules arrange in various solvents, mixing and aggregation of variety of small conjugated molecules for applications in organic photovoltaics using classical molecular dynamics, semi-empirical, and density functional simulation methods.[12, 13] We are also trying to understand how the presence of fullerene derivatives affects the mechanical properties of thin films of polymer-fullerene photoactive layer.

Drawing from my experiences and expertise, apart from collaborating and continuing my previous research I am interested to expand my research into: 1.) Fundamental understanding of how molecular scale properties and interactions (shape, size, dispersion and electro-static) drive aggregation in various solvents. 2.) Effect of ligands on the diffusion limitations on nano-particle catalysts. 3.) The interfacial phenomenon of water and organic solvents next to graphene oxide, and various clay and shale surfaces at atomic level using both quantum and classical simulations. And also, study the transport of solutes in such nano-pores for understanding the shale-solvent interactions at both molecular level and continuum level using a combination of techniques like non-equilibrium molecular dynamics simulations and Lagrangian tracking methods. I would like to investigate shale-surfactant and shale-polymer interactions to better understand the fundamental processes involved in hydraulic fracturing.

PhD Advisor: (2005 - 2010) Prof. Alberto Striolo (University of Oklahoma)

Post-Doctoral Advisor (2010 - 2011) Prof. Dimitrios Papavassiliou (University of Oklahoma)

Post-Doctoral Advisor (2011 - Present) Prof. Jean-Luc BrÃ©das (Georgia Institute of Technology)

[1] M. Suttipong, N. R. Tummala, B. Kitiyanan, A. Striolo, The Journal of Physical Chemistry C 2011, 115, 17286.

[2] N. R. Tummala, B. H. Morrow, D. E. Resasco, A. Striolo, ACS Nano 2010, 4, 7193.

[3] N. R. Tummala, A. Striolo, ACS Nano 2009, 3, 595.

[4] N. R. Tummala, A. Striolo, Physical Review E (Statistical, Nonlinear, and Soft Matter Physics) 2009, 80, 021408.

[5] M. Suttipong, N. R. Tummala, A. Striolo, C. S. Batista, J. Fagan, Soft Matter 2013, 9, 3712.

[6] N. R. Tummala, B. P. Grady, A. Striolo, Physical Chemistry Chemical Physics 2010.

[7] N. R. Tummala, L. Shi, A. Striolo, Journal of Colloid and Interface Science 2011, 362, 135.

[8] L. Shi, N. R. Tummala, A. Striolo, Langmuir 2010, 26, 5462.

[9] N. R. Tummala, A. Striolo, The Journal of Physical Chemistry B 2008, 112, 1987.

[10] D. Argyris, N. R. Tummala, A. Striolo, D. R. Cole, The Journal of Physical Chemistry C 2008, 112, 13587.

[11] R. Parthasarathi, N. R. Tummala, A. Striolo, The Journal of Physical Chemistry B 2012, 116, 12769.

[12] B. H. Wunsch, M. Rumi, N. R. Tummala, C. Risko, D.-Y. Kang, X. Steirer, J. Gantz, M. M. Said, N. R. Armstrong, J. L. BrÃ©das, D. Bucknall, S. R. Marder, Journal of Materials Chemsitry C 2013, Submitted.

[13] N. R. Tummala, Y.-T. Fu, S. Mehraeen, C. Risko, J. L. Bredas, Adv. Funct. Mater. 2013, DOI:10.1002/adfm.201300918.

(4cs) Nanoscale Semiconductors for Energy Conversion
Panthani, M. G., University of Chicago

Inorganic semiconductors have a proven track record for high performance electronic devices. However, they are difficult to manufacture inexpensively and the processes used to form them are do not scale well to large-area devices such as photovoltaics, displays, and sensors. Crystallization of inorganic semiconductors requires high temperature treatments which force trade-offs between device performance, fabrication costs and compatibility with flexible substrates amenable to roll-to-roll processing. Solution-based processes such as spin coating, spray-coating or inkjet printing combined with clever materials design can introduce lower cost alternatives to conventional technologyies but high perfomance must be achieved. Colloidal nanocrystals and molecular compounds have emerged as promising alternatives that can be deposited using inexpensive, high-throughput methods. Such semiconductor materials not only have the benefit of solution processability, they also offer optical and electronic properties that can be tuned based on size and structure. It has also been demonstrated that the performance of these materials can approaches that of traditionally manufactured technologies for electronic and optoelectronic applications.

A major concern in nanoscale materials is the increased role of interfaces. For example, can good charge transport be achieved in highly granular materials? Recent breakthroughs in surface passivation techniques for nanomaterials have opened a wide variety of techniques to facilitate charge transport in nanocrystalline thin films by using molecular compounds. These compounds can link grains electronically as well as modulate their mobility, carrier concentration, conduction type, and optical properties. These technological breakthroughs can be exploited to design interfaces in granular materials from the bottom up. Gaining control of interfaces in complex systems can enable the engineering of energy flows. This concept could be used to control electron and hole flow in photovoltaics or help isolate heat and charge carrier flows in thermoelectrics. By utilizing interface engineering, the performance of these materials could match or exceed the performance of traditionally manufactured counterparts.

Studies of nanocrystal-based solar cells, transport in solution processed films based on colloidal nanocrystals and inorganic molecular semiconductors, and future research plans will be discussed.

Selected Publications

Panthani, M. G.; Akhavan, V.; Goodfellow, B.; Schmidtke, J. P.; Dunn, L.; Dodabalapur, A.; Barbara, P. F.; Korgel, B. A. "Synthesis of $CuInS_2$, $CuInSe_2$, and $Cu(In_XGa_{1-X})Se_2$ (CIGS) Nanocrystal "Inks" for Printable Photovoltaics." *Journal of the American Chemical Society*, 130 (49), 16770-16777 (2008).

Panthani, M. G.; Hessel, C. M.; Reid, D.; Casillas, G.; Jose-Yacaman, M.; Korgel, B. A. "Graphene-

Supported High-Resolution TEM and STEM Imaging of Silicon Nanocrystals and their Capping Ligands." *J Phys Chem C,* 116 (42), 22463-22468 (2013).

Panthani, M.G.; Stolle, C.J.; Reid, D.K.; Rhee, D.J.; Harvey, T.B.; Akhavan, V.A.; Yu, Y.; Korgel, B.A. "$CuInSe_2$ Quantum Dot Solar Cells with High Open Circuit Voltage," *submitted.*

Panthani, M.G.; Khan, T.; Reid, D.K.; Hellebusch, D.J.; Rasch, M.; Maynard, J.; Korgel, B.A. "$CuInSe_XS_{2-X}$ Quantum dots for Biological Imaging,"*manuscript in preparation.*

Panthani, M. G.; Korgel, B. A. "Nanocrystals for electronics." *Annual review of chemical and biomolecular engineering,* 3, 287-311 (2012).

Holmberg, V. C.; **Panthani, M. G.**; Korgel, B. A. "Phase Transitions, Melting Dynamics, and Solid-State Diffusion in a Nano Test Tube." *Science,*326 (5951), 405-407 (2009).

Steinhagen, C.; **Panthani, M. G.**; Akhavan, V.; Goodfellow, B.; Koo, B.; Korgel, B. A. "Synthesis of Cu_2ZnSnS_4 Nanocrystals for Use in Low-Cost Photovoltaics." *Journal of the American Chemical Society,* 131 (35), 12554-12555 (2009).

Akhavan, V. A.; **Panthani, M. G.**; Goodfellow, B. W.; Reid, D. K.; Korgel, B. A. "Thickness-limited performance of CuInSe nanocrystal photovoltaic devices." *Optics express*,18 Suppl 3, A411-20 (2010).

Stolle, C. J.; **Panthani, M. G.**; Harvey, T. B.; Akhavan, V. A.; Korgel, B. A. "Comparison of the Photovoltaic Response of Oleylamine and Inorganic Ligand-Capped $CuInSe_2$ Nanocrystals." *ACS Applied Materials & Interfaces*, 4 (5), 2757-2761 (2012).

(4o) Interdisciplinary Approach to the Design, Synthesis, and Evaluation of Inhalable Therapeutics

Brenza, T., Iowa State University

Infections and diseases of the respiratory system contribute to morbidity and mortality in both the developing and the developed world. The therapeutic advantages of inhalable therapeutic treatment of respiratory diseases include the application of the therapeutic to the site of infection which enables a high localized concentration of therapeutic with low systemic exposure. However, the successful development of inhalable therapeutics requires the ability to successfully formulate the therapeutic into an inhalable form, the successful deposition of the material into specific respiratory regions, and controlled interactions of the material with extracellular and cellular environments. The solution to these challenges requires the expertise and interaction of multiple disciplines including: material science, engineering, molecular biology, immunology, pharmaceutics, and physiology.

My graduate and post-doctoral experiences in chemical engineering and pharmaceutics have uniquely prepared me to tackle these challenges. For drug delivery vehicle design, I have synthesized biodegradable polymers and used these to encapsulate antibiotics and proteins into particles on both the micron and nano scales. I have also used oligosaccharides from the surface of non-typeable *Haemophilus influenzae* and polyethylene glycol to modify the surfaces of polystyrene nanospheres to provide cell targeting and mucosal avoidance, respectively. The biodistribution, extracellular and cellular interactions of particles were evaluated utilizing *in vivo* murine models and *in vitro* models of the upper respiratory tract, mucosal secretions, and alveolar macrophages.

To supplement the *in vitro* experience, translational studies into particle aerosolization and aggregation of dry powder formulations have provided an understanding of deposition and distribution of inhalable therapeutic formulations. Combining this knowledge with my experience with human respiratory physiological measurements provides me with the tools required to address the challenge of designing inhalable therapeutic delivery systems for treatments of respiratory diseases.

(4dc) Advanced Components and Methodology in Microfluidic Systems
Rhee, M., Sandia National Labs

For almost three decades, microfluidic systems have shown a wide variety of advantages and possibilities for biological, medical, and analytical studies, yet their use in the life sciences is still limited because of insufficient functionality of components, the required level of operational knowledge by non-experts, and the level of fabrication expertise required for construction. Here, I present novel microfluidic components with their simplest operational forms and most advanced functionalities as well as futuristic concept microfluidic systems. All of the systems listed here were based on entirely new concepts at the time of publications, and my inventions brought about very active follow-up research on similar systems around the world.

First, a novel assembly approach for microdevice construction using prefabricated microfluidic components is presented. Microfluidic assembly blocks are basic building units to assemble custom devices for diverse applications. Each Microfluidic Assembly Block (MAB) has a unique fluidic function such as inlet/ outlets, valves, straight/curved/bifurcated channels, and chambers. The MAB methodology allows for full flexibility in planar configuration. The MAB system provides a simple way for non-fluidic researchers to rapidly construct custom, complex microfluidic devices.

Second, I present microfluidic digital pneumatic circuits that operate upon serially encoded pneumatic signals. Digital pneumatic components include most of electronic counterparts such as flip-flops, logic gates, and shift registers. These circuit components are integrated to form advanced microprocessors capable of performing complex parallel operations with a single input source. The proposed digital pneumatic microprocessors can be a potent universal on-chip platform as a 'self-thinking' fluid chip to autonomously manipulate microfluids in a high throughput manner.

Third, a novel design for efficient three-dimensional hydrodynamic focusing is presented. Hydrodynamic focusing in a microfluidic channel provides homogeneous reaction conditions that allow for various synthesis applications including polymeric nanoparticle synthesis by nanoprecipitation. Although microfluidic platforms enable rapid diffusive mixing and controlled precipitation resulting in homogeneous nanoparticle synthesis, control of unpredictable aggregation is a nontrivial issue since polymer particles tend to stick to channel walls and rapidly agglomerate inside, blocking the entire channel. Polymer aggregation can be efficiently avoided by three dimensional hydrodynamic focusing where the polymer stream is constrained both horizontally and vertically. The system not only enables stable and efficient focusing but has also a very straightforward structure, minimizing complicated device preparation process.

Fourth, I have designed a microfluidic technology that performs counter-current flow dilution where prepared NPs solution with impurities and washing deionized water are oppositely directed in the microchannel network. This counter-current flow maintains maximum concentration gradients of impurities that in turn facilitate diffusive dilution by washing flow. Since the organic solvent and non-encapsulated drug molecules have higher diffusivities than that of polymeric NPs, the impurities are washed out and NPs can be purified and concentrated with the specially designed microfluidic system where convection between two opposite flows is minimized by imposing high hydrodynamic resistance in connecting bridges. Such integration of purification functionality eliminates the need for manual collection, delivery, and filtration and thus accelerates the NP preparation and screening process by orders of magnitude.

Finally, I present a powerful microfluidic processor on a single chip, capable of performing (*i*) fluorescence-based flow cytometry and cell sorting, (*ii*) encapsulation of selected cells in picoliter droplets, and (*iii*) injection of amplification reagents into the droplets for single-cell genome amplification. Especially, to my best knowledge, this is the first report in the field that proposes on-demand droplet generation for single cell sorting. Encapsulation of bacteria in picoliter plugs in particular allows us to scale down conventional assays into much smaller reaction volumes better suited to the size of an individual microbe. By dramatically reducing the reaction volume, the effective concentration of template is also increased, reducing amplification artifacts that often arise in single-cell reactions carried out at a conventional scale.

(4do) Gas Transport and Sorption in Polymer Membranes for Energy-Efficient Separations
Smith, Z. P., The University of Texas at Austin

In the petrochemical industry, gas separations are traditionally accomplished using energy-intensive processes such as absorption and distillation. With proper synthesis and design, polymer membranes could provide an attractive alternative to these traditional separation techniques. However, there are three major challenges that need to be addressed to improve currently available polymer membranes. Polymers must have high permeability, high selectivity, and a resistance to plasticization. Permeability is the gas flux normalized by the film thickness and pressure driving force, and selectivity helps set the purity ratio of gas mixtures. Plasticization is the swelling of a polymer in the presence of condensable gases, which results in a decrease in gas selectivity during the operating lifetime of a polymer membrane.

In 2007, thermally rearranged (TR) polyimides were first investigated for separating CO_2 from CH_4. These polymers showed exquisite combinations of CO_2 permeability and CO_2/CH_4 selectivity while still maintaining a resistance to CO_2 plasticization under mild conditions. Because of these composite characteristics, this study aims to elucidate the mechanism of transport through TR polymers and to determine the relationship between polymer structure and gas separation properties.

TR polymers are formed via a solid state, thermally-induced decarboxylation reaction of polyimides containing reactive groups *ortho*-position to the polyimide diamine. Between temperatures of approximately 350°C and 450°C, these polymers are thermally reacted from fully soluble, processable polyimides to crosslinked, insoluble polybenzoxazoles. To understand the mechanism of transport for small molecules in TR polymers, the gas diffusivity and solubility was determined as a function of TR polymer conversion for H_2, N_2, O_2, CH_4, and CO_2. For CO_2, conversion of a TR polymer known as HAB-6FDA resulted in an over 20-fold increase in gas permeability and an over 10-fold increase in diffusivity. However, solubility only increased by a factor of approximately 2 during conversion. Therefore, TR polymers achieve their separation properties through largely increased gas diffusivity.

This increase in gas diffusivity indicates that TR polymers are potentially useful for separations typically governed by diffusion selective polymers such as olefin/paraffin separation. To investigate the utility of TR polymers for these separations, several polyimide structures with varying chemical backbone functionality were synthesized and tested for ethylene/ethane and propylene/propane separation. Similar to what has been observed for other gas pairs, conversion of TR polymers to their final polybenzoxazole structure resulted in increased permeability, a minimal loss in selectivity, and an increase in plasticization resistance.

(4bo) Plastic Electronics: Structure-Property Relationships of Polymer Semiconductors in Thin Film Transistors

Kim, F. S., University of Washington
Jenekhe, S. A., University of Washington

The era of plastic and organic electronics has emerged in the past decade towards the development and applications of low cost electronics and energy conversion systems.[1,2] Understanding of the charge transport properties of polymer semiconductors is critical to the future advances in plastic electronics. Our research aimed to understand the relationships between molecular structure, processing, morphology, and charge transport properties of polymer semiconductors. Organic field effect transistors (OFETs) are excellent platforms to study the charge transport properties of diverse conjugated polymers. The polymer thins were also studied by UV/Vis absorption and photoluminescence spectroscopies, cyclic voltammetry, X-ray diffraction, optical microscopy, atomic force microscopy, scanning and transmission electron microscopies. In addition, various molecular and interfacial engineering approaches were pursued to understand the physical factors that influence the performance of polymer transistors and to develop high-performance devices.

Our studies of structure-processing-morphology-property relationships in n-channel,[3,4] p-channel,[5,6] and ambipolar[2,7,8] polymer transistors revealed that variations in molecular structure and processing can be used to tune electronic energy levels, solid-state morphology and crystallinity, and thus charge-carrier mobility of the polymer semiconductors. The studies resulted in achievement of high-performance transistors with high mobility (1.5 cm^2/Vs).[5] We found that the long-term air-stability of n-channel and p-channel polymer transistors depended on the crystallinity (morphology) and electronic energy levels of the polymer thin films.[3,6]

Interface engineering was applied to reveal the effect of dielectric constant of a gate insulator layer on charge transport of n-channel polymer transistors.[9] The polymer transistors with a thin polymer dielectric buffer layer exhibited excellent charge transport properties, such as higher mobility, negligible hysteresis between forward and backward scans, and superior electrical stability under multiple cycles of gate voltage scans, compared to the device without the buffer layer. Electron mobility was found to increase exponentially with decreasing dielectric constant of the polymer dielectric buffer layer. These findings highlight the importance of engineering the dielectric/semiconductor interface for a better understanding of electron transport in polymer semiconductors and for developing high-performance OFETs.

We have also studied polymer semiconductor nanowires and their nanocomposites with an insulating polymer matrix as a means of controlling the solid-state morphology of active thin films in transistors.[10]

Well-dispersed polymer semiconductor nanowires with a high aspect ratio in an insulating polymer matrix resulted in high mobility (~0.02 cm^2/Vs) and DC conductivity (7 mS/cm) with a low percolation threshold (0.5 wt%). Intra- and inter-nanowire charge transport was proposed and examined as the origin of the observed high mobility/conductivity and composition dependence.

Selected Publications (30 total; total citation >1000).

(1) **Kim, F. S.**; Ren, G.; Jenekhe, S. A., *Chem. Mater.* **2011,** *23* (3), 682-732.

(2) **Kim, F. S.**; Guo, X.; Watson, M. D.; Jenekhe, S. A., *Adv. Mater.* **2010,** *22* (4), 478-482.

(3) Briseno, A. L.; **Kim, F. S.**; Babel, A.; Xia, Y.; Jenekhe, S. A., *J. Mater. Chem.* **2011,** *21* (41), 16461-16466.

(4) Guo, X.; **Kim, F. S.**; Seger, M. J.; Jenekhe, S. A.; Watson, M. D., *Chem. Mater.* **2012,** *24* (8), 1434-1442.

(5) Subramaniyan, S.; **Kim, F. S.**; Ren, G.; Li, H.; Jenekhe, S. A., *Macromolecules* **2012,** *45* (22), 9029-9037.

(6) Ahmed, E.; Subramaniyan, S.; **Kim, F. S.**; Xin, H.; Jenekhe, S. A., *Macromolecules* **2011,** *44* (18), 7207-7219.

(7) **Kim, F. S.**; Ahmed, E.; Subramaniyan, S.; Jenekhe, S. A., *ACS Appl. Mater. Interfaces* **2010,** *2* (11), 2974-2977.

(8) Wu, P.-T.; **Kim, F. S.**; Jenekhe, S. A., *Chem. Mater.* **2011,** *23* (20), 4618-4624.

(9) **Kim, F. S.**; Hwang, D.-K.; Kippelen, B.; Jenekhe, S. A., *Appl. Phys. Lett.* **2011,** *99* (17), 173303.

(10) **Kim, F. S.**; Jenekhe, S. A., *Macromolecules* **2012,** *45* (18), 7514-7519.

(4bm) Applications of High Performance Computing and Importance Sampling to Problems in Statistical Mechanics

Keys, A. S., University of Michigan

The field of statistical mechanics currently sits at the precipice of major breakthroughs in both its computational and theoretical components. Over the next decade, cloud computing and general-purpose GPU hardware will revolutionize the way molecular simulations are performed. Simultaneously, new importance sampling techniques will extend the reach of molecular simulation beyond typical equilibrium thermodynamics to far-from-equilibrium systems often encountered in clean energy, colloidal self-assembly, and biological applications.

In this poster, I present excerpts from my Ph.D. and post-doctoral research in the areas of soft matter and condensed materials. My Ph.D. work applies high-performance computing and importance sampling techniques to study the self-assembly of nano-and-micro scale target structures. In particular, my research focuses on assembling materials with a photonic band gap, which allows for the fabrication of tiny switches that can control the passage of light. My postdoctoral work applies importance sampling of non-equilibrium trajectory space to study the structure and dynamics of glassy materials. One important question my work addresses is how glasses differ from frozen liquids, and in particular, why a glass reacts very differently than an instantaneously-frozen liquid when melted. Important unifying concepts throughout my work are the ability of fluctuations and entropy to drive systems into interesting and unexpected structures, and the extension and application of importance sampling techniques to non-equilibrium phenomena.

(4cf) Synthetic Biology Tools for Protein and Genome Engineering

Mansell, T. J., Johns Hopkins University

Rapidly advancing technologies like multiplex oligonucleotide synthesis and next-generation high-throughput sequencing have opened the door to tremendous progress in synthetic biology. My experience in both protein engineering and strain engineering puts me in an excellent position to forge the **next generation of synthetic biology tools for protein and genome engineering**.

My research career started as an undergraduate at Johns Hopkins University under **Prof. Marc Ostermeier.** I helped engineer **allosteric molecular switches via directed evolution** (Guntas, Mansell, et al., PNAS, 2005). During my PhD (with **Prof. Matthew DeLisa**, Cornell), I developed a suite of tools for examining protein folding, protein-protein interactions, and post-translational modifications in *E. coli*. I developed a **protein folding reporter** for the bacterial periplasm (Mansell et al., Protein Science, 2010) using a Beta-lactamase (BLA) based selection. I showed that we could not only probe intrinsic protein folding, but **factors in the cellular environment** (e.g., chaperones, redox state) that affected protein folding. In addition, I optimized a system for detecting **protein-protein interactions** in the periplasm using split BLA. Finally, I refactored the protein folding reporter to function as a detector for **glycosylation in bacteria** (Mansell et al., Biotech. J., 2013) as a pipeline for engineering post-translational modifications. Basically, I was in the business of **selection design**.

My postdoctoral work with **Prof. Ryan Gill** at the University of Colorado Boulder gave me a broader view of both metabolic engineering and the design of directed evolution experiments. By combining **multiplex oligonucleotide synthesis and recombineering**, I developed a platform for directed evolution of orthogonal repressors for cellular circuits on both the DNA and RNA level. My postdoctoral work has given me a unique perspective on generating targeted libraries, generating **programmed diversity** to complement my expertise in selection design.

My future research directions will combine protein engineering and strain engineering to produce high-value products in bacteria. I am particularly interested in the effect of **post-translational modifications** on protein stability and function. Multiplex recombineering and high-throughput sequencing allow us to glean an unprecedented amount of information about structure-activity relationships, including the engineered placement of post-translational modifications in a protein. Multiplexed synthesis allows for heretofore unapproachable pathway engineering opportunities as well. By combining these molecular toolkits, we can develop platforms to engineer both native hosts and exogenous pathways for the production of therapeutic proteins, biofuels, and other high-value products using **programmed diversity and smart selections.**

Finally, I have a particular interest in teaching, especially in teaching biology to chemical engineering students. I have been trained in applying active learning techniques in the classroom to keep students engaged and involved in the material. **Active and cooperative learning** will be central to whichever courses I teach. As Niels Bohr once supposedly said: "Biology is too important to be left to the biologists."

(4db) Structural Characterization of Conjugated Polymer/Fullerene Dispersed and Thin-Film Composites for Organic Photovoltaic Applications

Richards, J. J., University of Washington
Pozzo, D., University of Washington

Understanding the structural morphology of conjugated polymer/fullerene composites is an important step toward improving the performance of photovoltaic devices. The efficiency of exciton dissociation and carrier transport to the working electrodes are both important factors to optimize with respect to thin-film morphology.

In thin-film composites, we have shown that generally the films can be viewed as composed of three main phases: a polymer aggregate phase, a fullerene aggregate phase, and a disordered polymer/fullerene phase. The concentration, distribution, and composition of these phases are directly tied to how devices are processed and the specific film treatment history. However, regardless of the specific thermal treatment we have shown that the performance of the thin-film composite can be understood by examining its instantaneous morphology. Coupled with spectroscopic and microscopy techniques, Grazing Incidence Small Angle X-Ray (GISAXS) scattering techniques provide detailed structural information about the structural morphology of these thin-films and provides a means to understand performance of thin-film composites.

Similar to their thin-film counterparts, the performance of photovoltaic devices derived from conjugated polymer/fullerene composite nanoparticles (CNPs) is a function of their internal structural morphology. However, whereas the structural morphology of the thin-film composite is a function of the detailed history of the coating process, the structure of CNPs is fixed as a result of their fabrication. Because of this fact, the structure can be identified in the dispersed phase and tied, using single particle characterization, to its performance before it is deposited into a device. Therefore, the optimization of devices derived from CNPs can be broken into two discrete problems: improving the **intrinsic** properties of a nanoparticle as a function of how it is produced and optimizing the **extrinsic** properties of devices through the deposition of optimized CNPs. Our work has addressed both of these challenges. First, using model conjugated polymer and fullerenes, poly(3-hexylthiophene) (P3HT)/[6,6]-phenyl-C_{61}-butyric acid methyl ester (PCBM), we show that internal phase segregation of P3HT/PCBM CNPs is a complex function of their production and composition. Second, we demonstrate the deposition of monolayers of these CNPs onto device relevant electrodes. The combination of these two separate approaches opens up the potential for novel roll-to-roll processes to produce high performance photovoltaic modules of arbitrary complexity with flexible form factors.

(4bg) Nanotechnology-Assisted, Mechanism-Based Phototherapy of Cancer

Huang, H. C., Massachusetts General Hospital, Harvard Medical School

Photomedicine involves the understanding and application of light and it's effect to advance health care and medical science. My post-doctoral research, with Prof. Tayyaba Hasan, focuses on the photochemical approaches for the treatment (photodynamic therapy, PDT) and diagnosis of pancreatic and ovarian cancers. The overall strategy is to develop nanotechnology-assisted, mechanism-based, imaging-guided combination treatment regimens that involve photo-activation of near-infrared-absorbing chemicals. My Ph.D. thesis, with Prof. Kaushal Rege, investigated the fundamental thermophysical and biological responses of the light activated-gold nanoparticles. We have developed multifunctional nanobioconjugates, and then applied them to improve prostate cancer treatment (combination drug delivery and photothermal therapy, PTT), and gastrointestinal anastomoses (photothermal tissue welding).

In this poster, I highlighted both the fundamental studies and the technological applications from my research, which demonstrate my ability in critical thinking, designing and executing experiments from theranostic nanoparticle synthesis to *in vivo* testing. My future research program will build upon my expertise in multifunctional nanoconstructs, photochemical (photothermal) approaches to (i) understand the fundamental therapy resistance mechanisms via photo-activation (or priming) of sub-cellular organelle, which will guide (ii) the development of advanced combination therapies for cancer (or other diseases) in a translational research setting.

Selected publications (out of 12):

Huang, H.C.; Walker, C.R.; Nanda, A.; Rege, K. *Laser welding of ruptured intestinal tissue using plasmonic polypeptide nanocomposite solders*. ACS Nano, 2013, 7(4), 2988-2998

Huang, H.C.; Nanda, A.; Rege, K. *Investigation of Phase Separation Behavior and Formation of Plasmonic Nanocomposites from Polypeptide-Gold Nanorod Nanoassemblies*. Langmuir, 2012, 28, 6645-6655

Huang, H.C.; Yang, Y.; Nanda, A.; Koria, P.; Rege, K. *Synergistic administration of Photothermal Therapy and Chemotherapy to Cancer Cells using Polypeptide-based Degradable Plasmonic Matrices*. Nanomedicine, 2011, 6(3), 449-458

Huang, H.C.; Heys, J.J.; Rege, K. *Spatiotemporal Temperature Distribution and Cancer Cell Death in Response to Extracellular Hyperthermia Induced by Gold Nanorods*. ACS Nano, 2010, 4(5), 2892-2900

Huang, H.C.; Barua, S.; Kay, D.B.; Rege, K. *Simultaneous Enhancement of Photothermal Stability and Gene Delivery Efficacy of Gold Nanorods Using Polyelectrolyte*. ACS Nano, 2009, 3(10), 2941-2952

(4ci) New Materials and Photon Management for Scalable Solar Energy Conversion

McPeak, K. M., ETH Zurich

The late Nobel Laureate, Richard Smalley, declared "Energy is the single most important challenge facing humanity today." The use of large scale, renewable energy sources is necessary to meet the demands of both population growth and CO_2 reduction in the 21st century. Solar energy is the only clean, renewable energy source available that is capable of supplying the 13 terawatts that are necessary to power the planet. I will develop a research program which focuses on harnessing this abundant resource for energy conversion with photoelectrochemical devices. The utilization of both advanced photon management techniques and new materials to enhance the performance of these devices will be the primary focus of my group.

During my PhD I studied the chemical bath deposition of nanomaterials for low cost photovoltaic devices. I developed a combinatorial approach for depositing oxide and chalcogenide nanomaterials which enabled rapid and unprecedented understanding of the effect of growth conditions on optical, electronic, and structural properties of the material. Furthermore, I pioneered the use of *in situ* x-ray absorption fine structure spectroscopy to better understand the fundamentals of chemical bath deposition.

Throughout my postdoctoral work at ETH Zurich I have developed novel, low cost nanofabrication techniques to construct large-area, complex, two dimensional metal grating structures for back reflectors in solar cell devices. I have used both finite element modeling and experimental results to verify the efficacy of these grating structures on improving solar cell efficiency.

As a faculty member, I will combine my bottom-up material synthesis, top-down nanofabrication and simulation expertise to establish a leading program in solar-to-fuel energy conversion. This application area is fundamental to energy sustainability and public health in the world.

(4cv) Ophthalmic Drug Delivery System: From Tear Dynamics to Contact Lenses

Peng, C. C., University of California, Berkeley

Traditional ophthalmic drug delivery via eye drops accounts for more than 90% of current ocular drug administration. Unfortunately, drop instillation is an inefficient route since only 1-5% of the applied drug enters the cornea while the remaining drains or is absorbed by the bloodstream leading to side effects. Nevertheless, topical ophthalmic drug delivery remains the primary application modality to the human eye. This research focuses on improving the pharmacokinetic (PK) accuracy of topical ocular drug delivery by a tear-dynamic-based continuum model, and on developing new silicone hydrogel (SiHy) soft contact lens materials as extended drug delivery vehicles. The aims of my research are to improve the ophthalmic drug delivery prescription so that FDA approval of formulation and dosage is expeditious and to develop a long term solution for extended drug delivery.

I. Tear-Dynamic-Based Pharmacokinetics Model for Topical Ophthalmic Drug Delivery

Classical PK models are limited to measuring concentrations in ocular tissues over time following single or multiple administration, which prevents quantitative bioavailability predictions for changes in formulation or dosage regimen. We propose a compartmental description of human tear dynamics that quantifies drug pharmacokinetics (PK) by topical administration including bioavailability in each compartment of the anterior eye. The tear supply/drainage system is divided into 5 compartments: upper and lower menisci, upper and lower conjunctival sacs, and tear film. A transient description is employed for each compartment for tear volume, salinity, and drug concentration. Compartments are coupled to each other and to drug delivery through kinetic routes.

This tear-dynamic-based PK model predicts the effect of drop size and dosing sequences on the drug loss to conjunctiva and tears drainage, which is directly related to the systemic side effects of drug delivery. Another case study predicts precorneal residence time of a fluorescent tracer mixed with commercial artificial tears. The viscosity of solution depends on the concentration of active agent. Although dilution reduces the initial high viscosity, thickness of precorneal tear film increases and tear drainage rate decrease, thus extending drug residence time. These predictions are consistent with clinical observations.

The proposed tear-dynamics model is the first attempt to develop a topical ocular drug-delivery model based on compartmental tear dynamics. The model is especially useful for optimizing drug formulations and dose regimens for both normal and dry-eye subjects and, thus, for streamlining FDA approval. Further, quantitative prediction of conjunctival uptake can be integrated with available PK models to improve drug uptake prediction accuracy in both the anterior and posterior segments of eye. The new model is suitable for quantifying ocular drug delivery by other vehicles including contact lens, and emulsion and nanoparticle suspensions.

II. Extended Ophthalmic Drug Delivery by Soft Contact Lenses

Contact lenses serve as an alternative ocular drug delivery vehicle due to higher drug bioavailability and potentially enhanced patient compliance. However, current commercial hydrogel contact lenses do not provide sufficient transport resistance, usually releases the included drug for no longer than a few hours. Our approach to achieving extended and controlled ocular drug delivery focuses on creating biocompatible transport barriers within commercial hydrogel contact lenses by using vitamin E as a diffusion-barrier material. Nanosized vitamin E barriers are created in a prefabricated contact lens by soaking the lens in a solution of vitamin E in ethanol, followed by rapid extraction of ethanol in water. Transport rates for several drugs with various hydrophobicity and surface activity were successfully modeled proving the validity of the approach. We also investigate the impact of the vitamin E barriers on transport of ions and oxygen, and on other properties relevant to the safety of contact lens wear. Finally we conducted the *in vivo* animal experiments in glaucoma model of Beagle dogs to deliver a glaucoma drug through the vitamin E loaded contact lenses. Our combination of *in vitro*, *in vitro*experiments along with mathematical models for transport prove that glaucoma and several other ophthalmic diseases can be treated effectively with vitamin E impregnated contact lenses.

-- **Peer-Reviewed Publications**

1. **C.-C. Peng**, C. Cerretani, R. J. Braun, C. J. Radke, ***Adv Colloid Interface Sci*** (2013), accepted
2. C. Cerretani, **C.-C. Peng**, A. Chauhan, C. J. Radke, ***Contact Lens & Anterior Eye***, 35 (2012) 260-265
3. **C.-C. Peng**, M.T. Burke, B.E. Carbia, C. Plummer, A. Chauhan, ***J Controlled Release***, 162 (2012) 152-158
4. **C.-C. Peng**, A. Chauhan, ***J Membrane Sci***, 399-400 (2012) 95-105
5. **C.-C. Peng**, A. Ben-Sholmo, E. MacKay, C.E. Plummer, A. Chauhan, ***Curr Eye Res***, 37 (2012) 204-211
6. **C.-C. Peng**, M. T. Burke, A. Chauhan, ***Langmuir***, 28 (2012) 1478-1487
7. **C.-C. Peng**, A. Chauhan, ***J Controlled Release***, 154 (2011) 267-274
8. **C.-C. Peng**, L.C. Bengani, J. Leclerc, C. Gupta, A. Chauhan, ***J Drug Deliv Sci Tech***, 21 (2011) 111-121
9. J. Kim, **C.-C. Peng**, A. Chauhan, ***J Controlled Release***, 148 (2010) 110-116
10. **C.-C. Peng**, Jinah Kim, Anuj Chauhan, ***Biomaterials***, 31 (2010) 4032-4047

(4cg) Electrokinetics for Analysis and Manipulation of Biological Systems
Marshall, L. A., Stanford University

Electrokinetics for Analysis and Manipulation of Biological Systems

Lewis A. Marshall

Stanford University

Stanford, CA 94305, USA

Electrokinetics effects, effects caused by the movement of ions under electric field, play an important role in biological systems. On the level of individual molecules, electrostatic attraction, repulsion, and shielding can dramatically influence binding between biomolecules. On the cellular level, selective ion transport can create potential gradients that affect cell behavior. At the scale of an individual organism, neurons achieve synchronized control of cells throughout the body by relaying electrochemical signals, and some organisms, such as sharks, can even sense the electric fields around them to navigate the world.

My most recent research activity has been in the area of sample preparation of nucleic acids using isotachophoresis, an electrophoretic technique in which shockwaves focus ions into highly concentrated zones. I have used this electrokinetic technique, in combination with control of sample chemistry, to selectively purify nucleic acids from highly complex samples, including whole human blood. At the same time, I have worked to design better devices to perform this technique by drawing on knowledge of capillarity and microfluidics.

I would like to develop a research program to bring a deep knowledge of electrokinetic phenomena to the study of living biological systems. By using an electrokinetic framework to the study of cell biology and by building microfluidic platforms for cell analysis and manipulation, I hope to bring together a research team that can contribute to our understanding of the basic physiology of neurons and other electrically active cells. Ultimately, I hope to contribute to the technological development of neural interfaces for control of computers and prosthetic devices, which rely on constructing a deep understanding the information that we can receive from electrically active cells.

(4em) Engineering Tailor-Made Microbial Cell Factories With Metabolic Engineering and Synthetic Biology

Xu, P., Rensselaer Polytechnic Institute
Koffas, M. A. G., Rensselaer Polytechnic Institute

Global energy demand, rising petroleum prices and environmental concerns have stimulated increased efforts to develop more sustainable and cost-effective chemicals that are derived from renewable resources. Metabolic engineering entails targeted manipulation of biosynthetic pathways to maximize yields of desired products. As an emerging discipline, synthetic biology is becoming increasingly important to design, construct and optimize metabolic pathways leading to desired overproduction phenotypes. Exploitation of the diverse microbial pathways based on metabolic engineering and synthetic biology frameworks provides a promising solution to synthesis of fuel and pharmaceutical molecules that are previously produced from the petroleum-based resources.

Harnessing cell factories for producing biofuel and pharmaceutical molecules has stimulated efforts to develop novel synthetic biology tools customized for modular pathway engineering and optimization. Here we present a versatile gene assembly platform ePathBricks supporting the modular assembly of multigene metabolic pathways and combinatorial generation of pathway diversities. Based on this gene assembly platform, we have combinatorially optimized the transcriptional levels of three genetic modules in the fatty acid pathway and identified conditions that balance the supply of acetyl-CoA and consumption of malonyl-CoA/ACP. Refining protein translation efficiency by customizing ribosome binding sites for both the upstream acetyl coenzyme A formation and the downstream fatty acid synthase module has enabled further production improvement. Fed-batch cultivation of the engineered strain led to a final fatty acid production of 8.6g/L. The modular engineering strategies demonstrate a generalized approach to engineering cell factories for valuable metabolites production.

As heterologous pathways become larger and more complicated, it becomes increasingly difficult to optimize them with static regulatory control. An ideal approach would be to use dynamic regulatory networks to control gene expression so that the host cells can adjust their metabolic functions when the environmental condition changes. Here we report engineering both the positive and negative feedback controls for dynamic tuning of metabolic flux in *E. coli*. We have identified a dual transcriptional regulator which can act either as an activator or a repressor for two different promoters. The level of activation or repression is dependent on the level of intracellular malonyl-CoA. As a proof of concept, we demonstrated that the expression of two reporter proteins can be exclusively switched between the on and off state. By engineering this synthetic malonyl-CoA controller, we envision that both the malonyl-CoA source pathway and the malonyl-CoA sink pathway can be dynamically modulated so that carbon flux can be efficiently redirected to synthesize our target compounds. Implementation of this dynamic control will maintain the intracellular malonyl-CoA at the optimal level and improve both the productivity and yield of value-added metabolites in *E. coli*.

(4dl) Molecular Simulations of Room Temperature Ionic Liquids - Novel Solvents for Sustainable Energy and Environment

Shah, J. K., University of Notre Dame

Ionic liquids – defined as substances formed due to combination of ions and existing as liquids near ambient conditions – have emerged as novel solvents over the last 15 years. Among the most beneficial properties of ionic liquids are their negligible vapor pressure and a large number of cations and anions providing the flexibility with which they can be designed for specific physicochemical properties. Additionally, ionic liquids exhibit a number of desirable properties such as low flammability, good solvation properties for both polar and nonpolar substances and large electrochemical window. Such favorable properties of ionic liquids are the main drivers for efforts directed at using ionic liquids as potential solvents as CO_2 capture media and gas separation agents, lubricants, reaction media, solvents in dye-sensitized solar cells, batteries and refrigeration cycles.

One of the challenges in the ionic liquid research is to rationally design an ionic liquid for desired physicochemical and biological properties. As one can envision nearly a billion ionic liquids, a simple brute force approach to synthesizing and measuring thermophysical properties of pure ionic liquids and phase equilibria with other substance is infeasible. My doctoral research and current efforts as a research assistant professor are geared towards unraveling molecular level interactions between ions and other substances with molecular simulation techniques.

In this presentation, I will provide key results from Monte Carlo and molecular dynamics studies for prediction of thermophysical, transport and phase equilibria properties of pure ionic liquids and their mixtures with water and a number of gases.

As ionic liquids are specialty materials, chemical processes carried out in bulk ionic liquids can be cost prohibitive. In such instances, nanoconfinement of ionic liquids offers an attractive alternative. To this end, I will discuss how confinement of ionic liquids at nanometer lengthscales alters CO_2 and H_2 gas sorption in ionic liquids for potential application in precombustion CO_2 capture.

My proposed research as a faculty will draw upon the ionic liquid and molecular simulation research for *in silico* screening and design of environmentally benign ionic liquids. I am also interested in establishing research in dye-sensitized solar cells (DSSCs) with focus towards increasing the solar-to-electric conversion efficiency. Efforts will be directed towards application and development of novel computational tools to gain molecular level understanding of processes responsible for conversion of solar energy into electricity.

(4eg) Accelerating Iterative Materials Design With Simulations and Experiments

White, A. D., University of Washington

The Materials Genome Initiative (MGI) is a national concerted effort to accelerate materials design. The MGI aims to accomplish this by leveraging advances made in information technology from genetics and bioinformatics. A key goal of MGI is to enable rapid integration of simulations and experiments for rational design of materials. However, the connection between models of materials and experiments is large and growing as nano-composites, meta-materials, and other exotic new material classes outpace innovation in simulations and modeling. Whereas other research efforts have focused on ab initio models, where material property predictions are made with little to no experimental data, my past and future research is the unification of experimental data and molecular models. It is uncommon to pursue materials design without experiments, and this should be considered when constructing molecular-leve models. My PhD research focused on designing self-assembling biomaterials for use in complex biological environments by iteratively combining bioinformatics, simulation, and QSAR models with high-throughput experiments. In addition to materials design, I combined experiments and simulations to better understand peptide self-assembly, hydration of charged molecules, and protein chaperone models. I plan to continue this new direction of molecular modeling where iterative materials design is accomplished by new methods of combining experiments and models.

(4c) Automated, Flowable Formats for Carbon Dioxide Sequestration and Tailored Manufacturing of Colloidal Nanomaterials

Abolhasani, M., University of Toronto
Guenther, A., University of Toronto

Since our demand for energy is increasing due to the growth of populations and development of countries, the emissions of greenhouse gases (mostly carbon dioxide (CO_2)) from fossil fuels will constantly increase. Currently, carbon capture and sequestration is the most efficient solution to control and reduce the concentration of the released greenhouse gases in the atmosphere. But the underground storage of the captured CO_2 is expensive and limited to some specific locations. On the other hand, the current rate of energy consumption and inevitable future demands means that energy will be one of the crucial issues of the near future. Therefore, there will be a huge demand for new sets of sustainable energy substitutes. In 2008, it was found that CO_2 in the presence of a frustrated Lewis pair and hydrogen can be converted to methanol and water. This chemical reaction has the potential to be the key solution to address both of the above-mentioned challenges; captured carbon storage and energy.

In my PhD, I was working on a multidisciplinary project, collaborating between Mechanical Engineering and Chemistry Departments. During this time I made significant contributions in flow chemistry and microfluidic (MF) research fields specifically related to screening gas solubilities in liquids, measuring carbon dioxide-physical solvent mass transfer characteristics associated with CO_2 sequestration and on-demand preparation of high quality colloidal nanomaterials (e.g. CdSe/CdS).

In my PhD project, I have designed and developed the first automated MF platform that allows the routine screening of solubility data for a wide range of CO_2-physical solvent mixtures. The platform allows screening of approximately 300 different conditions (temperature and pressure) during one day. I have also designed a MF platform for high temperature (240-330 C) synthesis of high quality (Fwhm ~30 nm) nanocrystals such as CdSe, CdS and ZnS. This automated platform can produce high quality core or core/shell quantum dots with the desired spectral characteristics at a throughput of 0.6 ml/min.

As an independent scientist, I will focus on integrating the carbon capture process with its fuel conversion step. To do this, I will use MF technologies as a tool for screening the reaction conditions and process development and optimization to develop a robust, integrated and high throughput system for capturing and converting CO_2 to fuel. I will also study the underlying transport phenomena in the presence of chemical reactions with experimental and numerical methods.

(4bh) Hierarchical Nanostructured Electrodes for Efficient Electrochemical Energy Storage Devices

Hyder, N., Massachusetts Institute of Technology

Next generations electrochemical energy storage devices for hybrid or plug-in-hybrid vehicles would require both high-power, high-energy density with lower emission for sustainability. To achieve the goals, it is essential to identify novel electrode materials that can accommodate efficient transport ions and electrons in addition to high surface area. Energy storage devices based on based on nanoscale materials (nanoparticles, nanotubes or nanofibers of electroactive polymers, metal oxides and carbon nanotubes) recently gained significant interest to afford high-power, high-energy density without compromising cycle life. During my postdoctoral research at MIT, I've developed expertise in the design of novel electrode assembly for energy storage devices. I focused on developing hybrid electrodes based on conjugated polymers (Polyaniline, PANi), metal oxides (TiO_2) and multi-walled carbon nanotubes (MWNT); these nanostructured hybrid electrodes performed exceptionally well (compared to graphitic electrodes) delivering high-power, high-capacity simultaneously for electrochemical pseudocapacitors that suitable for electric vehicles or plug-in electric vehicles. Along with conventional electrode development, I designed alternative layer-by-layer assembled electrodes based on electrostatic interaction of nanomaterials. The electrode architecture, assembly technique and post treatment were optimized to create high performance films. These electrodes were engineered to achieve high electronic conductivity and high porosity for improved ion transport of aqueous or organic electrolytes rendering higher capacitance than graphite electrodes with excellent mechanical and electrochemical stability over thousands of cycles. The development of hybrid electrodes of MWNTs and PANi nanofibers with controlled aspect ratio and morphology exhibits attractive potential in terms of improved electronic conductivity and porosity that are competitive with state-of-the-art RuO_2. However, these multilayer electrodes of nanoscale materials would reach a performance limit unless we optimize the electrolytes that can severely affect ion conduction and thereby overall electrochemical performance. Polymer-based gel electrolyte could a solution to address the challenge that could be part of the separator membrane in the overall electrochemical cell design. My Ph.D research expertise on polymeric membrane design and manufacturing at the University of Waterloo would allow me to develop such novel separator membranes.

In future, as an independent researcher, I would like to focus on developing hierarchical nanostructured electrodes and electrolytes for the fundamental study and development of energy storage and conversion technologies. The novel electrode materials along with the gel electrolyte have tremendous potential to impact the electrode architecture and performance for high performance energy storage devices like metal-air batteries and pseudocapacitors. I will apply the expertise and ideas of nanomaterial synthesis, self-assembly, structure-property-processing of materials electrochemistry and the study of electronic and ionic interactions to create efficient and robust energy storage device for a sustainable environment.

(4de) Unravelling Structure-Activity Relationships of Heterogeneous Catalysts

Runnebaum, R. C., University of California, Berkeley

As the world's sources of energy and hydrocarbons evolve, much opportunity is created in the fields of catalysis and catalytic materials. Several challenges for which I am particularly well-positioned to make important advancements include the following: developing fundamental understanding of selective removal of heteroatoms from hydrocarbons found in feedstocks such as shale, coal and biomass; elucidating structure-activity relationships for reactant adsorption and conversion via in-situ catalytic and spectroscopic characterization; synthesizing catalytic materials, including zeolites and enzyme-inspired supported-organometallic clusters, to advance the understanding of structure-activity relationships. Data such as these will lead to the development of catalysts and processes for the conversion of new energy and chemical feedstocks.

As a post-doctoral researcher at University of California—Berkeley, in conjunction with Dr. Alexander Katz and Dr. Stacey Zones, I am elucidating structure-activity relationships for new classes of catalytic materials—delaminated zeolites and supported metallocalixarenes—by using probe reactions. Mild delamination of layered zeolites, which retain high structural integrity and have acid sites available to bulky molecules, is expected to expand the range of molecules useable for zeolite-catalyzed reactions. Supported metallocalixarenes comprise bulky ligands on metal clusters to facilitate stability, allow access by small molecules, and create tunable metal catalysts by varying ligand electron donating or withdrawing properties. The supported enzyme-like cluster catalysts are characterized in-situ via catalysis, spectroscopic, and isotopic exchange experiments to determine structure-activity relationships. The combination of synthesis (by which the structure of the catalyst can be modified) and catalysis gives me tools to better elucidate fundamental understanding in the adsorption and catalytic conversion of new feedstock molecules.

My thesis work at the University of California—Davis, in conjunction with Dr. Bruce Gates and Dr. David Block, focused on developing a fundamental understanding of the catalytic conversion of lignin-derived compounds to fuels. Most researchers had focused on the conversion of sugars and polyols from cellulose degradation, and left the conversion of nearly 30 wt% of biomass—lignin—untouched. Our goal was to determine such a detailed reaction network for compounds representative of lignin-derived bio-oils and an important functional group characterizing compounds in them. Data characterizing these reactions determined conversions and yields of products, including trace products; an approximate reaction network for each reactant and generalization of the results to reactant types and reaction classes; conversion data sufficient to approximately quantify kinetics of the principal reactions in the networks. The data provide understanding to allow resolution of the roles of the separate catalyst functionalities (e.g. solid acid and metal) and provide a basis for prediction of reactivities of whole bio-oils. This research area was new for Professors Gates and Block. As a result, I now have experience in starting up an academic research lab including designing and constructing laboratory process equipment necessary for my research.

Future work will build upon my expertise in catalyst synthesis of zeolites and supported enzyme-inspired organometallic clusters and in heterogeneous catalysis to develop fundamental understanding of structure-activity relationships for the formation of useful chemicals and fuels and to provide a basis for understanding more complicated systems.

(4m) Nanostructured Materials for Solar Energy Conversion
Bohn, C., National Institute of Standards and Technology

Title: Nanostructured Materials for Solar Energy Conversion

Author: C.D. Bohn

The sun provides 10,000 times more energy to the earth than the current global consumption rate of 16 TW. Storing solar energy to overcome diurnal supply is critical for meeting growing energy demand. My future research lab will focus on solar energy technologies including (i) photoelectrochemical (PEC) hydrogen production, *i.e.* solar water splitting, and (ii) solar thermochemical cycles for fuel production.

Inexpensive metal oxides such as α-Fe_2O_3 with a bandgap of approximately 2.1 eV can theoretically exceed a photoelectrochemical solar to hydrogen efficiency of 10 %. In my poster, I show that activating n-type dopants in iron oxide by thermal annealing improves efficiency by three orders of magnitude to approximately 0.1 %. Additionally, I establish a correlation between conductivity of the metal oxide and photocurrent. Preliminary strategies to exceed the current efficiency record of approximately 0.4 % with α-Fe_2O_3 including, increasing the surface area of the photoanode by nanostructuring and optically manipulating incident light using metallic contacts, are discussed. Additional results are presented for another earth abundant photoanode material, SiC. The stability of n-type 6H-SiC in a pH=7 buffer and the effect of applying a Pt catalyst to the surface is presented. Finally, I will also discuss my future teaching and outreach plans at this poster session.

Education

- **Center for Nanoscale Science and Technology**, (2011-Present)

National Institute of Standards and Technology

Advisor: Dr. Veronika Szalai

- **University of Cambridge**, (2006-2010)

PhD with Gates Cambridge Scholarship

Advisor: Prof. J.S. Dennis

- **Princeton University**, (2002-2006)

BSE with honors

Recent Publications:

1. Bohn, C.D.; Agrawal, A.K.; Walter, E.C.; Vaudin, M.D.; Herzing, A.A.; Haney, P.M.; Talin, A.A.; Szalai†, V.A. Effect of tin doping on α-Fe2O3 photoanodes for water splitting. *J. Phys. Chem. C* 116, 15290–15296 (2012).
2. Bohn, C.D.; Scott, S.A.; Dennis, J.S.; Müller, C.R. Validation of a lattice Boltzmann model for gas–solid reactions with experiments. *J. Comput. Phys.* 231, 5334–5350 (2012).
3. Bohn, C.D.; Cleeton, J.P.E.; Müller, C.R.; Davidson, J.F.; Hayhurst, A.N.; Scott, S.A.; Dennis, J.S. The kinetics of the reduction of iron oxide by carbon monoxide mixed with carbon dioxide. *AIChE J.* 56, 1016–1029 (2010).
4. Bohn, C.D.; Cleeton, J.P.E.; Müller, C.R.; Scott, S.A.; Dennis, J.S. Stabilizing iron oxide used in cycles of reduction and oxidation for hydrogen production. *Energy Fuels.* 24, 4025–4033 (2010).

(4bx) Biomolecular Engineering for Renewable Fuels and Chemicals

Li, H., University of California

Fossil energy sources supply about 82% of US's total energy need. While the total reserve of fossil resources is finite, the energy consumption grows at a stunning rate. Moreover, CO_2 produced from fossil fuel combustion has been implicated to contribute to climate changes. These reasons argue for an eventual replacement of fossil energy sources, which means that the fuels and chemicals must ultimately be produced from CO_2, H_2O, and sunlight. To this end, my doctoral work in Dr. James Liao's group at UCLA was focusing on engineering of the biocatalysts to produce pharmaceutical intermediates, industrial commodities, and gasoline substitutes from a variety of renewable feed stocks.

i) A synthetic pathway has been designed to produce homophenylalanine, a core pharmaceutical intermediate, in *Escherichia coli*. To make this non-native chemical accessible to metabolic pathways, the key enzymes in carbon chain elongation and amination in *E. coli* were engineered by rational design and directed evolution. The carbon chain elongation platform demonstrated in this work could also be used to produce other non-native chemicals such as long chain alcohols. ii) Heterotrophic microbes such as *E. coli* utilize sugars as the feed stocks which are generated from plant biomass decomposition. To bypass plant biomass, photosynthetic chemical production was explored using the cyanobacterium *Synechococcus elongatus*. A synthetic pathway was installed in *S. elongatus* to produce 1,2-propanediol, an industrial commodity, directly from CO_2. The pathway was further tailored to match the special metabolic features in the photosynthetic organism. iii) The photosynthetic microbial process faces challenges in scaling-up due to the requirement of the light exposure surface area. To achieve light independent fuel and chemical production from CO_2, a lithoautotrophic microorganism *Ralstonia eutropha* was used. Metabolic engineering and electrochemical process design approaches were used to produce C4 and C5 alcohols, the gasoline substitutes, in an integrated electro-bioreactor using CO_2 as the sole carbon source and electricity as the sole energy input.

Taken together, biomolecular engineering may play a critical role in building of a sustainable society. Meanwhile, it is also one of my research interests to understand the universal chemical logic behind the versatile metabolic pathways, and the fundamental mechanisms of enzymatic catalysis. To further train myself in biochemistry and protein science, I will do my postdoc research in Dr. Peter Schultz's group in the Scripps Research Institute. The ultimate goal would be to address some of the most interesting scientific and engineering challenges in renewable biomolecule synthesis using metabolic engineering, protein engineering, synthetic biology, bioinformatic, and process engineering approaches.

(4ao) Tissue Engineering Tools for Modulating the Immune Response

Gower, R. M., Northwestern University

My research focuses on understanding and exploiting the immune system at the systems level. At the intersection of engineering and immunology, this program utilizes tissue-engineering tools including biomaterials, gene and protein delivery, and cell and tissue transplant to enhance regenerative strategies and cure disease. My overall goal is to define the fundamental control mechanims for directing the immune response to biological and artificial materials so that they may be used to modulate cellular and molecular behavior to favor immunity or tolerance, as desired. This presentation features the development and implementation of technologies to direct the trafficking, differentiation, and activation, of key immune cell populations in order to orchestrate the immune system's response to implanted biomaterials and transplanted tissue.

(4ad) High-Performance Phosphorescent Organic Light-Emitting Diodes (PhOLEDs) Achieved By Orthogonal Solution-Processing Strategy

Earmme, T., University of Washington
Jenekhe, S. A., University of Washington

Phosphorescent organic light-emitting diodes (PhOLEDs) are being extensively explored for scientific interest and for commercial applications in display technology and solid state lighting. High-performance PhOLEDs are currently fabricated by sequential thermal evaporation to obtain multilayered structures. Despite the considerable progress made in developing vacuum-processed PhOLEDs, studies of potentially low cost solution-based high-performance PhOLEDs are relatively few. One of the key challenges to fabricate high performance multilayered organic electronic devices by solution-processing is that the solvent used to deposit the subsequent layer can easily dissolve the underlying layer. A general approach to overcome this problem is to develop an orthogonal solution-processing strategy, whereby a second solvent is used to deposit the overlayer without dissolution of the underlying layers. We have developed a novel orthogonal solution-processing of electron-transport materials (ETMs) and fabrication of high-performance multilayered PhOLEDs with triplet emitter-doped polymeric emission layers followed by orthogonal solution-processing of new or commercial small molecule ETMs. Charge injection and transport of ETMs were enhanced by orthogonal solution-processing and thus led to significant increase of the overall device performance. We have achieved a luminous efficiency of 37.7 cd/A with an EQE of 19.0%, which is the best device performance observed to date in all-solution-processed blue PhOLEDs. We also show that the orthogonal solution-processed electron-transport layer (ETL) exhibits a unique rough surface morphology which facilitates efficient charge-injection and transport from the cathode metal. The orthogonal solution processing demonstrated here is a promising strategy for the fabrication of high performance multilayered PhOLEDs and other organic electronic devices.

Selected Publications:

1. **Earmme, T.**; Jenekhe S. A. "Solution-Processed, Alkali Metal-Salt-Doped, Electron-Transport Layers for High-Performance Phosphorescent Organic Light-Emitting Diodes," *Adv. Funct. Mater.* **2012**, *22*, 5126-5136.

2. **Earmme, T.**; Jenekhe, S. A. "High-Performance Multilayered Phosphorescent OLEDs by Solution-Processed Commercial Electron-Transport Materials," *J. Mater. Chem.* **2012**, *22*, 4660-4668. (Selected as *Hot Paper* by Editor)

3. Ahmed, E.; **Earmme, T.**; Jenekhe, S. A. "New Solution-Processable Electron Transport Materials for Highly Efficient Blue Phosphorescent OLEDs," *Adv. Funct. Mater.* **2011**, *21*, 3889-3899.

4. **Earmme, T.**; Ahmed, E.; Jenekhe, S. A. "Solution-Processed Highly Efficient Blue Phosphorescent Polymer Light-Emitting Diodes Enabled by a New Electron Transport Material," *Adv. Mater.* **2010**, *22*, 4744-4748.

5. Ahmed, E.; **Earmme, T.**; Ren, G.; Jenekhe, S. A. "Novel n-Type Conjugated Ladder Heteroarenes: Synthesis, Self-Assembly of Nanowires, Electron transport, and Electroluminescence of Bisindenoanthrazolines," *Chem. Mater.* **2010**, *22*, 5786-5796.

6. **Earmme, T.**; Ahmed, E.; Jenekhe, S. A. "Highly Efficient Phosphorescent Light-Emitting Diodes by Using an Electron-Transport Materials with High Electron Affinity," *J. Phys. Chem. C.* **2009**, *113*, 18448-18450.

(4cz) Molecules That Tangle: Polymer Entanglement and Melt Structure

Qin, J., Pennsylvania State University

Linear polymers are flexible chain molecules containing many chemical repeat units. The motion of individual polymers in a dense polymer liquid (molten plastic) is severely constrained by surrounding chains, and by the fact that chains cannot cut through one another. Effectively, polymers may be considered as being confined inside a tube-like region. The tube diameter, or the entanglement length, is the key parameter needed by the standard molecular theory for polymer rheology. But a molecular understanding of the origin of the tube diameter is still lacking. We approach this problem by closing polymers into rings, in order to obtain a system with well-defined, permanent topology, and using tools from the mathematical theory of knots to identify and count topological entanglements. For simulated polymer melts, this approach enables us to get a tube diameter value that is based on topological considerations alone, and that agrees with values obtained by more heuristic methods. We use this approach to study the effects of chain flexibility and addition of diluents upon the tube diameter.

Another consequence of the unique chain-like structure of polymers is the presence of composition fluctuations at the length scale of a polymer coil size. These composition fluctuations can be measured by small angle neutron and X-ray scattering. Understanding these fluctuation phenomena is important for understanding the structure and self-assembly behavior of multi-component polymeric systems. We show that composition fluctuations in systems of diblock copolymers can be accurately described by a statistical mechanical theory that we developed, by comparing its predictions with simulation results.

(4ah) Fabrication of Engineered, Monodisperse Particles for Respiratory Drug Delivery

Fromen, C. A., North Carolina State University

Drug delivery via the lung holds great promise for the local and systemic delivery of a wide variety of materials; however, current respiratory drug delivery systems are surprisingly inefficient, making delivery of expensive materials, such as biologicals, impractical. Using the top-down, micro-molding technique referred to as PRINT (Particle Replication In Non-wetting Templates), powders have been fabricated with predetermined individual particle properties which can improve dispersion, control deposition and tailor therapeutic responses. These aerosols have been characterized with geometric standard deviations well below 1.2 when nebulized, indicating a monodisperse aerosol. Shape factors have been characterized for all non-spherical geometries by correcting for the drag of shaped particles in the flow conditions of an aerodynamic particle size (APS) spectrometer. Dry powder PRINT aerosols have been sized from numerous aerosol devices, yielding high emitted doses and tailored deposition as a function of particle geometry. Particle shape was found to also be critical in downstream cellular interactions in the lung; particle uptake by alveolar macrophages was observed to differ by over two-fold between particles of similar aerodynamic diameters but different shapes, implying the potential to tailor avoidance of macrophage clearance in the lung. When dosed to mice airways via intratracheal administration, cargo-free PRINT particles were observed to remain immunologically inert in the lung for seven days, with minimal clearance, airway remodeling, or triggering of host immunity by either pro-inflammatory cytokine production or cellular recruitment. In total, these data suggest that PRINT aerosols hold promise for localized and sustained pulmonary therapeutics through designed particle characteristics which modulate dispersion, deposition, cellular interactions and therapeutic responses.

(4el) Predictive Models for Speeding Up Catalyst Discovery

Xin, H., University of Michigan

Catalytic materials are essential components in ~90% of the world's large-scale chemical processes. It is, however, astonishing that all industrial catalysts used today were discovered via costly and time-consuming empirical testing, and far from optimized for energy efficiency and product selectivity. Recent developments of density functional theory (DFT) and the emerging catalysis informatics project have led to an unprecedented atomic-scale understanding of the complex processes occurring on catalyst surfaces and an identification of potentially improved catalysts from first-principles. However, the immense phase space of catalytic materials spanned by structural and compositional degrees of freedom precludes thorough screening, even with combinatorial quantum-chemical calculations and high-throughput experiments. Development of predictive structure-reactivity models for rapid discovery of novel catalytic materials still remains a grant challenge in fundamental catalysis.

During my PhD studies under the supervision of Prof. Suljo Linic, we developed a general theoretical framework for understanding variations in the surface reactivity of transition metals upon a perturbation of their electronic properties. I will discuss this framework on three applications: 1) *rapid screening of Pt multimetallic electrocatalysts for the oxygen reduction reaction in PEM fuel cells*; 2) *understanding of alkali promotion mechanisms for oxidation reactions on Ag surfaces*; 3) *coupling of thermal energies and energetic electrons for chemical bond activation on optically-excited plasmonic Ag nanoparticles*. Each model system is characterized by one specific type of perturbation, introduced either by alloying with impurity elements, by doping of substrates with chemical promoters, or by exposing plasmonic nanoparticles to visible photons. Our study opened up a new path for designing highly efficient and selective chemical, electrochemical, and photochemical processes.

My post-doctoral research with Prof. Jens Nørskov and Dr. Frank Abild-Pedersen aims to obtain a detailed understanding of the surface reactivity of complex materials using DFT calculations and molecular dynamics (MD) simulations. One of such systems involves transition metal alloys that are outliers of the *d*-band model description of surface chemisorption. Another system of importance is associated with transition metal oxides that are computationally cumbersome to describe with reasonable accuracy due to strong electronic correlations. Establishing structure-reactivity relationships for these systems becomes tremendously important for the rapid discovery of highly-efficient and cost-effective materials for chemical manufacturing, renewable energy generation and storage, etc.

My future research interests focus on dynamic aspects of chemical reactions driven by energetic charge carriers. My knowledge and experience on multiscale modeling has provided me a unique opportunity to tackle those challenges in energy related catalyst design. The objective is to develop 1) *an ab-inito mechanistic understanding of chemical energetics and excited-state dynamics of molecule-surface interactions*, and 2) *predictive models that link easily accessible physical characteristics of constituent components of catalytic materials to their surface reactivity.*

Relevant publications:

1. **H. Xin**, N. Schweitzer, E. Nikolla, S. Linic, J. Chem. Phys., 132, 111101, (2010)
2. N. Schweitzer, **H. Xin**, E. Nikolla, J.T. Miller, S. Linic, Top Catal, 53, 348, (2010)
3. **H. Xin**, S. Linic, J. Chem. Phys., 132, 221101, (2010)
4. P. Christopher, **H. Xin**, S. Linic, Nature Chem., 3, 467, (2011)
5. **H. Xin**, A. Holewinski, S. Linic, ACS Catal., 2, 12, (2012)
6. **H. Xin**, A. Holewinski, N. Schweitzer, E. Nikolla, S. Linic, Top Catal, 55, 376, (2012)
7. P. Christopher, **H. Xin**, A. Marimuthu, S. Linic, Nature Mater. 11, 1044, (2012)
8. S. Linic, P. Christopher, **H. Xin**, A. Marimuthu, Acc. Chem. Res., accepted (2013)
9. A. Holewinski, **H. Xin**, E. Nikolla and S. Linic, Curr Opin Chem Eng., accepted (2013)

(4di) Microfluidics and Engineering Solutions for Systems Biology Studies in the Nematode Caenorhabditis Elegans

San Miguel, A., Georgia Institute of Technology

Microfluidics has recently emerged as a powerful tool for lab-on-chip research in a wide variety of areas such as soft matter, rheology, synthesis, crystallization, transport and reactions. Moreover, microfluidics is gaining importance as an essential tool in biological research since it offers many advantages that enable studies of live microorganisms and cells impossible to perform using conventional techniques. Microfluidic devices are especially beneficial for the study of the roundworm *C. elegans*. This soil dwelling nematode is an important model organism used to explore many biological questions, such as development, aging, neuroscience and even learning. This nematode is a highly studied multi-cellular organism that offers many experimental advantages, including a short life span, ease of culture, transparency and straightforward genetic manipulation. Additionally, its full genome and neuronal wiring diagram are known. These features make it an ideal model system for biological study, especially neuroscience. Microfluidics, combined with other engineering techniques, provide a powerful platform to perform high-content studies in a high-throughput fashion.

Here I show how we have applied an integrated approach to perform biological studies enabled by microfluidics, automation, mathematical algorithms and machine learning. I present how microfluidics is an excellent technique that allows otherwise unfeasible high-content studies of complex biological systems. Specifically, I discuss how integrated engineering platforms can be used to elucidate relevant questions in neuroscience, such as synaptic plasticity, learning and the effects of aging on neurological health.

(4t) Explorations of Synthesis and Application of Mesoporous Nanoparticles

Chung, P. W., University of California, Berkeley

Explorations of Synthesis and Application of Mesoporous Nanoparticles

Po-Wen Chung, Ph. D

Department of Chemistry, *Iowa State Uniersity, Ames, IA 50010*

Department of Chemical and Biomolecular Engineering,

University of California, Berkeley, CA 94720

In light of the development of MCM-41 mesoporous silica materials, a variety of mesoporous nanoparticles have been synthesized by either using soft-template as structure directing agents or hard-template nanocasting replication. Such mesoporous nanoparticles as mesoporous silica, mesoporous metal oxides, mesoporous organic-inorganic hybrid, mesoporous carbon and mesoporous polymer nanoparticles have drawn attention to their applications on drug delivery, catalysis, adsorption, separation, and energy storage. Herein, my doctoral research has demonstrated the novel synthesis of mesoporous organic-inorganic hybrid silica nanoparticles, highly ordered mesoporous carbon nanoparticles, and mesoporous metal oxides as well as the applications of drug delivery, separation and catalytic supports for sustainable chemistry by using aforementioned nanoparticles. Furthermore, my postdoctoral research has unveiled the unprecedented adsorption of glucans on mesoporous carbon materials, and is currently in pursuit of exploring efficient routes of depolymerizing lignocellulosic biomass.

Selected Publications:

1.) **Po-Wen Chung**,* Alexandre Charmot, Oz M. Gazit, and Alexander Katz,* "Glucan Adsorption on Mesoporous Carbon Nanoparticles: Effect of Chain Length and Internal Surface", *Langmuir* **2012,** *28*, 15222-15232. (ISI impact factor: 4.186)

2.) **Po-Wen Chung**,* Tae-Wan Kim, Yulin Huang, Weihua Deng, Brian G. Trewyn, Robert C. Brown and Victor S.-Y. Lin "Ordered Mesoporous Carbon Nanoparticles as a Catalyst Support for Syngas Reaction", *manuscript in submission.*

3.) **Po-Wen Chung**, Rajeev Kumar, Marek Pruski and Victor S.-Y. Lin,* "Temperature Responsive Solution Partition of Organic-Inorganic Hybrid Poly(N-isopropylacrylamide)-coated Mesoporous Silica

Nanoparticles", *Advanced Functional Materials*, **2008**, *18(9)*, 1390-1398. (ISI impact factor: 6.990)

4.) Tae-Wan Kim,* **Po-Wen Chung** and Victor S.-Y. Lin, "Facile Synthesis of MCM-48 Ordered Mesoporous Silica Nanoparticles with Control of the Particle Size", *Chemistry of Materials,* **2010**, *22,* 5093-5104. (ISI impact factor: 5.368)

5.) Tae-Wan Kim, **Po-Wen Chung**, Igor I. Slowing, Makoto Tsunoda, Edward S. Yeung, and Victor S.-Y. Lin,* "Structurally Ordered Mesoporous Carbon Nanoparticles as Transmembrane Delivery Vehicle in Human Cancer Cells", *Nano Letters*, **2008**, *8(11),* 3724-3727. (ISI impact factor: 9.991)

6.) Tae-Wan Kim,* Igor I. Slowing, **Po-Wen Chung**, and Victor S.-Y. Lin, "Ordered Mesoporous Polymer-Silica Hybrid Nanoparticles as Vehicles for Intracellular Controlled Release of Macromolecules", *ACS Nano* **2010**, *5(1),* 360-366. (ISI impact factor: 7.493)

(4f) Conjugated Polymer Semiconductor for Electronics, Optoelectronics, Diagnostics and Therapeutics

Ahmed, E., University of Washington

Conjugated polymer semiconductors have attracted significant interest due to their potential application in low cost plastic electronics and optoelectronics. They also represent a novel class of promising alternative nanomaterials for diagnostics and therapeutics. The ability to engineer the electronic structures of organic/polymer semiconductors using molecular design and synthesis and understand structure-property relationships are essential for developing novel materials for the next generation high-performance organic electronics, and for early detection and targeted delivery of therapeutics. During my graduate studies, the focus of my research was on engineering the electronic structures of organic semiconductors using molecular design and synthesis and understanding structure-property relationships that are essential for developing the next generation, high-performance, low-cost organic electronics including field-effect transistors, light-emitting diodes, and solar cells. Currently, the focus of my postdoctoral work with Timothy Swager at MIT is to develop a modular platform that combines fluorescent probe, targeting ligands and therapeutics in multicomponent conjugated polymer nanoparticles with tunable photophysical properties for early detection of biological analytes, *in vivo* dual imaging and multi-targeted drug delivery. This versatile platform allows for independent tuning of optical properties, targeting agents, and nanoparticle size. As an independent investigator, I am interested in developing a highly interdisciplinary research program at the interface of synthetic organic chemistry, materials science and engineering and biomedical diagnostics and delivery of therapeutic agents. My research interests broadly include: (i) synthesis of novel and functional organic/ polymer semiconductors; (ii) device engineering (field-effect transistors, photovoltaics, lasers, organic spintronics); (iii) self-assembly and morphology of organic/polymer semiconductors, block copolymers, nanowires and multicomponent nanostructures; and to (iv) develop multiblock copolymers for diagnostics and therapeutics.

(4cl) Assessment of Fouling in Native and Surface-Modified Water Filtration Membranes

Miller, D. J., University of Texas at Austin

Fouling is a major obstacle to implementation of polymeric membranes in water purification applications, causing deterioration of membrane performance and, consequently, increases in operational expenses. Surface modification is one way of alleviating membrane fouling. Most porous membranes are made of hydrophobic polymers, promoting the adhesion of foulants via strong hydrophobic-hydrophobic interactions. Hydrophilization of the membrane surface tends to reduce fouling because a layer of water molecules is tightly bound to the hydrophilic surface, acting as a buffer against foulant ingress. Dopamine was recently found to non-selectively deposit from buffered alkaline solution to form "polydopamine" on virtually any surface, rendering the surface very hydrophilic after only a few minutes of contact with dopamine solution. The chemical structure of polydopamine, which was previously ambiguous, was investigated by many techniques, including solid state NMR, FTIR, UV/vis and powder X-ray spectroscopies. Polydopamine was also used as a platform for the molecular conjugation of other anti-fouling materials, such as poly(ethylene glycol), to the membrane surface.

This work has focused on the evaluation of fouling behavior in native and surface-modified commercial membranes. Water filtration membranes can be operated in two ways: constant permeate flux with variable transmembrane pressure (TMP) or constant transmembrane pressure with variable permeate flux. Membrane fouling may be measured using either of these two methodologies. A constant permeate flux crossflow system was carefully constructed and used to challenge native and surface-modified membranes with an oil/water emulsion. The critical flux—the flux below which minimal fouling occurs—was determined by a flux-stepping technique during constant flux crossflow filtration. The membrane performance during fouling was evaluated at fluxes higher than and lower than the critical flux. Membrane resistance evolution during fouling is compared for constant-flux and constant-TMP operation. The efficacy of surface modifications in reducing biofouling was also studied. Many evaluations of anti-biofouling behavior rely on short-term tests of protein or bacterial adhesion. The results of such experiments were compared to those obtained for comparatively long-term operation under hydrodynamic conditions relevant for industrial module operation.

(4df) Advanced Materials for Catalysts, Membranes and Sorbents Applications and Renewable Energy Technologies

Rownaghi, A. A., Georgia Institute of Technology

My research focuses on the interface of chemistry, chemical engineering and materials science. Over the past six years, I have excelled in my research efforts related to the synthesis, characterization and application of organic/inorganic materials for catalysts, membranes and sorbents applications and renewable energy technologies. My PhD research experience at University Putra Malaysia involved development of a cost-effective method for synthesizing novel mixed-metal oxide catalysts being used in light alkane dehydrogenation and oxidation applications, in particular in n-butane to maleic anhydride process, whereby I adapted recent developments in reaction science to advance the development of novel catalysts. My PhD work resulted in 3 patents and 7 papers that have been published in high quality catalysis and chemical engineering journals.

During my previous research at Lulea University of Technology (LTU) in Sweden, I was involved in catalytic production of methanol, dimethyl ether and biofuel from black liquor synthesis gas. The work focused on catalyst synthesis (mainly mixed metal oxides and zeolites), characterization and evaluation into two different catalytic process; (1) conversion of black liquor synthesis gas to methanol (high-pressure hydrogenation reactor), and (2) methanol conversion to dimethyl ether and gasoline-range hydrocarbons (high-temperature dehydration reactor). As part of a multidisciplinary team of researchers and graduate students from LTU university and industry partners, for the first time in the world, we produced bio-methanol from black liquor synthesis gas.

Currently, I am a post-doctoral fellow at Georgia Tech, with Prof. Bill Koros working on membrane gas separations by tailoring membrane characteristics through investigation of new materials and structures to selectively control mass transport at the molecular level. In particular, the current research is focused on producing novel mixed-matrix hollow-fiber membranes for post-combustion CO_2 capture. This work addresses the economics of CO_2 capture with easier scalability.

My research interests in the near future will focus on the fundamentals and applied aspects of catalysis and separation employed in science and engineering for sustainability. The proposed research will seek to design, synthesize and evaluate advanced catalytic materials to allow for revolutionary advances in separation and reaction science to address critical issues in pollution control, waste-to-energy and renewable energy and fuels.

Selected Publications (15 total, 12 first author, h-index=7):

1. Rownaghi A.A., Hedlund J. "Methanol to gasoline-range hydrocarbons: Influence of uniform nanoparticle size versus mesoporosity on catalytic performance and product distribution of ZSM-5" *Industrial & Engineering Chemistry Research* 50 (2011) 11872-11878.
2. Rownaghi A.A., Rezaei, F., Hedlund J. "Catalytic cracking of n-hexane: influence of pore characteristics and particle size on catalytic performance of uniform ZSM-5 single crystals" *Chemical Engineering Journal* 191 (2012) 528-533.
3. Rownaghi A.A., Rezaei, F., Stante M., Hedlund J. "Selective catalyst for conversion of methanol to dimethyl ether" *Applied Catalysis B: Environmental* 119-120 (2012) 56-61.
4. Rownaghi A.A., Taufiq Yap Y. H. "High surface area vanadium phosphate catalysts for n-butane oxidation" *Industrial & Engineering Chemistry Research* 48 (2009) 7517-7528.

(4co) Understanding Catalysis Through Organic-Inorganic Hybrid Catalytic Materials

Nigra, M. M., University of California Berkeley

The design of metal cluster/ nanoparticle catalysts that are both sinter-resistant (stable) and have tunable electronic properties enables the understanding of structure-function relationships in heterogeneous and homogeneous catalysis. It also provides a method to tune the activity and selectivity of real-world catalysts by overcoming one of the major stumbling blocks for cluster catalysts – deactivation via sintering. I investigate the use of organic ligands, such as calixarenes, to control catalytically active noble metal sites on metal oxide supports and in homogeneous solution, utilizing an approach that is conceptually similar to that used by enzymes to control reactivity though organic side-chains acting as ligands. Noble metal catalysts have been synthesized with calixarene ligands bound to the metal nanoparticle yet still demonstrate reactant accessibility to the metal nanoparticle surface as measured by chemisorption probe experiments. Catalytic activity has been demonstrated for these using several different reaction systems such as reduction and oxidation reactions on these noble metal catalysts. Results provide a unique perspective into the structure-activity relationships present in catalysis.

(4bb) Replicating Signaling Cues to Restore Musculoskeletal Tissue Function Using Hydrogel-Based Materials

Holloway, J. L., Drexel University

Musculoskeletal injuries and diseases, which include meniscal tears and degenerative disc disease, are a significant health concern in the United States. Current treatments, however, typically rely on donor tissues (either allo- or autografts) and suffer from poor availability. Furthermore, donor tissue does not always adequately restore function (meniscus allografts), integrates poorly with surrounding tissue (bone grafts), and can have a high morbidity. Tissue engineering aimed at replicating and/or restoring the biological and mechanical cues required for tissue function offers an advantage to current treatments and prevents further musculoskeletal degeneration. My research has focused on the intelligent design of hydrogels towards this aim.

During my Ph.D., under the advisement of Anthony Lowman and Giuseppe Palmese, I developed and characterized a fiber-reinforced hydrogel composite for meniscal replacement aimed at replicating the mechanical properties of the native meniscus. The hydrogel composite was characterized in compression (0.1 - 0.8 MPa) and tension (0.1 - 250 MPa) over a wide range of formulations demonstrating fine control over mechanical properties within the range of the human meniscus. Furthermore, the fiber-hydrogel interface was tailored for optimal adhesion using a novel biocompatible grafting technique to form a direct covalent linkage between the fiber and hydrogel. Grafting resulted in significant improvements to interfacial shear strength from 11 kPa without any treatment to above 220 kPa following grafting.

Currently, I am a postdoctoral fellow in Jason Burdick's laboratory researching synergistic growth factor delivery for improved osteogenesis using proteolytically degradable hydrogels. In vitro cell migration assays indicate stromal cell-derived factor-1 alpha (SDF-1α) significantly increases human mesenchymal stem cell chemotaxis. This motivates the use of SDF-1α to recruit stem cells in conjunction with more traditional osteogenic growth factors (i.e. bone morphogenic proteins (BMPs)) in the development of therapies for improved tissue repair. Ongoing work is evaluating the synergistic effect of SDF-1α and BMP on osteogenesis in an in vivo rat cranial defect model.

In the future as a faculty member, I will use my experience with and knowledge of polymer science, biomechanics, and growth factor signaling to develop finely tunable materials that replicate and restore the mechanical and biological signals needed for function in musculoskeletal tissues.

(4bj) Design of Pharmaceutical Crystallizers Using Dual-Impinging-Jet Mixers, Multi-Phase Flow, and Controlled Dynamic Operations

Jiang, M., Massachusetts Institute of Technology

Crystallization is a key unit operation in the pharmaceutical industry. The crystal size distribution (CSD) can affect the efficacy of drug products (such as the amount of drug reaching the lungs from a nasal spray) as well as the efficiency of downstream processes such as filtration and milling. CSD control can be challenging when the crystallization simultaneously includes many phenomena, such as growth, particle attrition and breakage, agglomeration, and primary and secondary heterogeneous nucleation. This poster describes new crystallizer designs that greatly enhance the control of the crystal size distribution. One strategy is to exploit dual-impinging-jet mixers and/or multiphase flow to decouple nucleation and growth so that these phenomena can be individually controlled. Another strategy is to increase the degrees of freedom in the dynamic operation of the crystallizer. Experimental validation confirmed that the designs suppressed secondary nucleation, attrition, and aggregation/agglomeration—phenomena that worsen the ability to control the CSD but are dominant in most existing crystallizer designs.

The last portion of the poster will propose some new design strategies for the crystallization of biopharmaceuticals, especially proteins, which constitute a growing proportion of new therapeutic compounds. Strategies are proposed both for the generation of single protein crystals for use in determining the protein structure for use in drug design, and for the generation of large quantities of protein crystals as a route towards inexpensive purification, extended shelf-life, or alternative modes of drug delivery.

(4ay) Engineering Functionality in Nano-Materials

Hernandez, S. C., Naval Research Laboratory

Nanomaterials have been rigorously investigated due to their superlative mechanical, thermal, and electronic properties. Being composed entirely of surface atoms, they are incredibly amenable to surface modification and thus provide the opportunity towards extreme control over surface interactions.

Surface engineering of 1-D and 2D materials composed of carbon based materials (graphene, carbon nanotubes, conducting polymers), oxides, and metals is achieved by plasma modification and site-specific electrochemical deposition. Well controlled patterned surface functionalities provide the ability to manipulate the material attributes locally, offering a range of opportunities for the composite materials. Such precise nano-engineering opens up a wealth of opportunities in biosensing, plasmonics, catalysis, and device applications.

(4aq) Tailoring Ionic Block Copolymer Structure and Function for Therapeutics Delivery and Energy Applications

Green, M. D., University of Delaware

The tunable chemical composition, thermomechanical properties, and nanoscale morphologies of block copolymers enable access to myriad applications, including energy storage and harvesting devices as well as drug and gene delivery vehicles. My graduate research (with Prof. Timothy Long, Virginia Tech) focused on using controlled radical polymerization techniques to prepare designer triblock copolymers for electromechanical transducers. Electromechanical transducers have potential applications in energy harvesting devices, biomimetic materials, and sensors. However, the design of polymeric systems that maximize both ion transport and mechanical strength has been challenging, which has limited performance and marketability. Therefore, I developed systematic structure-property relationships for imidazole-containing homopolymers and block copolymers, which enabled the design of ionomeric membranes that optimized ionic conductivity and storage modulus. Specifically, I investigated the influence of (i) charge placement and density in the central and outer blocks; (ii) thermomechanical properties; (iii) ionic liquid incorporation; and (iv) chemical composition of the cation and anion on the properties of imidazole-containing polymers. Ultimately, optimization of the ion content and thermomechanical properties in a triblock copolymer led to my preparation of the first successful cationic electromechanical transducer.

My postdoctoral research (with Prof. Thomas Epps, III and Prof. Millicent Sullivan, University of Delaware) focuses on solution assemblies of diblock copolymers for drug and gene delivery. Amphiphilic block copolymers form solution assemblies of a variety of structures based on the molecular weight and volume fraction of each component; these structures include spherical micelles, cylindrical micelles, and vesicles, which can be used to prepare drug delivery vehicles. Specifically, we are interested in understanding the solution assembly of drug-loaded micelles and the assembly of light-activated nucleic acid delivery structures. Our first set of studies is addressing the influence of the targeting ligand location on the solution assembly, cellular internalization efficiency, and *in vivo*circulation time; we have also analyzed the efficiency of hydrophobic drug encapsulation and delivery. The novel series of amphiphilic block copolymers studied herein present a methodology for decoupling prolonged circulation and targeted therapeutic delivery, which could prove pivotal for effective drug delivery strategies. Second, a series of light-activated diblock copolymers for gene delivery were prepared. These block copolymers contain a non-fouling poly(ethylene oxide) block and a cationic block containing pendant photoactive nitrobenzyl functional groups. The cationic block enables the condensation of anionic nucleic acids, and undergoes photocleavage upon irradiation with light. This unique release mechanism enables controlled, externally stimulated delivery of nucleic acids, which presents exciting opportunities for gene delivery.

The research directions of the program I plan to develop will explore bulk and solution assemblies of novel, ionic block copolymers for drug or gene delivery and energy applications. This draws upon my graduate experience with polymerizable ionic liquids and incorporation of ionic liquids into membranes, while also utilizing my postdoctoral experience characterizing novel assemblies for drug and gene delivery. Gaining inspiration from ionic liquids, I will design and prepare novel, ionic block copolymers with tunable chemical structures, thermomechanical properties, and targeted morphologies for energy storage devices. Additionally, I will develop ionic block copolymers with tailored solution properties to optimize their performance as therapeutic delivery vehicles. Ionic block copolymers present exciting opportunities for energy and therapeutic delivery applications, and developing novel structures with optimized performance will help revolutionize their utility and availability.

(4ba) Rational Design of (Electro)Catalytic Systems for Energy Conversion
Holewinski, A., University of Michigan

Electrochemical transformations permit the direct interconversion of electrical energy and the energy of chemical bonds. Such processes can be carried out at high efficiencies and are of particular interest for distributed, mobile, and auxiliary power generation, where energy demands fluctuate rapidly and there is limited opportunity for heat integration. Electrochemical reactions can also be coupled with renewables such as wind and solar energy to efficiently convert low-value feedstocks to fuels. Herein, a multi-faceted approach combining molecular level insights from detailed kinetic analysis, spectroscopic observations of reactive intermediates, and quantum chemical calculations, along with advanced synthetic techniques and thorough characterization is used to design optimal (electro)catalyst structures. This research methodology is discussed in the context of the development of stable and economical electrode catalysts for low-temperature hydrogen fuel cells, which are advantageous in transportation applications but currently suffer from the requirement of costly Pt-based catalysts.

Insights from quantum chemical calculations (Density Functional Theory) and microkinetic modeling were used to target ideal catalyst materials for electrochemical oxygen reduction ($O_2+4H^++4e^-=2H_2O$), which is a major source of lost efficiency in low temperature fuel cells. Materials based on alloying Ag with 3d transition metals (Co,Ni,Fe) were identified, and a novel synthesis technique was developed to force alloying between these ordinarily immiscible metals. The Ag-alloys were demonstrated to exhibit oxygen reduction activity on the same order of magnitude as Pt, making them competitive on a cost basis. A variety of techniques, including cyclic voltammetry, x-ray diffraction, aberration-corrected electron microscopy, and electron energy loss spectroscopy were used to characterize the materials and confirm the presence of alloyed surface structure and its contribution to the activity enhancements.

In contrast to the trial-and-error approach that has historically dominated catalyst research, the discovery of the Ag-alloy electrocatalysts was enabled by a relatively firm understanding of the oxygen reduction reaction mechanism. This mechanistic understanding was derived from the development of a new theoretical formalism for electrochemical kinetics. The relationship between applied electrical potential and observed current was shown to be a predictable fingerprint that can be determined from elementary-step reaction free energies, activation barriers, and the availability of catalytically active sites under reaction conditions. This result was generalized to describe arbitrary electrochemical reactions, permitting analysis of a variety of important energy conversion reactions.

This work illustrates the efficacy of combining theoretical analysis, synthetic chemistry and characterization tools to rationally design (elecro)catalytic materials for energy conversion processes with great potential impact. Future work using this approach in the areas of advanced lithium-air batteries, electrochemical fuel production from CO_2, and small-molecule oxidation in fuel cells will also be discussed.

(4as) Computational Studies of Phase Transition in Systems of Anisotropic Building Blocks With Applications to in Silico Design of Functional Materials

Haji-Akbari, A., Princeton University

Phase transitions are ubiquitous in nature. The very existence of life on earth would not have been possible without the multitude of thermodynamic and kinetic phase transitions that take place both *in vitro* and *in vivo*. Computational studies of phase transition are not only important from a theoretical perspective, but can also help us gain a better understanding of profoundly important natural processes such as climate change or protein folding. Studies of phase transition have gained further momentum in light of recent advents in nanotechnology and material sciences, which has made the bottom-up assembly of complex functional materials possible. Most notable are the wide range of anisotropic nano- and colloidal building blocks that can spontaneously assemble into complex structures with interesting mechanical, electrical, magnetic and/or optical properties. Bottom-up assembly of such complicated structures from the corresponding building blocks usually involves one or several phase transitions. Computational studies of such transitions can help us devise engineering guidelines for designing new building blocks and new structures. The main focus of my research is to study the phenomenology, thermodynamics and kinetics of phase transitions in systems of anisotropic building blocks or molecules, or in systems with confined geometries. In the following paragraphs, I mention different flavors of phase transition problems that are of particular interest to me.

1- **Entropy-driven disorder-order transitions in hard particle sysrems**: From a theoretical perspective, hard particles are suitable model systems for studying the exclusive effect of building block geometry on the thermodynamic phase behavior, because the thermodynamics of a hard particle system is purely determined by its shape. In addition, hard particles closely mimic colloidal particles that interact via short-range repulsive interactions, and can be used for *in silico* prediction of the experimental behavior of such colloidal systems. Theoretical and computational studies of hard particles are also intimately related to the problem of packing in discrete geometry, since the densest packing of a given shape is the most stable structure of the corresponding hard particle system in the limit of infinite pressure. Since it is generally difficult to obtain analytical solutions for the packing problem, numerical approaches- based on hard particle simulations- are becoming popular heuristic ways solving packing problems. Although entropy-driven transitions can happen in energetic systems as well, all disorder-order transitions in hard particle systems are entropy-driven [1].

Hard tetrahedra, the main focus of my PhD dissertation, are the simple yet perplexing examples of geometrically anisotropic building blocks that self-assemble into complicated structures. The regular tetrahedron is the simplest of the five Platonic solids. In my research, Monte Carlo simulations revealed that these geometrically simple building blocks assemble into a dodecagonal quasicrystal, at sufficiently high densities due to entropy alone [2]. These quasi-periodic arrangements are surprisingly dense. An approximant of the quasicrystal that we constructed manually is an even denser packing and constituted the world record at the time of its discovery. Quasicrystals are an exotic class of solids with long-range order that lack periodicity. In 2011, the Nobel Prize in Chemistry was awarded to Danieh Schechtman for the discovery of the first quasicrystal [3]. We found that the quasicrystal forms robustly from hard tetrahedra, and subsequently found the phase again in a system of hard triangular bipyramids (TBPs), which are dimers of hard tetrahedra [4]. We also confirm that in both systems, the quasicrystal approximant is more stable at intermediate densities than the now known densest packing of these polyhedra [4, 5]. This research suggests the rich and subtle role that entropy plays in stabilizing complex structures and the potential that geometric anisotropy has in inducing complex forms of order.

2- **Homogeneous Nucleation of Ice and Gas Hydrates**: I am also interested in phase transitions in systems of small anisotropic molecules such as water, CO_2 and small organic molecules such as hydrocarbons. I use advanced sampling techniques such as umbrella sampling and forward-flux sampling for calculating the free energy profiles and nucleation rates for these crystallization processes. I am particularly interested in studying homogeneous nucleation in confined geometries such as droplets, films and porous media since such confined environments are usually more relevant to what really happens in nature, and are therefore more interesting to atmospheric, oceanic and geological sciences. The knowledge obtained from these studies will be extremely helpful in designing materials for energy storage and sequestration.

1. Daan Frenkel. *Entropy-driven phase transitions*. **Physica A**, 263:26–38, 1999.

2. Amir Haji-Akbari, Michael Engel, Aaron S. Keys, Xiaoyo Zheng, Rolfe Petschek, Peter Palffy-Muhoray, and Sharon C. Glotzer. *Disordered, quasicrystalline and crystalline phase of densely-packed tetrahedra.* **Nature**, 462:773-777, 2009.

3. D. Shechtman, I. Blech, D. Gratias, and J. W.Cahn. *Metallic phase with long- range orientational order and no translational symmetry*. **Phys. Rev. Lett.**, 53:1951–1953, 1984.

4. Amir Haji-Akbari, Michael Engel, and Sharon C Glotzer. *Degenerate quasicrystal of hard triangular bipyramids*. **Phys. Rev. Lett.**, 107:215702, 2011.

5. Amir Haji-Akbari, Michael Engel, and Sharon C Glotzer. *Phase diagram of hard tetrahedra*. **J. Chem. Phys.**, 135:194101, 2011.

(4du) Novel Thermosets and Their Nanocomposites

Throckmorton, J. A., Drexel University
Palmese, G. R., Drexel University

Thermosetting materials and thermoset composites are widely used for structural applications, adhesives, insulation materials, electronic encapsulants, and coatings in a wide array of industries including wind energy, aerospace, sporting goods, and automotive. Recent research has shown that properties of these materials, including CTE, modulus, fracture toughness, electrical conductivity, thermal conductivity, and volatile perfusion, can be enhanced by the addition of nano-scale fillers with various chemistries and morphologies. Nanocomposites offer exciting ways to modify polymers and conventional composites by providing a moldable, flowable, polymer-like composite with advantageous properties taken from embedded materials with desirable properties over small length scales.

Achieving effective macro-scale properties from these nanofillers requires good solvent chemistry and processing for nanoparticles dispersion, effective choice of nanoparticle material and geometry for the desired property improvements, and strong interfaces between the embedded particles and the polymer matrix. Our recent work has demonstrated that well-chosen ionic liquids can simultaneously disperse nanoparticles within an epoxy matrix and initiate epoxy cure reaction. This results in materials with potentially unique interphase properties resulting from interactions among the nanoparticle surface, ionic liquid, and thermoset matrix.

This study explores the structure and properties of nanocomposites made with novel ionic liquid-thermoset chemistries and a unique high-shear processing method to embed graphite nanoplatelets into various thermosetting resins. Experimentally and conceptually, various methods of measuring dispersion for differing particle geometries, the relationship between nano-scale dispersion and macroscopic properties, and the effect of a percolated nanofiller network are explored.

(4u) Multi-Scale Approaches in Nanoscale Systems and Phenomena

Chung, P. S., Carnegie Mellon University

During the past few decades, along with remarkable scientific and engineering achievements, the world has faced immense challenges in efficiency and system size, which motivates the investigation of novel materials and nanoscale systems. Analytic approaches have been widely studied to satisfy the demand by tuning physical properties, chemical structures, and functionalities of the new materials as a part of the eye-opening progress. The broad progress in nanotechnology provides us the understanding of nanoscale physics to control the system properties. However, advances in nano materials and analysis methods direct us to the critical issue of the interaction to meso/macroscopic scales since most systems comprise phenomena at different time and length scales, and often described via a hierarchy of scale-specific models. Consequently, the multi-scale approach based on a clear understanding of nanoscale phenomena becomes an essential multidisciplinary analysis paradigm in science and engineering fields in order to realize the new nano materials and systems. Via this approach, the mechanism of information communication can be obtained from one scale and passed to another scale allowing one to predict the system scale behavior from first principles in a bottom-up approach. The device scale properties, nonetheless, can be obtained with nanoscale resolution using the top-down approach.

My research interest covers the broad area of both theoretical and experimental multi-scale approaches in novel materials and nanoscale phenomena including tribology, energy, and nano electronics applications by using novel atomistic/molecular and mesoscale/continuum theory and developing methodologies for hierarchical integration of the systems at the different scales, which will be expanded to multi-phenomena/physics related issues. The device scale properties will be coupled with optimization for the process system integration. I also intend to perform experiment and will vigorously collaborate with experimentalist to complement the full-scale simulation. In this poster session, I will introduce and discuss the details of my research plan.

(4eu) Sustainable Nanocomposites Towards Electrochemical Energy Storage and Environmental Remediation

Zhu, J., Lamar University

Energy shortage and environmental pollution are two most concerns right now for the long term sustainable advance of human society. New technology developments are the key solutions to these challenges, which strongly rely on the continuous upgrading of advanced material performance. Sustainable nanocomposites targeting to large energy/power density electrochemical energy storage and highly efficient environmental remediation have been developed. In structure design, materials with different dimensionalities (0D, 1D, 2D and 3D) will be integrated to achieve high specific surface area, multi-functionality and to reveal the structure-property-performance relationship. In composition optimization, materials with unique properties including carbon, metal/metal oxide, conductive polymer and others will be integrated into one unit to achieve the targeting performances. Most importantly, interfacial property is the key factor to realize the carefully designed functionality in a composite, which largely depends on the developed synthetic methods.

Targeting a high energy density composite material, an interfacial polymerization method has been developed to fabricate an integrated 3-dimensional (3D) structure comprising 2D graphite oxide (GO) sheet and 1D polyaniline nanofibers. This facile method can be used as a general method to synthesize similar hybrid structures, which provides large specific surface area, high capacitance and satisfactory energy density. Besides, microwave energy has been utilized to fabricate carbon/metal oxide nanocomposites that show better nanostructure control, interfacial interaction and energy storage performance than the nanocomposites obtained from conventional approaches. Hybrid materials synthesized from these methods exhibit much higher energy density and power density than most of the currently available carbon based composites.

Targeting the environmental remediation, a comprehensive and sustainable waste-free coupled process of polymer nanocomposite recycling is introduced to synthesize magnetic carbon nanocomposites (MCNCs) and to produce useful chemical radicals simultaneously. The embedded NPs serve as catalysts for the waste polymer decomposition. At the same time, the coke produced on the NP surface (poison the catalyst in conventional process) during pyrolysis serves as a carbon source for the in situ formation of a compact carbon shell structure. The MCNCs demonstrate an extremely fast heavy metal removal from the polluted water within 10 min (conventional adsorbents takes hours even days). The large saturation magnetization of these novel magnetic carbon nanocomposites allows fast recycling of both the adsorbents and the adsorbed heavy metals from liquid suspension in a more energetically and economically sustainable way by simply applying a permanent magnet. The significantly reduced treatment time required to remove the heavy metals makes these MCNCs promising for efficient polluted water treatment.

SELECTED PEER-REVIEWED PUBLICATIONS (15 out of 51) & BOOK CHAPTERS (2).

visit http://dept.lamar.edu/zhanhu/icl/Jiahua Zhu.htm for complete publication list.

1. **J. Zhu,** M. Chen, H. Qu, Z. Luo, S. Wu, H. A. Colorado, S. Wei and Z. Guo, Magnetic Field Induced Capacitance Enhancement in Graphene and Magnetic Graphene Nanocomposites. ***Energy & Environmental Science***, 6, 194-204 (2013).
2. **J. Zhu**, M. Chen, S. Pallavkar, N. Yerra, Z. Luo, T. C. Ho, S. Wei, Z. Guo, Microwave Synthesized Wearable Microtubular Carbon/Co-Co_3O_4 Nanocomposites toward Electrochemical Energy Storage. ***Nanoscale,*** 5, 1825 (2013).
3. **J. Zhu**, S. Pallavkar, M. Chen, N. Yerra, Z. Luo, H. A. Colorado, H. Lin, N. Haldolaarachchige, Thomas C. Ho, D. P. Young, S. Wei and Z. Guo, Magnetic Carbon Nanostructures: Microwave Energy – Assisted Pyrolysis vs Conventional Pyrolysis. ***Chemical Communication*****,** 49, 258-260 (2013)
4. **J. Zhu,** M. Chen, H. Qu, X. Zhang, Z. Luo, H. A. Colorado, S. Wei and Z. Guo, Interfacial Polymerized Polyaniline/Graphite Oxide Composites toward Electrochemical Energy Storage. ***Polymer***, 53(25), 5953-5964 (2012)
5. **J. Zhu,** S. Wei, H. Gu, R. Sowjanya, Q. Wang, Z. Luo, N. Haldolaarachchige, D. Young, Z. Guo, One-pot Synthesis of Magnetic Graphene Nanocomposites Decorated with Core@Double-shell Nanoparticles for Fast Chromium Removal, ***Environmental Science & Technology,*** 46 (2), 977-985 (2012)
6. **Zhu,** H. Gu, Z. Luo, N. Haldolaarachchige, D. P. Young, S. Wei and Z. Guo, Carbon Nanostructures Derived Polyaniline Metacomposites: Electrical, Dielectric and Giant Magnetoresistive Properties, ***Langmuir*****,** 28(27), 10246-10255 (2012)
7. **J. Zhu,** Z. Luo, S. Wu, N. Haldolarachchige, J. He, D. P. Young, S. Wei and Z. Guo, Magnetic Graphene Nanocomposites: Electron Conduction, Giant Magnetoresistance and Tunable Negative Permittivity, ***Journal of Materials Chemistry*****,** 22, 835-844 (2012).
8. **J. Zhu,** S. Wei, Y. Li, L. Sun, N. Haldolaarachchige, D. Young, C. Southworth, A. Khasanov, Z. Luo and Z. Guo, Surfactant-free Synthesized Magnetic Polypropylene Nanocomposites: Rheological, Electrical, Magnetic and Thermal Properties, ***Macromolecules*****,** 44(11), 4382-4391 (2011)
9. **J. Zhu,** S. Wei, J. Ryu and Z. Guo, Strain Sensing Elastomer/Carbon Nanofibers "Metacomposites", ***Journal of Physical Chemistry C***, 115, 13215-13222 (2011)
10. **J. Zhu,** S. Wei, N. Haldolaarachchige, D. P. Young and Z. Guo, Electromagnetic Field Shielding Polyurethane Nanocomposites Reinforced with Core-Shell Fe-Silica Nanoparticles, ***Journal of Physical Chemistry C***, 115, 15304-15310 (2011)
11. **J. Zhu,** S. Wei, Y. Li, S. Pallavkar, H. Lin, N. Haldolaarachchige, Z. Luo, D. P. Young and Z. Guo, Comprehensive and Sustainable Recycling of Polymer Nanocomposites, ***Journal of Materials Chemistry*****,** 21, 16239-16246 (2011)
12. **J. Zhu,** S. Wei, L. Zhang, Y. Mao, J. Ryu, A. B. Karki, D. P. Young and Z. Guo, Polyaniline-Tungsten Oxide Metacomposites with Tunable Electronic Properties, ***Journal of Materials Chemistry*****,** 21, 342-348 (2011).
13. **J. Zhu**, S. Wei, M. J. Alexander, T. D. Dang, T. C. Ho and Z. Guo, Enhanced Electrical Switching and Electrochromic Properties of Poly(p-phenylenebenzobisthiazole) Thin Films Embedded with Nano-WO_3, ***Advanced Functional Materials***, 18, 3076-3084 (2010).
14. **J. Zhu**, S. Wei, J. Ryu, M. Budhathoki, G. Liang and Z. Guo, In-Situ Stabilized Carbon Nanofibers (CNFs) Reinforced Epoxy Nanocomposites, ***Journal of Materials Chemistry***, 20, 4937-4948 (2010).
15. **J. Zhu**; S. Wei; M. J. Alexander; D. Cocke; T. C. Ho; Z. Guo, Electrical Conductivity Manipulation and Switching Phenomena of Poly(p-phenylenebenzobisthiazole) Thin Film by Doping Process, ***Journal of Materials Chemistry***, 20, 568-574 (2010).
16. "**Nanostructured Ceramic Oxides: Preparation, Property Analysis and Applications**"

Encyclopedia of Semiconductor Nanotechnology, Edited by Ahmad Umar, in press (2010).

17. S. Wei, **J. Zhu,** P. Mavinakuli, Z. Guo, **Chapter 4: "Magnetic Polymer Nanocomposites: Fabrication, Processing, Property analysis and Applications"** *Multifunctional Polymer Nanocomposites*, CRC Press, Edited by Jinsong Leng and Alan K. T. Lau, (2010).

AWARDS/HONORS:

1. Scholarship from Thermoplastic Elastomers Special Interest Group, Society of Plastics Engineers Foundation, 2013.
2. Chinese Government Award For Outstanding Self-Financed Students Abroad. March, 2012.
3. NSF travel Grant, *2012 NSF-CMMI Research and Innovation Conference*, Boston, MA, July, 2012.
4. 3rd Research Excellence Award in the *2011 Chinese-American Chemical Society Southwest Chapter Poster Competition,* Houston, TX, October, 2011.
5. 1st place award, Environmental Division*, AIChE Annual Meeting,* Minneapolis, MN, October, 2011.
6. 3rd place award, Materials Engineering & Sciences Division, *AIChE Annual Meeting,* Minneapolis, MN, October, 2011.
7. 2nd place award, *International Polyolefins Conference,* Houston, Texas, February, 2011.
8. NSF travel Grant, *2011 NSF-CMMI Research and Innovation Conference*, Atlanta, GA, January, 2011.

(4eh) Designing Nano-Scaled Micro-Environments for Molecular Catalysis and Synthetic Biology

Walker, D. A., Northwestern University

One of the canons of organic chemistry has been that a molecule's properties are inherently 'coded' by its chemical structure. This perspective, however, is an over simplification as there are a multitude of examples demonstrating that environmental factors contribute to a molecule's properties equally to, if not more than, the chemical structure. This is best exemplified by living systems – the complexity of human life is created out of just 21 amino acids which are strung together to create proteins and other sophisticated molecular machines. By controlling the environments around this handful of chemical species, and as such controlling their nanoscale interactions, nature has developed an endless list of highly efficient molecular machines.

Nanoparticles (NPs) provide a useful platform through which we can begin to probe the effects of molecular environments and various types of inter- or intra-molecular interactions. Electrostatic interactions tend to be one of the more versatile interactions at this length-scale as they can be attractive or repulsive, short- or long-ranged, and their magnitudes readily controlled. Here we demonstrate that on-particle self-assembled monolayers can be used to create artificial microenvironments which control the chemical properties of organic molecules. By controlling geometric orientations and electrostatic interactions one can essentially program the pK_a of carboxylic acids and redox potentials of organometallics in a manner analogous to that of proteins and enzymes. Additionally, using NPs functionalized with mixed self-assembled monolayers (mSAMs) one can elucidate important mechanistic information for catalytic diads or triads.

The ability to create artificial micro-environments offers a fundamentally new approach, as inspired by biology, for synthetic organic chemists. Furthermore, in the field of synthetic biology artificial microenvironments show great promise for tuning the properties of existing enzymes and proteins while, simultaneously, allowing for their stability and function in non-physiological chemical environments.

(4eq) Heterogeneous Catalysis: Producing Energy While Minimizing Environmental Footprint

Yi, N., Tufts University

Increasing global demand for non-renewable sources of energy has resulted in higher energy prices and emissions, but positive spillover effects have also ensued such as innovation in renewable energy and novel research into the consumption of conventional energy that is more efficient and less deleterious. Technological innovations must be sufficient in scale to supplant existing technology. This requires us to elucidate the reaction mechanism and active sites, as well as, to identify and eliminate the bottlenecks in catalytic processes. My research explores the role of heterogeneous catalysis in the production of fuels and chemicals, while, at the same time, decreasing our carbon footprint. I aim to integrated fundamental mechanistic studies, controllable synthesis methods, and techniques for improving practices implementation.

My Ph.D. projects investigated the efficacy and practical application of nanoscale precious metal clusters (e.g., gold and platinum) for hydrogen production through methanol and formic acid reactions, which could be applied to fuel cells. The strategy of selecting suitable supports and promoters as ligands to synthesize atomically dispersed precious metals was systematically explored. We designed catalysts that used trace amounts of precious metals, yet still resulted in high activity, selectivity, and stability. These catalysts were mindful of industrial economic realities, and were designed with low cost but with robust performance.

My first postdoctoral project investigates novel approaches to control the acid strength/sites over solid acids. The primary application is to catalyze light alkane cracking at low temperature to produce value-added chemicals. My second project explores the interaction between metal and multi-walled carbon nanotubes (MWCNTs) through structure sensitive probe reactions. Different approaches allow us to tune the size and shape of the metal and to control different types of functional groups on MWCNTs, which will identify the parameters that determine the products' selectivity. The significance of this research is that it could guide us to design desired materials while maximizing those promoting factors.

(4ak) Multifunctional Nanomaterials Design and Applications in Micro/Nanoelectronics, Sensing and Catalysis

Gao, F., University of Massachusetts Lowell

The overarching goal of my research work is to develop nano-joining materials for micro/nanoelectronics assembly and packaging applications and as templates with which to investigate the fundamental joint/substrate interactions. This poster gives an overview of my past and current research efforts on the design and development of nanostructured materials for various applications, including (i) fundamental research on the investigation and applications of lead-free nanosolders for micro/nanoelectronics assembly and device packaging; (ii) multifunctional nanowires and nanoparticles for electrochemical/bio-sensing and gas sensing applications; (iii) synthesis of metallic and metal oxide nanoparticles and use nanoparticles as catalyst for environmental application such as dye degradations.

Building on my past experience, I will develop a research program that includes, but are not limited to, the following three main research areas: (1) design and development of hybrid nanostructures with multiple functionalities, (2) novel nanodevice assembly and packaging techniques, and (3) integration of the nanomaterial synthesis methods and engineering processes, such as scale up into nanomanufacturing processes. Behind the diverse range of research topics that I have explored is the determination to explore the most unconventional ideas about the creation, manipulation, and use of materials, with the ultimate objective of contributing to solving the most difficult and important scientific and engineering problems such as sustainability and green engineering.

(4n) Utilizing Geometrically and Chemically Anisotropic Particles for Biomedical Applications

Bong, K. W., Harvard Medical School, Massachusetts General Hospital, and Department of Chemical Engineering in Massachusetts Institute of Technology
Doyle, P. S., Massachusetts Institute of Technology
Toner, M., Massachusetts General Hospital
Irimia, D., Harvard Medical School

Flow lithography (FL) pioneered by the Doyle group in MIT has been a powerful synthesis technique that enables the mass-production of hydrogel microparticles with geometrical and chemical patterns. The technique combines photolithography with microfluidic methods, providing precise control over shape and chemical patchiness in particles. Much of my PhD works has been focused on the development of advanced flow lithography that can achieve much higher degree of geometrical and chemical complexity in particles than before. Briefly, my techniques can be used to create hydrogel particles with 3D morphologies and chemical patterns. Also, I have extended the process of FL to non-PDMS based channels (ex. glass or plastic channels), and prepared novel class of functional particles using water insoluble monomers, organic solvents, and hydrophobic functional entities such as quantum dots, ruthenium dyes, and single-walled carbon nanotubes.

One of the powerful advantages of the anisotropic multifunctional hydrogel particles prepared by FL is that the particles are very well-suited for the biomedical applications. Typically, the gel particles are synthesized from monomers based on polyethylene glycol which is bio-friendly, highly tunable, and can be functionalized with a variety of biomolecules. Also, the controlled shapes and chemical patterns in particles can provide unique function for applications in diagnosis, treatment, and tissue engineering. For example, multifunctional barcoded particles have been designed for the rapid screening of target biomolecules in a complex biological mixture. Another example can be the artificial red blood cells (RBCs). Recently, soft disk particles with a few micron dimensions were synthesized to mimic RBCs, exhibiting the reversible deformability which is the vital property to allow RBCs to function.

In this presentation, I summarize the previous works, introduce current works that use the particle technology to do research on cell motility with applications to clinical cancer research as well as the separation of circulating tumor cells, and emphasize the importance of future works to bridge the particle technology to the biomedical area.

Selected Publications (* denotes co-first author status)

Bong KW, Xu JJ, Kim JH, Chapin SC, Strano MS, Gleason KK, Doyle PS. Non-polydimethylsiloxane

Devices for Oxygen-Free Flow Lithography. *Nature Communications*, **2012**, 3, 805.

Bong KW*, Chapin SC*, Pregibon DC, Baah D, Floyd-Smith TM, Doyle PS. Compressed-air Flow Control System. *Lab Chip*, **2011**, 11(4): 743-747.

Suh SK*, **Bong KW***, Hatton TA, Doyle PS. Using Stop-Flow Lithography to Produce Opaque Microparticles: Synthesis and Modeling. *Langmuir*, **2011**, 27(22):13813-13819.

Bong KW, Bong KT, Pregibon DC, Doyle PS. Hydrodynamic Focusing Lithography. *Angew. Chem. Int. Ed.,* **2010**, 49(1): 87-90.

Bong KW, Chapin SC, Doyle PS. Magnetic Barcoded Hydrogel Microparticles for Multiplexed Detection. *Langmuir*, **2010**, 26(11): 8008-8014.

Bong KW, Pregibon DC, Doyle PS. Lock release lithography for 3D and composite microparticles. *Lab Chip*, **2009**, 9(7): 863-866.

(4ae) The Thermodynamics and Chemistry of Organic Atmospheric Particulate Matter

Epstein, S. A., Carnegie Mellon University

Research Interests

Atmospheric particulate matter (PM) is a global pollutant with adverse effects on human health, visibility, and climate. In the United States, 75 million people live in counties that fail to meet government standards for concentrations of fine PM (<2.5 nm in diameter). In order to design effective and efficient regulation, we must understand the sources, formation, and evolution of PM in the atmosphere. This is a complex problem that involves modeling upwards of 100,000 compounds in three phases (gas, aerosol, and aqueous) under constant attack by atmospheric oxidants and sunlight. PM is composed of inorganic and organic compounds. The inorganic fraction is relatively well-characterized. However, due to its complexity, the organic fraction (25 to 75% of the total mass) remains to be adequately described. My work is aimed at understanding the formation and evolution of organic aerosol (OA) in the atmosphere.

OA is either directly emitted by anthropogenic and biogenic sources or is formed by reactions of volatile organic precursors in the atmosphere. The chemistry of these volatile organic precursors governs their eventual fate. My research experience focuses on the formation of organic particulates from oxidation by atmospheric oxidants. Ozonolysis oxidation reactions—a significant source of biogenic PM—are complex, containing several intermediates and branching points. I will summarize Temperature Programmed Reaction Spectroscopy experiments, performed at cryogenic temperatures to elucidate previously unknown barrier heights that govern the fundamental behavior of these reactions.

Organic aerosol consists of thousands of distinct chemical species in equilibrium with the gas phase. A large fraction of the species found in OA have vapor pressures that lead to repartitioning between the particle and gas phase in response to changes in temperature or total condensed organic mass. An aspect of my research experience and interests involves characterizing the thermodynamics that govern the interactions between gas and particulate phase species.

Along with oxidation, photolysis by sunlight is an important process governing the evolution and fate of organic compounds. The majority of photolysis studies have been focused towards gaseous compounds. However, photolysis can occur inside cloud and fog droplets, where hydrophilic gas and aerosol compounds may partition. In addition, particulate phase compounds can be photolized directly. Another major aspect of my research is aimed at determining the significance and quantifying the rates of these non-gaseous photolysis reactions with novel experiments and theoretical models.

Teaching Experience

I have had many opportunities to acquire teaching experience during my tenure at Carnegie Mellon. For several years I served as the AIChE Chemical Engineering Car team advisor for undergraduate students at Carnegie Mellon. I had the opportunity to mentor several student groups as they developed and tested their vehicles. I was also a chemical engineering instructor for the Carnegie Mellon Summer Academy for Math and Science. In this position, I gave lectures on fundamental chemical engineering concepts and led discussions sessions as students pursued their own design projects. I also had the opportunity to serve as a teaching assistant for four courses at Carnegie Mellon and give several guest lectures for both undergraduate and graduate students.

Service

I am actively involved within and outside the University community. I currently serve as the Vice-Chairman of the Laguna Beach Environmental Sustainability Committee, established to advise the city council on environmental policy decisions. I also regularly give educational air quality seminars to school children and adults throughout Orange County and coordinate the weekly AirUCI Institute seminar series. I am also an active member of the Heavenly Volunteer Ski Patrol in South Lake Tahoe, CA.

Educational Background

- Bachelors of Chemical Engineering from University of Delaware
- Ph.D. in Chemical Engineering from Carnegie Mellon University working with Prof. Neil Donahue
- Currently a postdoctoral researcher at the University of California, Irvine working with Prof. Sergey Nizkorodov in the Department of Chemistry.

Selected Publications

Epstein, S.A. and N.M. Donahue, *The Kinetics of Tetramethylethene Ozonolysis: Decomposition of the Primary Ozonide and Subsequent Product Formation in the Condensed Phase.* The Journal of Physical Chemistry A, 2008. **112**(51): p. 13535-13541

Epstein, S.A., I. Riipinen, and N.M. Donahue, *A Semiempirical Correlation between Enthalpy of Vaporization and Saturation Concentration for Organic Aerosol.* Environmental Science & Technology, 2009. **44**(2): p. 743-748

Epstein, S.A. and N.M. Donahue, *Ozonolysis of Cyclic Alkenes as Surrogates for Biogenic Terpenes: Primary Ozonide Formation and Decomposition.* The Journal of Physical Chemistry A, 2010. **114**(28): p. 7509-7515

Epstein, S.A. Shemesh, D., Tran, V.T., Nizkorodov, S.A., Gerber, R.B, *Absorption Spectra and Photolysis of Methyl Peroxide in Liquid and Frozen Water.* The Journal of Physical Chemistry A, 2011. **116**(24): p. 6068-6077

Epstein, S.A. and S.A. Nizkorodov, *A Comparison of the Chemical Sinks of Atmospheric Organics in the Gas and Aqueous Phase.* Atmospheric Chemistry and Physics, 2012. **12**(17): p. 8205-8222

Epstein, S.A., Tapavicza, E., Furche, F., and Nizkorodov, S.A, *Direct Photolysis of Carbonyl Compounds Dissolved in Cloud and Fog Droplets.* Atmospheric Chemistry and Physics Discussions, 2013. **13**(4): p. 10905-10937

(4bf) Soft Materials Mechanics: Instabilities, Hierarchies, and Anisotropies

Hutchens, S. B., California Institute of Technology

My main scientific interest lies in understanding fundamental mechanisms governing deformation and failure in soft materials, e.g., hydrogels, elastomers, and carbon nanotubes, for the advancement of socially impacting technologies such as soft robotics, biomaterials, and low energy devices. To this end, I have acquired experience in multiple areas of expertise including small-scale mechanics of materials, finite element mechanical modeling, polymer science, and mean-field theory. I am currently a post-doctoral researcher at the University of Massachusetts–Amherst working under Prof. Alfred Crosby. Previously, I completed my Ph.D. in Chemical Engineering with Prof. Julia Greer in 2011 and M.S. with Prof. Zhen-Gang Wang in 2006 at the California Institute of Technology. During this time, I gave numerous technical talks and posters and afterward was recognized with two thesis awards: the Demetriades-Tsafka-Kokkalis Prize for best thesis in "Nanotechnology and Related Fields" at Caltech and the 2013 international Quadrant Thesis Award presented in Zurich, Switzerland. I obtained my B.S. in Chemical Engineering in 2004 from Oklahoma State University. Given both my research and professional experience, I believe I would make a good colleague and enthusiastic addition to any department looking to start or expand their participation in soft materials research.

Research Background

Postdoctoral Research

I play a role in several projects in the Crosby lab at UMass-Amherst, with my main focus on two areas: understanding and utilizing a soft materials characterization technique developed in that lab called Cavitation Rheology, and using nanoparticle assembly to enhance performance in organic photovoltaic active layers. In the former, I am combining analytical theory, finite element calculation, and a simple experimental technique to determine the instability mechanism, either cavitation or fracture, for rapid void growth at the tip of an embedded, pressurized needle. This knowledge is then utilized for the localized characterization of soft materials, e.g., hydrogels and biomaterials as well as for probing length scale dependent properties ranging from microns to millimeters depending on the size of needle used. In addition to providing spatial resolution for investigation of graded materials, this localized, size dependent mechanical probe presents a promising method to evaluate hierarchical materials. In the latter project, I am exploring processing methods by which to pattern 2-D nanoscopic assemblies of functionalized quantum dots. The aim of these assemblies is to provide a flexible platform, e.g., controllable geometry and readily interchangeable component materials, from which to study the complex charge transfer between photoactive organic materials while providing a percolated, but still pliant, pathway for enhanced charge extraction. These projects have given me expertise in mechanical instabilities in soft materials, particularly polymers, a potential characterization technique for hierarchical material properties, and the fabrication of mesoscopic polymer-like composite structures.

Graduate Research

My past studies have ranged from the mechanical characterization of vertically aligned carbon nanotubes (VACNTs) to an analytical theory for charged macromolecule solutions. While often associated with a high tensile modulus and toughness, VACNTs are as soft as many elastomers when responding collectively in as-grown structures. Additionally, VACNTs possess many of the excellent electrical and thermal properties of their carbon nanotube constituents, making them potential candidates for applications requiring multiple, extreme functionalities and properties such as energy dampers in space. These foam-like materials undergo a distinctive layer-by-layer, localized folding response when compressed uniaxially. For my thesis research, I both observed this response *in situ* and reproduced key elements of it numerically using the combination of finite element calculations and a proposed constitutive model that included a material instability. In my earliest studies, I explored a mechanism for the condensation of charged macromolecule solutions, e.g., proteins, which have been observed to undergo a two-step nucleation mechanism that including small intermediate clusters before crystalizing. I utilized thermodynamics, specifically density functional theory, to develop an analytical prediction for metastable cluster intermediates as a function of ion concentration and relative dielectric constant. This work gave me experience in applying fundamental thermodynamic principles to generate and numerically solve complicated sets of equations. These projects have given me expertise in an atypical soft-material, VACNTS, numerical calculation with finite elements, the development of constitutive relations, and theoretical thermodynamics.

Proposed Research

As an independent researcher, my aim is to leverage my experimental and numerical expertise in large strain, large displacement, and instabilities in soft materials to open and explore additional subfields of study. In particular, I am interested in two fundamental areas:

1. *Systems that achieve large deformation using small strain.* What are the necessary geometric and material attributes for such systems? How can the response timescale of such systems be tuned? Answers to these questions will enhance design practice in soft robotics and explore paths leading to the adoption of phyto-inspired, low energy input movement. I aim to elucidate these answers using phyto-inspired, artificial cellular materials with well-controlled geometries and common, non-toxic solid-fluid elements. Plant tissues are known to undergo nastic movements on timescales ranging from milliseconds to hours using only a narrow range of materials properties, varying their microstructure, and employing instabilities, diffusion, and phase change as driving forces.
2. *Materials that are both conformal/connective, while being low stress/non-damaging.* What intensive and inhomogeneous properties lead to the optimization of this type of behavior? Can such materials be made in a manipulable and processable way? One challenge for which soft materials are ideally suited is conformal adoption of a high aspect ratio from an initial 2-D geometry. An example of an application for this behavior is at an artificial interface with neurons. Current contacts achieve only short lifetimes before cells become inactive, which is thought to be due to an unfavorable mechanical environment. I will approach the transmittance of stress upon large deformations through the study of materials with controlled spatial and material anisotropy, e.g., graded elastomers, graded cellular solids, VACNTs, and CNT/polymer composites. Hierarchical materials in which conformability can vary with length scale will also be fabricated using nanoparticle assembly techniques to create flexible, mesoscale network super-structures.

Teaching and Professional Activities

In addition to my scholarly interests, I have spent time broadening my professional experience through teaching assignments, lecture organization, and outreach to under-represented demographics. I was a teaching assistant twice each for two graduate level courses at Caltech, statistical thermodynamics and polymer physics. As a result of this experience and in conjunction with my research expertise, I can teach courses in several subject areas including thermodynamics, polymer physics, and mechanics of materials. As a graduate student, I organized and hosted seven speakers for the Materials Science Department departmental lecture series as well as serving on the Everhart Lecture Series Committee which yearly selects three outstanding graduate student researcher/communicators. At both Caltech and UMass-Amherst, I have served as a mentor for women-in-science through a mentoring program and as a participant on two panels discussing success in graduate school.

(4x) Understanding Mechanically Robust Materials: An Important Issue in the Energy Problem

Declet-Perez, C., University of Minnesota

Increasingly important challenges such as energy needs and utilization of a growing global population can be approached from different perspectives. Materials with novel properties often play a crucial role in new technology that aim to address these challenges. Accordingly, emerging technologies will show augmented structural complexity and functional richness. Regardless of any specialized or application-driven property, e.g. ionic conductivity for batteries, the mechanical integrity of new materials will greatly influence their overall performance as consumer products. Simply put, if the new technology is not mechanically robust to be handled, it will remain useless in a greater scheme. Traditionally, mechanical performance has been the primary focus in structural composite applications. The transportation industry provides excellent examples where lightweight yet robust materials permit reductions in energy requirements and footprint. However, understanding the relationship between structure and mechanical response must be considered in a broader sense. As novel materials are designed with increasingly complex structures there is a growing need for fundamental understanding of the mechanical properties of these often multi-component materials. The example presented here shows how these challenges can be addressed by combining different techniques. Simultaneous small angle x-ray scattering (SAXS) was used to probe local deformation processes of nanostructured epoxies under tension. These events were correlated to the macroscopic mechanical response and possible toughening mechanisms. With recent advances in SAXS, this approach can be easily extended to study the mechanical interaction of complex structures and multi-length scale phenomena within the material. The goal of the research program I will establish is not only to obtain materials with controllable structures, but also to understand the fundamental processes that govern the mechanical interaction of components or structural features and how they determine deformation and failure.

(4er) Modeling Adaptive Materials and Nanoengineered Surfaces for Energy Applications

Yong, X., University of Pittsburgh

The incentives for conducting energy-centric research stem from the pressing need to improve energy efficiency, reduce greenhouse gas emissions, and to define renewable energy sources. Nanoengineeringed adaptive materials, such as core/shell nanoparticles, polymer gel nanocomposites, and nanoporous scaffolds, are attractive materials for energy applications because of their unique properties, which arise from interface dominated behavior, size dependent effects and an ability to respond to multiple stimuli. The goal of my proposed research is to design adaptive stimuli-responsive soft materials and nanoengineered functional surfaces that will be used in various energy applications. The proposed projects are highly interdisciplinary in nature, and spans fluid dynamics, polymer science, and material science and engineering.

Specifically, I aim to use coarse-grained simulations to design liquid-infused nanostructured or nanoporous surfaces in order to enhance electroosmosis flow over these surfaces. Increasing the throughput of electroosmosis pump could dramatically drop the parasitic load of water management in polymer electrolyte membrane (PEM) fuel cells. I also propose to model membrane-mediated self-assembly of nanoparticles through controlled vesicle fusion. Guiding these nanoscale building blocks to self organize into regular and precise patterns could turn the self-assembly into a cost-effective alternative pathway of nanolithography for manufacturing organic photovoltaics, energy-saving transistors and LEDs. My third project is focused on designing photoresponsive polymer-coated nanoparticles, which could have potential in increasing the long-term stability of dye-sensitized solar cells (DSCs). In carrying out the above studies, I will be able to make significant contributions in the fields of active materials, electrokinetics and renewable energy, which are of particular interest to the chemical engineering, mechanical engineering and materials science and engineering communities.

(4dp) Engineering Microbial Production Platforms for Efficient Carbon Utilization

Solomon, K., University of California, Santa Barbara
Prather, K. L. J., Massachusetts Institute of Technology
O'Malley, M. A., University of California, Santa Barbara

The diverse chemistries of cellular metabolism make microbial systems an attractive solution to grand challenges in bioenergy, sustainability and human health. Among these are successes in the production of high-valued bulk and specialty chemicals such as isoprene and artemisinic acid (an antimalarial agent). The full economic potential of such systems, however, remains unrealized due to challenges inherent in driving carbon flux towards the desired product. Drawing on tools from synthetic biology, metabolic engineering, systems biology, and next generation sequencing and synthesis, my research has focused on the development of molecular strategies to increase the viability of these processes, and the elucidation of cellular responses to such intervention. My overall goal is to elucidate fundamental cellular strategies for the design of efficient metabolic pathways in next generation microbial production platforms which exploit robust and versatile features of microorganisms.

Here, I present my work towards the engineering of more efficient bacterial and eukaryotic systems. In my doctoral research (with Kristala Prather, Chemical Engineering, MIT), I developed transcriptional and post-transcriptional tools that dynamically divert carbon from central metabolism into a pathway of interest in *E. coli*. These tools pave the way for the development of new classes of metabolic pathways and compounds and allow for the optimization of existing pathways by reducing carbon waste. As a postdoc (with Michelle O'Malley, Chemical Engineering, UCSB), I am studying the effects of compartmentalized expression of metabolic pathways on productivity. This approach has tremendous potential to both locally concentrate metabolites and enzymes leading to greater productivity, and limit carbon loss to metabolic side reactions localized elsewhere. While these examples pursue more efficient carbon utilization, there still remains the challenge of sustainable, cost-effective feedstocks for these processes. Towards that end, I am also analyzing next generation sequencing data to identify novel biomass-degrading enzymes from anaerobic gut fungi and to elucidate their hydrolytic strategies against diverse lignocellulosic substrates. Such research strives to secure sustainable feedstocks for biotechnology and presents a unique opportunity to examine novel design strategies in a non-model organism.

(4dq) Fundamental Studies of Nanoscale Bio-Interactions

Sorci, M., Rensselaer Polytechnic Institute

Nanoscale bio-recognition is an essential requirement in many biological processes (e.g. enzyme-subtract interaction, immune response, and signaling pathways). Although our knowledge of the molecular and structural basis of these interactions is limited, scientists have found a way to apply these interactions to several applications like affinity chromatography, bio-sensors, and drug delivery. My Ph.D. work focused on affinity membranes for bio-separations. It centered around the development (selection of ligands, modification of membranes) and testing (batch and dynamic adsorption experiments, modeling, estimation of binding capacity, dissociation and kinetic constants, effect of ionic strength and flux) of affinity membranes for the purification of different classes of biomolecules: monoclonal antibodies, lectins and MBP-fusion proteins. More recently, during my post-doctoral studies, I have continued to work on several projects with the common focus of protein-protein interactions, from separation to aggregation. The poster will give an overview of the research conducted in the following topics: (i) specificity of nuclear transport using QCM-D; (ii) insights into the lytic mechanism of antimicrobial peptide using supported lipid bilayer to mimic bacterial membranes; (iii) the use of AFM in force mode to measure the binding forces between different split-protein partners; (iv) biophysical characterization of amyloid aggregates, from dimers to fibrils; (v) charged membranes for the separation of proteins with similar molecular weight and charge. I have broadened my background as a chemical engineer, integrating it with a variety of bio-tools. I plan to continue my career maintaining this interdisciplinary approach in studying protein-protein recognition both as basic science (e.g. better understanding of biological mechanisms) and applied science (e.g. developing bio-sensors).

(4en) Design and Development of Responsive Advanced Biomaterials for Regenerative Medicine
Xue, C.,

Regenerative medicine has been emerging as a new therapeutic strategy aiming to understand the principles of tissue development, and translate this fundamental knowledge into clinical restoration, repair, maintenance, and even improvement of tissue and whole organ function for the health of human beings. This innovative strategy has been performed by combining new regenerative medicines such as protein, DNA, cells, and engineered biomaterials. The engineered biomaterials have special significance.

The main goal of this research is to develop polymer-based biomaterials such as hydrogels and nanoparticles with the ability to mimic the natural extracellular matrix, to regulate the functions of its loaded elements, and to function as a theranostic carrier for regenerative medicine. Also, the research aims to understand the fundamental sciences underlying these problems by engineering the biointerfaces with highly-controlled architectures, chemistries and mechanical properties.

(4ar) Emerging Applications of Molecular Programming in Biology and Nanotechnology

Green, A. A., Harvard University

RNA and DNA are information-bearing polymers that underlie all life on Earth and that interact with one another predictably through Watson-Crick base pairing. Researchers have learned to exploit these interactions in a programmable way to establish molecular circuitry that can carry out many of the same operations as electronic circuits. They have also learned to construct nanometer scale structures with precisely defined sizes and shapes. These efforts in the field of molecular programming are opening up exciting new directions in biology and nanotechnology that I plan to explore as a faculty member. Since RNA and DNA, the basic tools of molecular programming, are also the information carriers of life, biomolecular networks composed of these polymers offer a powerful pathway through which to control the behavior of cells. I plan to develop sophisticated new nucleic acid based systems, such as riboregulators and riboswitches, to program the decision making of cells so that they can increase production rates of desired chemicals for metabolic engineering, and to better understand and enhance their behavior for synthetic biology. From the nanotechnology perspective, I plan to use nucleic acid structures as scaffolds that will enable us to position functional nanomaterials at precisely defined locations over micron scale areas. These efforts will make use of my combined expertise in self-assembly, nanofabrication, and biochemical separation processes. For example, I plan to assemble large area arrays of carbon nanotubes with well-defined electronic behavior. These nanotube arrays will have great potential as field effect transistors and transparent conductors composed solely of earth abundant materials. Furthermore, large scale assembly of metal nanoparticles will lead to plasmonically active structures with novel optical properties enabled through the controlled nanometer scale positioning afforded by nucleic acid scaffolds.

Biography. Alexander Green obtained his B.A.Sc. in Engineering Science from the University of Toronto in 2005. He then pursued graduate studies in Materials Science and Engineering from Northwestern University with Mark Hersam. At Northwestern, Dr. Green developed some of the first commercially scalable methods for producing carbon nanotubes and graphene with well-defined electronic and optical properties, and studied the performance of these novel materials in transparent conductors and thin film field-effect transistors. Since obtaining his PhD in 2010, Dr. Green has been a postdoctoral researcher at the Wyss Institute for Biologically Inspired Engineering at Harvard in the labs of Peng Yin and Jim Collins. There he has applied concepts from molecular programming to design programmable riboregulators of gene expression with unprecedented orthogonality and dynamic range. In addition, he has developed new methods for self-assembling RNA nanostructures with prescribed geometries, which have implications in siRNA delivery and metabolic engineering. Dr. Green has co-authored 50 peer-reviewed journal articles that have accumulated over 2800 citations. His h-index is 24 and several of his scientific discoveries have been successfully translated to industry.

(4ee) Understanding the Chemistry of Graphene and Two-Dimensional Nanomaterials

Wang, Q. H., Massachusetts Institute of Technology

Two-dimensional (2D) materials have unique quantum-confined properties in a form that is highly processable. Graphene, the prototypical 2D nanomaterial, is a single atomic layer of carbon with extraordinary properties including exceptionally high electronic carrier mobilities, thermal conductivity, and mechanical strength. Graphene-based applications have been demonstrated in high-speed electronics, chemical and biological sensing, optoelectronics, and energy storage and conversion. In addition, there is a wide-ranging library of 2D nanomaterials including transition metal oxides and dichalcogenides with diverse physical properties that are attracting great interest from researchers. In applications of 2D nanomaterials, the interaction of these atomically thin sheets with other materials plays a crucial role, particularly in influencing the electronic and chemical properties. There is a need to understand and control these interactions at nanometer scales. The chemical functionalization of graphene via organic chemistry is actively being pursued for modifying the electronic structure and doping level of graphene, and for altering its affinity toward various organic, inorganic, and biological materials.

I have pursued the chemical functionalization of graphene via both covalent and non-covalent methods. First, I have explored the self-assembly of organic semiconducting molecules into ordered monolayers on graphene, and their characterization using atomic resolution scanning tunneling microscopy (STM). These organic layers can serve as high quality seeding layers for the atomic layer deposition (ALD) of oxides that act as gate dielectrics in graphene electronics. Second, I have studied the direct covalent attachment of chemical groups onto the graphene lattice from aryl diazonium salts. The chemical reactivity of graphene toward this reaction was found to depend strongly on the nature of the underlying substrate, which influences the Fermi level in graphene, as shown by Raman spectroscopic mapping. Finally, I have developed new methods in the spatial patterning of graphene chemistry via scanning probe lithography and reactivity imprint lithography, which will enable new applications in electronic devices and chemical sensors. The insights into the interactions of graphene with organic molecules will be applied toward newly emerging transition metal dichalcogenide nanomaterials.

References:

1. **Q. H. Wang** and M. C. Hersam, "Room temperature molecular-resolution characterization of self-assembled organic monolayers on epitaxial graphene", ***Nature Chemistry***, **1**, 206-211 (2009).
2. **Q. H. Wang** and M. C. Hersam, "Sub-5 nm Nanopatterning and Templating of Non-Covalent Organic Monolayers on Graphene", ***Nano Letters***, **11**, 589 (2011).
3. **Q. H. Wang**, Z. Jin, K. K. Kim, J. Kong, A. J. Hilmer, G. L. C. Paulus, C.-J. Shih, J. D. Sanchez-Yamagishi, K. Watanabe, T. Taniguchi, P. Jarillo-Herrero, and M. S. Strano, "Understanding and Controlling the Substrate Effect on Graphene Electron Transfer Chemistry via Reactivity Imprint Lithography", ***Nature Chemistry***, **4**, 724 (2012).
4. **Q. H. Wang**, K. Kalantar-Zadeh, A. Kis, J. N. Coleman, and M. S. Strano. "Electronics and Optoelectronics of Two-Dimensional Transition Metal Dichalcogenides", ***Nature Nanotechnology***, **7**, 699 (2012).

(4bc) The Impact of Biomass Fuels On Flame Structure and Pollutant Formation During Biomass Cofiring Combustion

Holtmeyer, M., Washington University in St.Louis

Pulverized coal flames have been optimized for large-scale power generation to minimize criteria pollutant emissions and attain maximum fuel conversion and boiler efficiency. This traditional way of producing power is facing new challenges due to rising CO_2 emissions, criteria pollutant emissions regulations, and renewable portfolio standards (RPS). Biomass cofiring is a technology that can maintain this traditional method of power generation, while reducing emissions and potentially meeting RPS. However, biomass fuels are widely variable and different from coal, which can cause changes in the overall flame structure.

The volatile flame zone, which is the part of the flame that is dominated by the combustion of the gaseous volatiles, is strongly affected by changes in fuel characteristics. The effects on volatile flame length in pulverized fuel flames, especially with biomass cofiring and oxyfuel combustion, have been studied recently. An important contribution of these works is understanding the impact of particle size and volatility on volatile flame length, and the potential that particles can break through the volatile flame zone prior to releasing all of their volatiles.

In thi work, a model is employed to predict the volatile flame size of pulverized fuel flames, and scaling laws are used to predict particle breakthrough, particularly for conditions associated with biomass cofiring. Time and length scales of the flame and particle devolatilization processes are calculated and compared to determine to the overall participation of volatiles towards the volatile flame length. A dimensionless number is developed to characterize conditions of particle breakthrough.

(4bk) Effects of DNA Methylation Pattern On the Chromatin Structure

Jimenez-Useche, I., Purdue University
Yuan, C., Purdue University

My research focus has been on understanding how DNA methylation affects chromatin structure and gene expression. This information is instrumental in the identification of DNA methylation biomarkers for early diagnosis of cancer.

Chromatin's degree of compaction regulates gene expression by restricting the accessibility of cellular machinery to genes. Such regulation mechanism is often controlled by epigenetic modifications. DNA methylation, the most commonly found epigenetic modification in DNA, refers to the addition of a methyl group to the 5th carbon of a cytosine base in the context of a CG sequence. The presence of methyl groups in the DNA leads to a more rigid DNA structure, affecting the overall conformation of the chromatin fibers. Increased levels of DNA methylation are associated with chromatin compaction and gene silencing. The detailed molecular mechanism as to how DNA methylation directly modulates chromatin structure and function remains unknown.

Using multiple molecular biology techniques, we have constructed chromatin-like molecules i.e. nucleosome and nucleosome arrays, to study the effects of DNA methylation in the conformation and dynamics of chromatin, *in vitro*. These *in vitro* systems allow for the evaluation of the direct effect of DNA methylation on the variables of interest without the influence of other proteins found in the cell nucleus. In addition, we have been able to introduce different DNA methylation patterns to our system to study the effects of the position of the methylation sites in the chromatin structure.

We measured the conformational and dynamic changes of nucleosomes and nucleosome arrays using various biophysical assays, in particular fluorescence spectroscopy techniques. Our results show a dynamic behavior of nucleosomes i.e. the fundamental chromatin unit, despite of their methylation status or binding to other proteins e.g. linker histone proteins. We also observed that nucleosome and nucleosome arrays conformation can be modulated by specific DNA methylation patterns, depending on the buffer conditions.

The results from this research allow us to identify methylation pattern(s) that are functionally important in regulating chromatin conformation. Besides being instrumental in understanding the molecular mechanism by which DNA methylation directly modulates chromatin compaction, this information is useful in the identification of DNA methylation biomarkers for the early detection of cancer.

Aberrant DNA methylation patterns are present in early stages of carcinogenesis. Either hypomethylation of oncogenes or hypermethylation of tumor repressor genes can lead to the development of cancer cells. The potential of DNA methylation as biomarker of early onset of the disease has been explored in many different cancers. The traditional approach for screening of DNA methylation biomarkers is based on comparing the overall methylation level of a particular gene in healthy and disease tissue. A statistically significant difference between the methylation levels in both tissues makes this gene a good candidate for DNA methylation biomarker. Such approach can be bias or mislead by differential DNA methylation levels due to other factors different from cancer onset, such as age and gender. The DNA methylation patterns that we are evaluating can be use as a tool for assessment of DNA methylation status in biomarker genes. Besides the overall methylation level, the functional DNA methylation pattern associated with gene silencing discovered in this study can be identified in biomarker genes and then be use as a criteria for determining cancer risk, classification or prognosis.

(4h) Engineering Systems to Study and Control Tumor Progression and Metastasis

Azarin, S. M., Northwestern University

Most cancer deaths result from metastasis, which often goes undetected until multiple organs have been affected, at which point the disease is incurable. The cell microenvironment plays a critical role in every stage of cancer progression, but the mechanisms through which the microenvironment drives various events in the metastatic cascade remain poorly understood. The goal of my research is to develop novel techniques for elucidating the role of the microenvironment during various stages of cancer progression in order to identify potential targets for early intervention and detection of metastatic disease. During my doctoral research at the University of Wisconsin-Madison, I developed strategies to engineer the human pluripotent stem cell microenvironment *in vitro* in order to regulate cell growth and differentiation to various cell types, including cardiomyocytes and brain-specific endothelial cells. As a postdoctoral researcher at Northwestern University I have pursued novel methods to control and study the cancer cell microenvironment though utilization of a cellular array that provides high-throughput analysis of dynamic transcription factor activity and techniques to create controlled environments *in vivo* using biomaterial scaffolds that can be modified to include transplanted cells, extracellular matrix proteins and gene delivery vectors. I have applied these tools toward understanding the role of the extracellular matrix in driving colon cancer initiation and progression as well as developing biomaterial scaffolds to attract metastatic breast cancer cells *in vivo* that can be coupled with label-free non-invasive imaging techniques to provide a method for early detection of metastatic disease. This poster highlights the application of both *in vitro* and *in vivo* tools to control the cell microenvironment in order to elucidate mechanisms of tumor progression and metastasis and to develop novel therapeutic strategies and detection techniques to enable intervention while the disease burden is low.

(4bt) Design of Advanced Organic Materials: Block Copolymers and Photoresists

Lawson, R. A., Georgia Institute of Technology

Modern materials require modern, interdisciplinary approaches to their design and development. This is especially true of highly functional materials such as organic electronics and photoresists. Design space can be very broad, yet the property specifications can be very narrow. This means that large amounts of time and effort can be spent investigating materials that will eventually fail to meet specifications in the end. An improved alternative to this is the development of predictive models and heuristics which can rapidly shrink design space to improve the likelihood of success when experimentally investigating these materials. This is a daunting task even at the level of simply predicting the bulk properties of a material, but unfortunately in many emerging applications (e.g. organic electronics) the manner in which the material is processed and used can strongly affect its physicochemical properties. Thus, intelligent material design must have both structure-property relations, but also processing-property relations. My Ph.D. and Post-Doctoral work has led to the development of a number of design and simulation tools for the rational design of block copolymers, photoresists, and organic electronics. In the realm of structure-property relations, models have been developed that successfully predict the glass transition temperature of organic glasses and the full range of dissolution properties of small molecule photoresists including intrinsic solubility and dissolution rate. Further work has been done to develop design heuristics for high χ block copolymers and for directly photodefinable guiding layers for directed self assembly of block copolymers. Investigating the processing effects required the development of sophisticated simulation tools due to the multiple variables at play in processing beyond just the material's intrinsic properties. On-lattice kinetic Monte Carlo simulations were developed to investigate photoresist design strategies and the effect of additives on the patterning process in order to improve material performance. Off-lattice molecular dynamics simulations of block copolymers were developed to better understand their use in directed self-assembly. These simulations are especially useful to investigate the advantages and disadvantages of potential design schemes that are time and cost intensive to investigate experimentally such as the use of block copolymer or homopolymer blending to obtain a specified domain size and the use of tri-block, quad-block, and higher multiplicity block copolymers. These structure-property and processing-property relations have already been used to develop improved experimental photoresists and high χ block copolymers. Similar approaches could be used for the design of any number of advanced organic materials.

(4ct) Charge Transport Properties of Electronic and Ionic Conducting Block Copolymers for Lithium Battery Electrodes

Patel, S. N., University of California, Berkeley

Materials with nanostructured conducting domains are essential for a wide range of applications related to alternative energy. Active materials in battery and fuel cell electrodes such as LiFePO4, graphite, and platinum, are either electronic or ionic insulators. Nanoscale electron- and ion-conducting domains are necessary for enabling redox reactions in these materials. For example, a traditional porous lithium battery electrode consists of a redox-active material, carbon black for electronic conduction, and non-conductive binder that holds the particles in place. The pores are backfilled filled with organic electrolyte for ionic conduction. In some cases such as $LiFePO_4$, electronic and ionic conductivities are so low that the active materials must be in nanoparticle form, and addressing such particles requires the transport of both kinds of charges to occur on nanometer length scales. Materials such as block copolymers can self-assemble and form co-continuous nanoscale domains. The primary focus of my Ph.D. work at UC Berkeley under the advisement of Prof. Nitash Balsara involved the synthesis and characterization of the poly(3-hexylthiophene)-*block*-poly(ethylene oxide) (P3HT-*b*-PEO) as a nanostructured electrode binder where the P3HT-domains conduct electronic charges and PEO-domains conducts ionic charges. The results of ac impedance spectroscopy and dc measurements provided the first evidence of the simultaneous conduction of electronic and ionic charges in a block copolymer. In addition, my work was the first attempt to study the relationship between morphology and simultaneous electronic and ionic transport. The application of this material in a lithium battery with $LiFePO_4$ showed specific capacity reaching the theoretical limit and with minimal capacity fade, thus demonstrating the practical application of P3HT-*b*-PEO as conductive binder material. Furthermore, the ability of our conductive binder to switch between electronically conducting and insulating states in the positive electrode provides an unprecedented route for automatic overdischarge protection within the battery. This is in stark contrast to traditional lithium ion batteries where external electronics is used to provide overdischarge protection.

My current work as a postdoctoral researcher at UC Santa Barbara under the advisement of Michael Chabinyc and Ed Kramer focuses on fundamental charge transport properties of semiconducting polymers related to organic transistors and thermoelectric devices.

I believe the combination my Ph.D. and postdoctoral research provides a unique background spanning all solid-state batteries, organic electronics (e.g. transistors and photovoltaics), and organic thermoelectrics research fields. The experience of my Ph.D. and postdoctoral research will be used define the foundation of my research group where conductive polymers will be used a tool for a variety of energy storage and conversion applications.

(4br) Sustainable Energy for the 21st Century and Beyond

Konda, N. V. S. N. M., U.S. Department Of Energy, National Energy Technology Laboratory

Just over a decade into the 21st century and new millennium, it is becoming increasingly evident that energy security and climate change are two of the most defining challenges of our time and generations to come. Unfortunately, there is no *silver bullet* and we need to nurture range of technologies to address these challenges. To accomplish this, motivated by my own diverse research experiences, several ***clean and green technologies*** need to be developed further. These include:

.

- *Bioenergy (including advanced biofuels)*
- *Carbon capture, transport and sequestration (technologies and infrastructure)*
- *Hydrogen driven transportation (technologies and infrastructure)*
- *Clean coal technologies based on gasification (including integrated gasification combined cycle)*
- *Renewable energy sources (including solar and wind)*
- *Energy systems of the future (e.g., urban energy systems, rural electrification, zero-carbon homes)*
- *Water and healthcare (technologies and infrastructure)*

Together, all these technologies have the potential to alleviate both the energy security and climate change concerns. This presentation will discuss effective use of various *Systems Engineering* tools (including process modeling, design, control, optimization and dynamic simulation) to *foster scientific innovation and facilitate the acceleration of commercial development and deployment of these clean and green energy technologies. Special emphasis will be on large-scale real world energy systems.*

(4ep) In Situ Grafting and Growing of Organic Functional Layers Via Initiated Chemical Vapor Deposition

Yang, R., Massachusetts Institute of Technology
Gleason, K. K., Massachusetts Insitute of Technology

Initiated chemical vapor deposition (iCVD) is a very benign solventless process that produces conformal and thin (<100nm) coatings for a variety of technological applications. It enables the room temperature modification with functional polymers on virtually any substrate: organic or inorganic, rigid or flexible, planar or three-dimensional, dense or porous. During iCVD polymerization, monomers are delivered to the surface in vapor phase and undergo polymerization and thin film formation simultaneously. Free radical polymerization is initiated by filament heating with a typical current value of 1-2 Amp for a total power of 0.0086-0.0343 Watts cm^{-2}. This energy-efficient process thus enables the cost-effective passivation of silicon for semiconductor, photovoltaic and biosensor applications. By eliminating the need to dissolve monomers or macromolecules, iCVD prevents solvent damage to delicate substrates such as reverse osmosis membranes. CVD film growth proceeds from the substrate up, allowing for interfacial engineering, real-time monitoring, and thickness control. Ultra-thin coatings (~30nm) have been fabricated for membrane-based separations to obtain high permeate flux. iCVD has shown successful results in the rational design of micro- and nano-engineered materials to control molecular interactions at material surfaces.

However, during the iCVD process, monomers physisorb to the substrate surface and polymerize; the lack of covalent bonding between coating and substrate results in delamination and loss of function when the coated device is subject to stress or heat. Here, we report on the new chemistries we developed for iCVD to control interactions at material surface down to the molecular level. Both traditional iCVD polymerization and novel interface reactions have been utilized. We report for the first time a novel antifouling coating with excellent chlorine-resistance to mitigate fouling on membranes used in sea water desalination. *In situ* interface grafting of the coating directly to the surface of commercial membranes has been developed to enhance stability and durability. A second example involves a novel interface reaction on silicon, which enables the organic vapor passivation of silicon at room temperature. Passivation is essential for the fabrication of high-quality thin film photovoltaic devices. Excellent passivation quality with surface recombination velocities less than 10cm s^{-1} has been obtained with the iCVD passivation scheme and the passivation is stable for >200 hours in air, corresponding to the highest passivation quality reported to date for air-stable organic passivation. Through these two examples, we demonstrated the ability of iCVD to graft and grow polymer thin films on both organic and inorganic substrates for better control of interface interactions at the molecular level and greater robustness and durability of as-fabricated devices.

(4z) Understanding of Polymers in Confinement: Fluorescence Based Approach

Dishari, S., The Pennsylvania State University

Nanoscale investigation of functional polymers has significant importance in a wide range of applications, such as fuel cells, antifouling, chemobiosensing, optoelectronic materials and so on. When the film thickness approaches to several tens of nanometers, the optical, mechanical and transport properties are dominantly controlled by the interfaces. The mobility of the polymer chains is constrained due to severe confinement effect. The strength of interaction of water with polymer and substrate governs the film wetting, density and stiffness. Tuning these interactions with a view to approaching to desired material properties is a unique challenge. The key step to improve materials design is to understand the structure-property relationships. There is a large database available for ionomer membranes. However, very little is known about thin films. The challenges start when the conventional bulk characterization techniques (e.g. NMR, TGA, FTIR, standard tensile techniques) fail to predict the nanoscale changes in materials properties and sample preparation for a specific measurement becomes too tough (e.g. making free standing ultrathin films, stringent and specific substrate requirement). Fluorescence based techniques can overcome many of these issues and can be used as an important tool in both energy related research and chemobiosensing.

The knowledge of hydration behavior is truly beneficial for molecular level understanding and advanced level designing of polymer-catalyst interface in fuel cells. Fluorescent probes sensitive to local viscosity and proton concentration, when incorporated into ionomer thin films, gives valuable insight into mechanical and proton transport properties of thin films as a function of polymer structure, hydration and film thickness when they emit. The application of fluorescence based techniques in fuel cell research is quite new, has already started offering interesting insights and has a lot of scope of work.

One of the most fascinating applications of fluorescence based techniques is solid state nanobiosensing. The fluorescent conjugated oligomers/polymers and/or dyes with high quantum yield, when incorporated into nanoscale self-assembled thin films, respond to subtle changes in the microenvironment (pH, ionic strength) of the films and senses biomolecules (DNA, protein), metal ions and so on. Fluorescence resonance energy transfer (FRET), photoinduced charge transfer (PCT) etc. which requires fluorescent partners within ~10 nm scale distance shows superior response in ultrathin films with unique sensitivity and selectivity to the solid state sensing platform. The cost of detection can be reduced due to simplicity of techniques (e.g. steady state fluorescence), but not at the cost of signal efficiency.

Tailoring the polymer structure, film density and using suitable fluorescent moiety might solve a lot of mysteries hidden inside confined geometries.

Publications:

1. Liu, B.; **Dishari, S. K. "Synthesis, Characterization, and Application of Cationic Water-Soluble Oligofluorenes in DNA-Hybridization Detection"** *Chem. Eur. J.* **2008**, *14*, 7366-7375.
2. **Dishari, S. K.**; Kan-yi, P.; Liu, B. "**Combinatorial Energy Transfer between an End-Capped Conjugated Polyelectrolyte and Chromophore-Labeled PNA for Strand-Specific DNA Detection"** *Macromol. Rapid Commun.*, **2009**, *30*, 1645-1650 (**cover paper**).
3. **Dishari, S. K.**; Hickner, M. A. "**Antiplasticization and Water Uptake of Nafion Thin Films"** *ACS Macro Lett.*, **2012**, *1*, 291−295 (**cover paper**).
4. **Dishari, S. K.**; Hickner, M. A. "**Confinement and Proton Transfer in Nafion Thin Films"** *Macromolecules*, **2013**, *46* (2), pp 413–421.
5. Modestino, M. A.; Paul, D. K.; **Dishari, S.**; Petrina, S.; Allen, F. I.; Hickner, M. A.; Karan, K.; Segalman, R. A.; Weber, A. Z. "**Self-Assembly and Transport Limitations in Confined Nafion Films"** *Macromolecules*, **2013**, *46* (3), pp 867–873.

(4au) Electrochemically Structured Materials for Energy

Hangarter, C., NIST

My research aims to understand mechanisms by which materials transform energy and pathways to improve upon these systems. I am particularly interested in how materials form, degrade and interact with various components of the system, contacts, electrolytes and photons. I will present recent progress on Pt alloy electrocatalysts, assesing the role of synthesis through dealloying on the resulting structure and electrocatalytic performance. Additionally I will discuss development of three dimensionally structured thin film solar cells, exciting insights from photocurrent mapping and gains obtained through device engineering. Finally, I will describe future directions and my approach to cultivating an independent research group with a unique perspective in materials for energy conversion and storage.

(4cj) Energy Storage: Li-Ion and Beyond

Meduri, P., Pennsylvania State University

In the last two decades, there has been tremendous interest in energy storage and use of nanoscale materials for energy storage due to their high surface area, strain relaxation and shorter diffusion lengths inside the material. A major challenge is to understand the fundamental properties of these materials in energy storage applications. Nanoscale material surfaces play an important role both in electronic properties and as interfaces with other materials / electrolytes. As an example, the interaction of solid surfaces with organic electrolytes can lead to the formation of solid-electrolyte interphase layer (SEI) with thickness of few ten nanometers. For energy storage, an in-depth understanding of the SEI is necessary to better design materials.

Materials design and application for energy storage will be presented in this poster.

(4cb) Tailoring the Membrane Structure: the Challenging Path From Nano- to Process Scale
Marchetti, P., Imperial College London

In recent years the application of membrane technology to molecular separation processes has stimulated interest and showed great potential in a number of industrial fields. In the pharmaceutical industry, as an example, spending on R&D into new medicines has been enormously increased over the past decades and nanofiltration (NF) has been proposed as an emerging technology for concentrating and purifying Active Pharmaceutical Ingredients, with a number of advantages, compared to conventional downstream techniques. During my PhD at Imperial College London, and simultaneous enrollment as Early Stage Researcher at Lonza AG (Switzerland), my research in the field focused on two main aspects of membrane technology: the design and development of membrane processes for purification/separation of peptides, and the fundamental understanding and modeling of transport processes. The application of NF membranes was investigated for model peptides and the separation mechanism was analyzed from a fundamental point of view. An improved model for solvent permeation through nanofiltration and ultrafiltration membranes was developed and the importance of molecular affinity effects on the nanofiltration of peptides in aqueous and organic solvent mixtures was identified. The advantages of Design of Experiments methods and statistical modeling were combined with the description of basic filtration processes for concentration and diafiltration. Finally, an efficient strategy for the fragment condensation of peptides was developed, by integrating membrane technology into the synthesis step. This work demonstrated that NF is a solid and competitive technique for peptide applications.

One of the still challenging tasks of membrane technology is the fabrication of porous materials, with tunable intrinsic structure and controllable performance. Recently, I have been working for the development of new membranes from pre-fabricated nanostructures, to be used in both aqueous and organic solutions, and for the understanding of correlated phenomena of adhesion, wetting and spreading.

From a more general point of view, my research interest covers the area of both theoretical and experimental multi-scale approaches in novel materials and nanoscale phenomena. I am interested in the fabrication of functional materials from organized particles, on strategies for colloidal assembly and potential applications of micro- and nanoparticles structures in materials. The challenge of this research is in the development of efficiently controlled fabrication procedures, solving the issues of scalability and precision in controlling porosity and strength of interactions. Furthermore, I am interested in understanding the physico-chemical and structural properties of such materials and the transport mechanism through the matrix. This research opens to various fields, apart from membrane applications, since there is a broad spectrum of uses and applications of nanostructured materials in medicine, manufacturing and materials, energy and environment.

(4cq) Development of a Reactive Semi-Empirical Potential for the Atomistic/Molecular Simulations of Damage and Failure in Polymeric and Polymer/Metal Material System

Nouranian, S., Mississippi State University

Molecular dynamics (MD) simulation is a powerful tool for exploring various materials phenomena at the nanoscale. However, successful prediction of these phenomena and associated material properties relies on the availability and use of an interatomic potential that correctly captures the energetics of the atomic interactions. Recently, a great deal of research interest has been expressed towards the MD simulation of damage initiation and evolution in various materials, specifically polymers and polymer/metal systems. Most interatomic potentials do not allow dynamic bond making and bond breaking during simulation, which is a prerequisite for studying the damage phenomena associated with the polymer failure at the molecular scale. Moreover, a great number of potentials have not been parameterized for multicomponent material systems. The objective of this work is to develop an interatomic potential for hydrocarbon-based polymers based on the modified embedded-atom method (MEAM), a reactive semi-empirical *N*-body potential based on density functional theory and pair potentials. To achieve the objective, an initial parameterization of the MEAM potential was performed for *saturated* hydrocarbons. The potential was parameterized by fitting to a large experimental and first-principles (FP) database consisting of bond distances, bond angles, and atomization energies of a homologous series of alkanes and their select isomers from methane to *n*-octane, and various potential energy curves of H_2, CH, and C_2 diatomics, hydrogen, methane, ethane, and propane dimers, and select pressure-volume-temperature (PVT) data of alkanes. The atomization energies and geometries of a range of linear alkanes, cycloalkanes, and free radicals calculated from the MEAM potential were compared to those calculated by other commonly used potentials for hydrocarbons, i.e., second-generation reactive empirical bond order (REBO) and reactive force field (ReaxFF). MEAM reproduced the experimental and/or FP data with great accuracy, comparable to or better than REBO or ReaxFF. The experimental PVT data for a relatively large series of methane, ethane, propane, and butane systems with different densities were predicted reasonably well by MEAM. This work is the first attempt to expand the parameter database of the MEAM potential beyond atomistic systems, i.e., metals and metallic alloys, to covalently bound molecular systems. The details of potential parameterization and results of various atomistic calculations are presented.

(4cn) Effect of Conformation in Charge Transport for Semiflexible Polymers

Noriega, R., UC, Berkeley
Salleo, A., Stanford
Spakowitz, A. J., Stanford University

Current models for the electronic properties of semiconducting conjugated polymers do not include the hierarchical connectivity between charge transport units that results from the physical makeup of the materials. Concepts like on-chain vs. interchain mobility anisotropy have been known for a long time, yet they must be artifically incorporated into simulations. Models that can achieve remarkable predictive power but provide limited physical insight when applied to this new class of materials are of limited use for the rational design of new conjugated polymers. Here we present a new model in which the morphology of individual polymer chains is determined by well-known statistical models and the electronic coupling between units along a polymer chain and on different molecules is described using Marcus theory. Combining knowledge from polymer physics and semiconducting materials into an analytical and computational model that realistically incorporates the structural and electronic properties of conjugated polymers, it is possible to explain observations that previously relied on phenomenological models. The multi-scale behavior of charges in these materials (high mobility at short scales, low mobility at long scales) can be naturally described with our framework. Additionally, the dependence of mobility with electric field and temperature is explained in terms of conformational fluctuations and correlations.

(4ej) Energy Transfer in Molecular Photovoltaics, Carbon Nanotubes, and Nanowires - a First-Principles Perspective

Wong, B.,

The ability to tune electronic properties in molecular photovoltaics and nanomaterials holds great promise for incorporating these materials in next-generation transistors, circuits, and nanoscale devices. In particular, the use of predictive first-principles calculations plays a vital role in rationally guiding experimental efforts to optimize energy harvesting in nanoscale and mesoscale materials. In this presentation, I will highlight my recent work in using various quantum-mechanical approaches for understanding and predicting the electronic properties in light-harvesting molecules, functionalized carbon nanotubes, and heterostructure nanowires. First, I will demonstrate that both the optical properties and excitation energies in photovoltaic molecules can be accurately predicted by constructing new exchange-correlation functionals for time-dependent density functional theory (DFT).[1] Next, the use of large-scale DFT calculations is presented to understand optical detection mechanisms in chromophore-functionalized carbon nanotubes. Through joint experimental-theoretical studies, I will show that a single-walled carbon nanotube functionalized with light-sensitive chromophores can function as a sensitive nanoscale color detector, where the chromophores serve as photoabsorbers and the nanotube operates as the electronic read-out.[2-3] Finally, a new theoretical approach is presented to understand electron localization effects in heterostructure nanowires. At nanoscale dimensions, the formation of mobile electron gases in AlGaN/GaN core-shell nanowires can lead to degenerate quasi-one-dimensional electron localization, in striking contrast to what would be expected from analogy with bulk heterojunctions. The reduction in dimensionality produced by confining electrons in these nanoscale structures results in a dramatic change in their electronic structure, leading to novel properties such as ballistic transport and conductance quantization.[4]

1. M. E. Foster and B. M. Wong*, *Journal of Chemical Theory and Computation*, **8** 2682 (2012)
2. X. Zhou, T. Zifer, B. M. Wong, K. L. Krafcik, F. LÃ©onard, and A. L. Vance, *Nano Letters* **9**, 1028 (2009)
3. C. Huang, R. K. Wang, B. M. Wong, D. J. McGee, F. LÃ©onard, Y. J. Kim, K. F. Johnson, M. S. Arnold, M. A. Eriksson, and P. Gopalan, *ACS Nano* **5**, 7767 (2011)
4. B. M. Wong, F. LÃ©onard, Q. Li, and G. T. Wang, *Nano Letters* **11**, 3074 (2011)

(4v) Global Optimization for System Identification and Green Engineering Applications

Conner, J. A., University of California, Los Angeles
Manousiouthakis, V., University of California Los Angeles, Los Angeles

My Ph.D. research with Dr. Vasilios Manousiouthakis explored interval analysis-based branch-and-bound techniques for globally optimal parameter identification, a novel method for the assessment/synthesis of sustainable systems, graphical/analytical optimization techniques for the global solution of the TAC minimization problem for a series of compressors, and the introduction of the novel concept of the attainable region for separator networks. Throughout the course of my work, I have experienced first-hand the "curse of dimensionality" prevalent in non-linear optimization algorithms. Thus, I have applied analytical techniques, dimensionality reduction, convexification strategies, and functional analysis to identify global optima and to enhance the efficiency of their computation.

Interval Analysis-Based Branch-and-bound Methods for System Identification and Optimal Control

Motivated by the lack of model validation and guarantee of global optimality in traditional non-linear regression techniques, I developed a method for parameter identification that couples finite-dimensional global optimization strategies with interval forms of numerical model approximations (e.g. Euler, 4th order Runge Kutta). This method is guaranteed to identify the global optimum of the corresponding numerical non-linear regression problem. The method was employed in two case studies: (1) an ODE model aiming to identify the kinetic parameters of a Trambouze reaction scheme in a batch reactor based on experimental data related only to the reactant, and (2) a DAE model exhibiting trajectory discontinuities and infinite solutions. It was shown in both cases that the globally optimum objective function value was identified within the desired level of accuracy, and intervals were also identified, within which the globally optimal kinetic parameters must lie. These intervals were shown to be loose even though the associated collection of ODE trajectories tightly follows the experimental data, suggesting overparameterization with respect to the available data.

Infinite-Dimensional LPs for Process Synthesis and Attainable Region Identification

Recently, I expanded the applicability of the Infinite DimEnsionAl State-space (IDEAS) framework to attainable region (AR) identification for separator networks. While extensive research has already been done on AR for reactive distillation (see [1] for a formal introduction, and [2], [3], [4], [5] for extension of the framework to more complex reaction schemes and azeotropic mixtures), and the application of IDEAS to separation network synthesis has been demonstrated ([6], [7], [8], [9], [10]), the AR concept for Separation networks has not even been defined in the literature. I have constructed the AR for a water/methanol/acetone mixture involving one network feed stream with known species molar fractions and three network outlet streams at 1 atm pressure. The binary methanol/acetone system exhibits a minimum-boiling azeotrope at 79.07% mole fraction of acetone at 328.5 K. The IDEAS-generated AR successfully identified that the binary methanol/acetone azeotrope was outside the AR.

Hydrogen Economy

In my exploration of hydrogen as an alternative energy source, I formulated the minimum operating cost and minimum capital cost problems for a system of compressors and coolers in series that bring an ideal gas from a specified initial state to a specified final state. Through mathematical proofs, I reduced the dimensionality of the optimization problems and established analytical properties of the compressor outlet temperatures when either operating costs or capital costs dominate. This allowed the development and Excel implementation of graphical/analytical solution strategies involving minimization of operating cost and capital cost for a given number of compressors and coolers. A case study involving hydrogen compression was done to illustrate the methods.

Sustainable System Assessment and Synthesis

In my work, I proposed a novel global optimization approach that transforms basic indicator data in the form of intervals containing all possible basic indicator values, to an interval containing all possible values of the sustainability index, termed the "Sustainability Index Interval" (SII). The fuzzy-logic-based framework detailed in [11] was adapted for this work, since fuzzy logic is well-suited for simulating public opinions. Despite the nonconvexity of the underlying global optimization problems, I have proven that they possess a number of properties which can be utilized to reduce the burden associated with SII's computation. Based on these properties, I developed a branch and bound algorithm exactly quantifying the SII in a finite number of iterations. The framework and results of case studies have been published in the AIChE Journal [12]. Further, to test the validity of the public opinion simulation of the framework, I supervised three high school students in conducting surveys of the public pertaining to the sustainability efforts of six leading aluminum-producing corporations.

References:

1. Nisoli, A., Malone, M. F., & Doherty, M. F. (1997). Attainable regions for reaction with separation. *AIChE journal*, *43*(2), 374-387.
2. Agarwal, V., Thotla, S., & Mahajani, S. M. (2008). Attainable regions of reactive distillation—Part I. Single reactant non-azeotropic systems. *Chemical Engineering Science*, *63*(11), 2946-2965.
3. Agarwal, V., Thotla, S., Kaur, R., & Mahajani, S. M. (2008). Attainable regions of reactive distillation. Part II: Single reactant azeotropic systems. *Chemical Engineering Science*, *63*(11), 2928-2945.
4. Amte, V., Nistala, S., Malik, R., & Mahajani, S. (2011). Attainable regions of reactive distillation—Part III. Complex reaction scheme: Van de Vusse reaction.*Chemical Engineering Science*, *66*(11), 2285-2297.
5. Amte, V., Gaikwad, R., Malik, R., & Mahajani, S. (2011). Attainable region of reactive distillation-part IV: Inclusion of multistage units for complex reaction schemes. *Chemical Engineering Science*.
6. Justanieah, A. M., & Manousiouthakis, V. (2003). IDEAS approach to the synthesis of globally optimal separation networks: application to chromium recovery from wastewater. *Advances in Environmental Research*, *7*(2), 549-562.
7. Holiastos, K., & Manousiouthakis, V. (2004). Infinite-dimensional state-space (IDEAS) approach to globally optimal design of distillation networks featuring heat and power integration. *Industrial & engineering chemistry research*, *43*(24), 7826-7842.
8. Drake, J. E., & Manousiouthakis, V. (2002). IDEAS approach to process network synthesis: minimum utility cost for complex distillation networks.*Chemical engineering science*, *57*(15), 3095-3106.

9. Drake, J. E., & Manousiouthakis, V. (2002). IDEAS approach to process network synthesis: minimum plate area for complex distillation networks with fixed utility cost. *Industrial & engineering chemistry research*, *41*(20), 4984-4992.
10. Ghougassian, P. G., & Manousiouthakis, V. (2012). Globally Optimal Networks for Multi-Pressure Distillation of Homogeneous Azeotropic Mixtures. *Industrial & Engineering Chemistry Research.*
11. Phillis, Y. A., & Davis, B. J. (2009). Assessment of corporate sustainability via fuzzy logic. *Journal of intelligent & robotic systems*, *55*(1), 3-20.
12. Conner, J. A., Phillis, Y. A., & Manousiouthakis, V. I. (2012). On a sustainability interval index and its computation through global optimization. *AIChE Journal*, *58*(9), 2743-2757.

(4aa) Multi-Scale Modeling and Optimal Design Of Particulate Processes With Application To Solar Grade Silicon Production

Du, J., Carnegie Mellon University

My research interest in particle technology and particulate processes was motivated by an industrial application of solar grade silicon production in a fluidized bed reactor. We completed a multi-scale model by integrating three modules to capture the complex dynamics along different temporal and spatial coordinates. The integrated model has the capability to predict the silicon yield in the bed. By collaborating with our industrial partners from Renewable Energy Corporation, we validated the model against actual experimental data. The model was used to scale up and optimal design the demonstration scale reactor and is continued to be used for operation of the commercial system. A multi-scale model predicts silicon production yield and powder loss in a fluidized bed reactor for solar silicon production. The reaction module calculates the silicon vapor deposition and powder scavenging rates used by a population balance model that calculates the particle mass distribution functions. The model results are validated against industrial data. Furthermore, we conduct a sensitivity analysis to investigate the effect of gas flow rate and inlet Silane concentration.

New findings of my work show that an inherent instability exists in population balance as long as small particles are present in the particulate processes. An analytical expression was derived to explain how the critical size for stability of the particle size distribution function depends on physical properties and control parameters of the process. The mass distribution function of particles at steady state is derived and serves as the stationary equilibrium point of the population balance. I developed a discretization scheme which requires negligible computing effort. The discretized population balance is formulated in a network representation and a modified Tellegen's theorem is developed to provide an equivalent stability condition for the discrete form.

I pursued research in the control structure design in the context of reactive systems involving population balance. A systematic approach to choose process measurements and controlled outputs for nonlinear control systems was developed. The exponential stability and strict passivity of closed-loop control systems are guaranteed if the control structure is selected properly. I applied it to the abovementioned fluidized bed reactor, which only has a few degrees of freedom to manipulate. The proposed control strategy is proposed to maintain the process at the desired operating condition.

(4eb) Fundamentals of Self-assembly and Contact-Line Motion

Van Lengerich, H., Cornell University

I am interested in studying two topics: Self-assembly and Contact-line motion. These are both fundamental problems in engineering, which have a range of applications, and exhibit very rich and interesting behaviour. I plan on investigating these theoretically through chemical/mechanical engineering tools such as statistical mechanics, thermodynamics and fluid mechanics, using computer simulations to aid the theory, and performing simple experiments to test the results. My interest in these problems, and my contributions to them thus far, are described below. My other interest is teaching. I enjoyed being a teaching assistant in graduate school and instructor of a class as a post-doc, and I am proud of the good student reviews I have received, and my rapport with the students.

Self-assembly is the future for creating ordered nano and micrometer scaled objects. It is also the method by which viruses assemble, and hence it has significant applications in the health industry. However, no simple theory can predict when or how self-assembly works. Which kind of subunit structures will assemble into the desired macro-structure? How long will the assembly take? How can assembly be hindered? What are the classes of structures that can be formed? These are questions I seek to answer through modeling, computation, and simple experiments. In my post-doc (with Richard James, Aerospace Engineering and Mechanics, University of Minnesota), I have modeled self-assembly using a Smoluchowski equation with rotation, compared these results with Langevin simulations and constructed a macroscopic experiment on which these results can be tested. We have identified certain symmetric structures, which are natural to objects that self-assemble. Future research will focus on understanding the dynamics of assembly.

The second topic I am interested in is the relation between contact-line motion, solid-liquid slip, and contact-angle. Contact-line motion is important for inject printing, printed electronics, pesticides application (ones that spread more thinly and can be used more sparingly), and water-repelling glass. The fundamentals of the problem are still poorly understood. In my PhD (with Paul Steen, Chemical Engineering, Cornell University) I added a quasi-equilibrium thermodynamic balance to this fluids problem, which allowed for the first prediction of contact-line motion, based on the slip-length and equilibrium contact-angle. The theory allowed us to construct a simple experiment from which a slip-length was measured, which compared favourably with those in the literature. This theory works only for low Reynolds number flows, and there is interesting and industrially applicable work involving faster flow rates, which may become unstable or violate the quasi-equilibrium thermodynamic balance.

(4b) A Vector-Free Antigen Delivery Platform for Cancer Vaccines

Sharei, A., MIT

Mao, S., MIT

Basto, P., MIT

Szeto, G., MIT

Hartoularos, G., MIT

Jhunjhunwala, S., University of Pittsburgh

Irvine, D. J., Massachusetts Institute of Technology

Langer, R., Massachusetts Institute of Technology

Jensen, K. F., MIT

A vector-free antigen delivery platform for cancer vaccines

Authors: Armon Sharei[a], Shirley Mao[a], Pamela A. Basto[b], Gregory Szeto[b], George C. Hartoularos[a], Siddharth Jhunjhunwala[b], Darrell J. Irvine[b], Robert Langer[a,b], Klavs F. Jensen[a]

Affiliations: [a]Chemical Engineering Department, [b]Koch Institute for Integrative Cancer Research, Massachusetts Institute of Technology. 77 Massachusetts Ave, Cambridge, MA 02139.

Introduction

Cell-based vaccines against cancer, composed of tumor antigen-loaded antigen presenting cells (APCs), hold great therapeutic potential due to their ability to activate the patient's immune system and drive a long-lasting, tumor-specific, CD8 T cell response. These therapies, such as the recently approved Provenge® for prostate cancer, have minimal side-effects relative to chemotherapies and radiation treatment. However, one of the main barriers to developing such vaccines has been achieving efficient intracellular loading of APCs with protein antigens for presentation to CD8 T cells. Since the mechanism of cross-presentation (a process wherein endocytosed proteins can induce a CD8 response) within antigen presenting cells remains elusive, one must develop a reliable method of delivering the desired antigen directly to the cell cytoplasm to advance therapies utilizing the potent effector CD8 response. In this work, our team has adapted our recently developed microfluidic delivery system (Sharei, A. et al. *PNAS, 2013.* 110, 1082-2087) to induce selective MHC class I antigen presentation by direct delivery of the target protein to the cell cytosol. Compared to endocytotic methods, this system has demonstrated a 10-100 fold improvement in antigen-specific CD8 T cell activation and can potentially be translated to more challenging systems, such as B cells.

Materials and Methods

This work was performed using the ovalbumin (ova) antigen presentation model. In our experiments, dendritic cells were derived from the bone marrow of healthy B6 mice and treated by the aforementioned microfluidic delivery system to facilitate intracellular delivery of the ovalbumin protein. Antibody staining, ELISA, flow cytometry, and proliferation assays using OT-I and OT-II transgenic mice were used to assess antigen presentation efficiency and T cell cytokine secretion.

Results and Discussion

In our studies, we have demonstrated enhanced MHC-I presentation of the SIINFEKL epitope by antibody staining and verified that the delivery process does not interfere with DC maturation as measured by up-regulation of CD80 and CD86 in response to stimulation by lipopolysaccharide. As illustrated in **Fig. 1**, this method has also successfully induced the activation/proliferation of antigen-specific CD8 T cells (derived from OT-I mice) as measured by T cell proliferation and cytokine secretion. Moreover, preliminary results indicate that this method can induce functional antigen presentation in primary B cells ? a significant challenge for existing technologies.

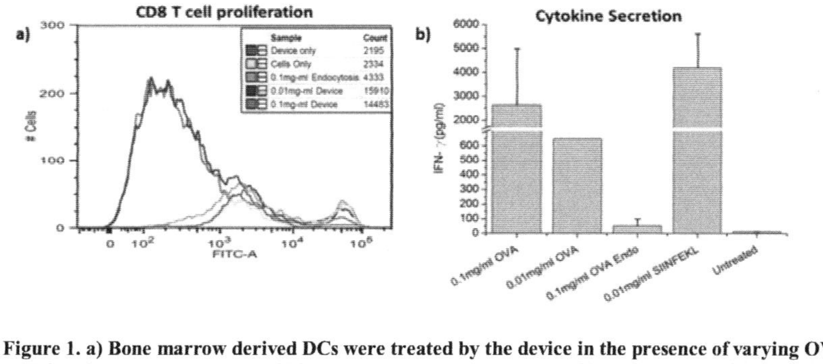

Figure 1. a) Bone marrow derived DCs were treated by the device in the presence of varying OVA concentrations and T cell proliferation was measured by CFSE staining and flow cytometry. b) Secretion of IFN-γ from CD8 T cells stimulated by treated DCs. The ?OVA Endo' condition corresponds to an endocytosis only case.

Conclusion

The described intracellular delivery technique has demonstrated enabling potential as a platform for high efficiency antigen presentation in cancer vaccine applications.

(4a) Polydopamine - Coated PCL Shape Memory Polymer Foams for Bone Regeneration

Munoz-Pinto, D., Texas A&M University

Polydopamine-Coated PCL Shape Memory Polymer Foams for Bone Regeneration

Monday, November 4, 2013: 2:57 PM
Franciscan D (Hilton)

Dany J. Munoz-Pinto[1], Andrea C. Jimenez-Vergara[1], Dawei Zhang[2], Melissa Grunlan[2] and Mariah S. Hahn[1,3], (1)Biomedical Engineering, Rensselaer Polytechnic Institute, Troy, NY, (2)Biomedical Engineering, Texas A&M University, College Station, TX, (3)Department of Chemical Engineering, Texas A&M University, College Station, TX

Introduction: Scaffolds employed in bone regeneration must promote new bone formation while forming a strong physical interface with (conforming with and bonding to) adjacent bone tissue to promote osteointegration. In the present study, the osteoinductivity of a "self-fitting", polydopamine-coated shape memory polymer (SMP) scaffold was evaluated. This SMP foam was fabricated by photopolymerization of diacrylated poly(e-caprolactone) (DAcPCL) followed by polydopamine coating. When treated with warm saline ($T > T_m$ of PCL), the compressed DAcPCL foam not only expands but is in a softened state, which permits its manipulation and conformability along irregular bone defect "boundaries" thereby promoting osteointegration. A polydopamine coating was selected due to its ability to maintain its integrity under repeated scaffold deformation (in contrast to calcium phosphate coatings) and due to recent literature indicating that polydopamine stimulates osteoblast adhesion and matrix mineralization. Results from the current work demonstrate that polydopamine coatings increase the osteogenic capacity of these SMP foams.

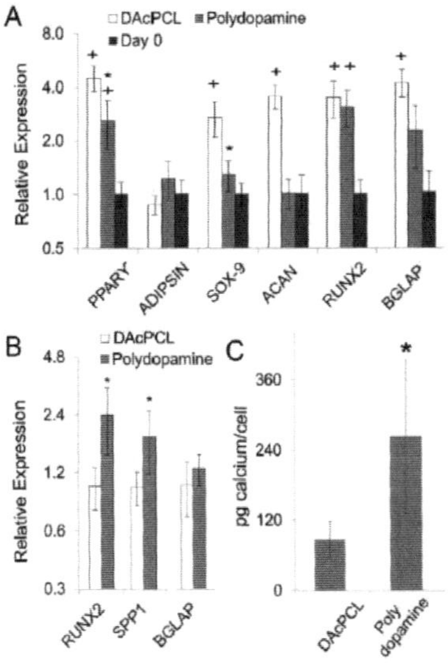

Fig. 1. (A) Relative gene expression of adipogenic, chondrogenic and osteogenic markers using GM; (B) Relative gene expression of adipogenic, chondrogenic and osteogenic markers using OM; (C) Specific calcium deposition using OM. +, significantly different from day zero; *, significantly different from DAcPCL, $p < 0.05$, n = 5-7.

Materials and Methods: *Fabrication of DAcPCL foams*. Photosensitive DAcPCL macromers were formed into open porous foams via solvent casting/particulate leaching (SCPL) technique which increases pore interconnectivity versus traditional salt-leaching. Following salt leaching, subsets of foams were coated with an aqueous dopamine chloride solution for 16 h to allow spontaneous oxidative polymerization of dopamine on pore walls. *Construct Culture and Analysis*. Coated and uncoated DAcPCL foams were seeded with human bone-marrow derived MSCs (Lonza) at 1×10^6 per cm^3. The resulting constructs were cultured for 2 weeks in Growth Medium (GM; DMEM with 10% MSC-qualified FBS) or Osteogenic Medium (OM) (GM supplemented with 0.1 μM dexamethasone, 50 μg/ml L-ascorbic acid-2-phosphate, 10 mM β-glycerolphosphate). Gene expression was then analyzed relative to β-actin using qRT-PCR. Total calcium deposition was quantified using the CPC liquid color kit (Stanbio).

Results and Discussion: The capacity of polydopamine-coated foams to support MSC osteogenesis was analyzed following 2 weeks of culture. In the absence of osteogenic media supplements, general upregulation of adipogenic (pparγ), chondrogenic (sox9, aggrecan [ACAN]), and osteogenic (runx2, osteocalcin [BGLAP]) markers was observed in the uncoated scaffolds relative to initial (day zero) expression levels ($p<0.05$; **Fig 1A**). In contrast, polydopamine-coated scaffolds did not promote chondrogenesis, although they still stimulated adipogenesis in addition to osteogenesis. When foams were cultured under conditions of "enforced" osteogenic specificity (under OM conditions), an increase in MSC expression of osteogenic markers (runx2, osteopontin [SPP1]) was observed on polydopamine-coated scaffolds relative uncoated controls ($p< 0.023$, **Fig. 1B**). Polydopamine-coated foams were also associated with a 3.2-fold increase in calcium deposition relative to uncoated scaffolds ($p = 0.04$, **Fig. 1C**). Cumulatively, these results support the osteogenic potential of polydopamine-coated DAcPCL foams.

Conclusions: DAcPCL SMP foams have significant potential as bone regeneration scaffolds due to their high porosity, elastomeric (non-brittle) nature, and capacity to be inserted in conformal contact with bone defect boundaries (promoting osteointegration). The present studies indicate that coating these SMP foams with polydopamine increases the osteoinductive capacity of these novel scaffolds.

Acknowledgements: MH and MG acknowledge NIH funding.

(30a) An Immersive Computer Game to Support ChE Learning

Liauw, M. A., RWTH Aachen University

Originally presented on: 11/3/2013 17:30:00 - 17:31:00

A virtual chemical factory, "Zink & Co.", is implemented as an online Flash computer game. Second- and fourth year's students register while they are taking Reaction Engineering I and II classes. In that factory, they have to solve problems with the newly acquired information (second year, non-graded activity) or design new, challenging problems (fourth year, graded activity). Reportedly, the interactivity and the link to industrially relevant settings are beneficial for the learning. It is noteworthy that the game was mainly created by students themselves, including nonsense details and gimmicks, giving it much more the student flavor than instructor-created games. This is confirmed by the students playing the game. It is well received, even if it cannot compete with professional computer games as far as graphics or dynamics are concerned.

At the workshop, the game will run on a computer so it can be experienced hands-on. In a handout, the creation and implementation of the game will be outlined. A focus will be the obstacles along the way, and how they were overcome.

(30b) Improving Student Attitudes Toward Capstone Laboratory With Gamification

Anastasio, D., University of Connecticut
Burkey, D. D., University of Connecticut
Suresh, A., University of Connecticut
Originally presented on: 11/3/2013 17:31:00 - 17:32:00

A gamified method of presenting capstone laboratory material and evaluating student performance was developed in order to improve student attitudes toward the course. The game split the class into teams, and these teams were encouraged to compete with one another to amass the most points. Students gained experience, or grade, points from required assignments, the distribution of which was delineated in a detailed rubric. Other optional tasks only increased their team's point score. Some of these tasks required students to work together as a group to gain an edge against the competition. The team with the highest point total "won" the game and received a prize. Other prizes were given for outstanding participation in the game.

The game was run for two subsequent semesters: Fall 2012 and Spring 2013. While participation in the game was optional, all students in the course participated in some of the optional activities each semester. Students who earned higher grades on required assignments tended to complete a large amount of the optional tasks. However, the remaining students all participated in optional tasks at equal amounts across the grade distribution, meaning the game was not exclusively appealing to the highest-achieving students. After each semester, students were polled as to their opinion of the course and the way the game was run in an attempt to improve the game in future semesters. Students reacted positively to most aspects of the game and enjoyed the class for the unique way the material was presented. Some students went on to say the game made them want to think about the class when they otherwise wouldn't. In both polls, students overwhelmingly voted to run the course as a game in future semesters, an attitude shared by equal percentages of both male and female participants.

(30c) Games and Game Mechanics in the Cheme Curriculum

Vigeant, M., Bucknell University
Liauw, M. A., RWTH Aachen University
Originally presented on: 11/3/2013 17:32:00 - 17:57:00

Interested in games? Want to know more about how you might use computer-based games to teach chemical engineering? Want to know how you can use game mechanics to make your courses more engaging? Want to meet potential collaborators for educational game development?

This 90 minute workshop will give participants hands-on experience with several sample computer-based games that can be used in ChemE courses. Computer-based games for a variety of scales will be shared - from 5-minute homework support to multi-day or week- simulations. Game designers will be present to offer advice on development of new games and insight into how their games may be used in the classroom. Participants will also discuss and observe how game-mechanics can be applied to make courses more engaging for students. As a final product, participants will have time to brainstorm their own game / game mechanics and have a chance to have their ideas constructively discussed with facilitators and peers.

(62a) Session Honoring Don Woods

Faucher, S., Hatch
Keith, J. M., Michigan Technological University
Originally presented on: 11/4/2013 8:30:00 - 10:35:00

Session Honoring Don Woods

Abstract

This session honors Professor Don Woods who passed away in April 2013. Dr. Woods was a chemical engineering professor at McMaster University from 1964 to 2000 where he used innovative teaching methods and won many teaching awards (and three honorary doctorates from Queens, Guelph and McMaster). He wrote many books and papers, and gave workshops in various parts of the world. What you may not know about Dr. Woods is that he was also an artist, banjo player, square dancer, badminton player, builder, author, genealogist, and great reader. He enjoyed spending time at Kinlough, the family's 200 acre vacation place in southern Bruce County, as well as travelling to many interesting destinations in retirement.

This session will highlight reflections on the impact of Dr. Woods career. Speakers in this session include:

- Angie Perna, NJIT

- Phil Wood, McMaster University

- Rich Felder, North Carolina State University

- Scott Fogler, University of Michigan

- Charles Petty, Michigan State University

- Suzanne Kresta, University of Alberta

- George Chase, The University of Akron

(62b) Cross Flow Filtration Hydrocyclones

Petty, C. A., Michigan State University

Originally presented on: 11/4/2013 10:35:00 - 11:00:00

A cross flow filtration hydrocyclone (CFFH) combines the desirable attributes of a vortex separator and a filter barrier. By selecting an appropriate hydrophilic filter medium, a CFFH clarifier has the potential to produce a filtrate phase with a low concentration of a dispersed organic phase in a single stage. The CFFH concept addresses three traditional problems associated with current hydrocyclone separators for oil/water separation: 1) the loss of separation performance due to core flow reversal; 2) the loss of separation performance due to entrained oil in the sidewall boundary layer; and, 3) the loss of separation performance due to hydraulic turndown. The paper will discuss the efficacy of using a forward-flow CFFH and a reverse-flow CFFH for clarifying produced water on offshore platforms and elsewhere. Computational fluid dynamic (CFD-) methods are used to identify practical prototype designs and operating characteristics.

"The Ubiquitous Hydrocyclone: Theory and Practice, C. A. Petty, AIChE Webinar, February 11, 2009, http://www.aiche.org/webinars.aspx.

C. A. Petty and S. M. Parks , 2001, "Flow Predictions within Hydrocyclones", Filtration+Separation, Volume 38, Number 6, 28-34.

Ali, S.K., K. Weispfennig, J.F. Foss, and C. A. Petty, 1992, "The Separation Performance of Long Hydrocyclones with an Annular Feed Entry", BHRA 4th Int'l Conf. on Hydrocyclones, Southampton, UK, 23-25 September 1992.

Dabir, B. and C. A. Petty, 1986, "Measurements of Mean Velocity Profiles in a Hydrocyclone Using Laser Doppler Anemometry", Chem. Eng. Commun., 48, 377-388.

Bahram Dabir, 1983, "Mean Velocity Measurements in a 3"-Hydrocyclone Using Laser Doppler Anemometry, Michigan State University, Ph.D. Dissertation, 1983.

Knowles, S. R., D.R. Woods, and I.A. Feinstein, 1973, The Velocity Distribution within a Hydrocyclone Operating without an Air Core", CAN J CHEM ENG, 51, no.3, pp.263-271.

(69a) Hydrodynamic Resistance of a Train of Confined Droplets
Suteria, N.,
Originally presented on: 11/4/2013 8:30:00 - 8:48:00

Microfluidic devices have become increasingly popular in the fields of science and engineering due to their ease of fabrication. These devices have significant potential for a wide range of applications including biomedical diagnostics and biosensing. Many of these applications involve the transportation, sorting and storage of droplets. When a single-phase fluid flows in a confined channel, there is a resistance to flow. Two-phase fluid flow, which includes droplets, introduces even more resistance to flow, known as hydrodynamic resistance of the droplets. The hydrodynamic resistance of droplet is the main key factor that controls the behavior of the droplets in a channel. Researchers have tried to determine the parameters that may affect the hydrodynamic resistance of droplets in a microfluidic channel. However, their studies have limited domain of parameters. The purpose of this study is to measure the hydrodynamics resistance of droplet over a wider range of parameters, which include capillary number, drop size, intra-drop spacing, viscosity ratio and surface tension between the inner and outer phase. We measure the resistance of the droplet using a microfluidic comparator; it directly compares a reference channel with known resistance to a parallel test channel containing droplets of unknown resistance. These water-in-oil droplets are generated upstream from the comparator using a cross-junction where droplet size and spacing can be easily controlled. Quantifying the resistance of droplet can help researchers better design their microfluidic devices for droplet trafficking, sorting or trapping.

(69b) An Improved in Vitro Model for the Study of Cancer Cell Adhesion to Endothelial Cells Using Micropatterned Surfaces

Gray, L., University of Kentucky
Fischer, J., University of Kentucky
Trinkle, C., University of Kentucky
Eitel, R., University of Kentucky
Anderson, K. W., University of Kentucky
Originally presented on: 11/4/2013 8:48:00 - 9:06:00

Adhesion of cancer cells to the endothelial lining of blood vessels is an integral step in metastasis. Because of shear stress produced by blood flow, the endothelium undergoes a distinct morphological change resulting in a more elongated and unidirectional morphology. It has recently been suggested that this morphology can affect surface expression, in turn affecting cell-cell binding to the endothelium. *In vitro* methods of studying cancer cell adhesion to endothelial cells include flow models and static well-plate cultures. Static well-plate cultures result in a cobblestone morphology and more random orientation than what is observed *in vivo*. While flow adapted endothelial cells are more representative of physiological conditions, flow systems can take up to 24 hours to obtain a valid morphology. In this study, the use of micropatterned surfaces to statically culture human vein endothelial cells (HUVECs) with desired morphology was investigated. The HUVECs were grown on glass slides containing grooves of 0.5 mm width and depth created using UV lithography. Both morphology and surface molecule expression of these cells were compared to cells grown statically and in a flow-adapted system. HUVECs cultured on micropatterned grooves demonstrated the desired morphology which was proved statistically more elongated and unidirectional than the control.

(69c) The Effect of Mechanical Shear On the Growth and Aggregation Dynamics of Taxus Suspension Cultures

Vilkhovoy, M.,
Bevacqua, S., University of Massachusetts Amherst
Wilson, S., Rowan University
Roberts, S. C., University of Massachusetts, Amherst
Originally presented on: 11/4/2013 9:06:00 - 9:24:00

Paclitaxel, derived from the *Taxus* Pacific Yew Tree, is a mitotic inhibitor used in cancer chemotherapeutics. Currently, paclitaxel is being produced using plant cell suspension cultures, but the process is limited by low and variable product yields. One possible cause for this variability is the tendency of plant cells to grow in a range of aggregates sizes, creating subpopulations throughout the culture due to varying oxygen and nutrient diffusion profiles. The Roberts' lab has found that cultures with smaller aggregate sizes have increased paclitaxel production levels. As a result, this project aims to investigate the effect of mechanical shearing on growth, viability and aggregation of *Taxus* suspension cultures. Different shearing methods (pipetting and vortexing) were investigated in order to determine the most efficient and effective way to break aggregates. Manual pipetting of cultures through a glass pipet was found to be the most effective method to decrease the aggregate size. The aperture tip, where most of the breakage occurs, plays an important role in determining the final culture aggregate size. Along with the aperture tip, the flow rate of pipetting affects the rate of particle breakage and determines the terminal diameter of the aggregates. The culture density had no effect on determining the terminal particle size when sheared through pipetting, which is important for shearing of cultures throughout a growth cycle. The effect of mechanical shear was then investigated on the culture growth and aggregation over multiple generations. The culture mean aggregate size was effectively reduced without inhibiting culture growth over three generations. Controlling aggregate size through mechanical shearing by pipetting could lead to cultures with reduced variability and increased paclitaxel production levels.

(69d) Biomimetic Significance Of The Nanofeatures and Nanomechanical Properties Of Organic Thin Films In Nacre

Parratt, K.,
Yao, N.,
Originally presented on: 11/4/2013 9:24:00 - 9:42:00

Many of important advances in materials engineering have come from transitioning to thinking of materials and their properties on the nano-scale rather than on the macro- or even micro-scale. Improvements in imaging technology have allowed us to see the nanofeatures that directly impact the chemical and mechanical properties of natural and man-made materials. This has been especially transformative to the study of the structure and failure modes of nacre, one of nature's most impressive composites which provides inspiration to the field of biomimetics. Now that these characteristics can be quantified and these features can be imaged, for the first time scientists have begun to determine the importance of nanoasperities and structure of the organic layer in nacre. This project investigates the structure of the organic component of nacre; specifically its mechanical properties, self-assembly, and interactions with the inorganic layer. Plasma-etching has only recently begun to be applied to nacre and has not been previously used to investigate the organic layer. The plasma-etching allowed me to selectively remove organic components and document their relative placement in the organic layer. I then applied these principles to explain the existence of the relatively less discussed "dome" platelets, to rationalize their role in the organic layer, and to analyze their potential benefits for biomimetic materials.

(69e) Manufacture and Characterization of Carbon Nanotube Microcable

Kienzle, N., University of Cincinnati
Ochmann, T.,
Shanov, V., University of Cincinnati
Alvarez, N. T., Rice University
Originally presented on: 11/4/2013 9:42:00 - 10:00:00

Carbon Nanotubes (CNTs) have gained attention over the past two decades because of their extraordinary strength and electrical properties. CNTs are a good candidate to replace copper wire for electric power distribution because of their corrosion resistance, low density, and high cycle fatigue. At the University of Cincinnati's Nanoworld Labs, CNTs are grown on a silicon wafer substrate by a Chemical Vapor Deposition process. CNTs are then "spun" in a similar fashion as silk into a macroscopic thread (CNT thread). This CNT thread is about half the width of a human hair. The objective of the reported research efforts was to determine the most effective way to improve the electrical properties of CNT thread. Currently, the conductivity of "as spun" CNT thread is around three orders of magnitude lower than copper. By post treatments, we have been able to achieve conductivity of 10^5 S/m, which is about two orders of magnitude lower than that of copper. In addition, experiments have been conducted to coat the CNT thread with a polymer making its surface insulating as that of a metal electric cable. We expect further improvement in the electrical conductivity of CNT thread which will allow using it in electromagnetic devices, thus replacing copper wire and reducing the weight of the devices.

(69f) Synthesis and High Temperature Thermoelectric Characterization of YCo1-XRhxO3 ($0 \leq x \leq 1.0$)

Dickens, C., Oregon State University

Li, J.,

Subramanian, M.,

Originally presented on: 11/4/2013 10:00:00 - 10:18:00

Polycrystalline samples of $YCo_{1-x}Rh_xO_3$ ($0 \leq x \leq 1$) were synthesized by solid state reaction, and found to be phase pure (with the exception of a minute amount of Y_2O_3) by x-ray diffraction. Co containing compositions required reaction in O_2 at 925 or 1000 °C while $YRhO_3$ required reaction in an evacuated quartz ampoule at 1100 °C. All compositions were shared the same distorted perovskite crystal structure, space group Pbnm. A linear relationship between Rh content and unit cell volume (calculated via Lebail fit) was found, confirming direct substitution of Rh for Co. Electrical resistivity, thermal diffusivity (converted to conductivity utilizing experimental density and published $LaCoO_3$ heat capacity), and Seebeck coefficient were measured from 450 – 750 K. All compositions exhibited semiconductor-like behavior over the temperature range tested. Conduction was facilitated either by mixed valence state of Co/Rh or by excitation of Co^{3+} from LS to HS. The activation energy of polaron conduction was found to increase near 600 K in all compositions besides $YRhO_3$, corresponding with a distortion of the lattice induced by the Co^{3+} spin state transition. In the high temperature limit, the Seebeck coefficient was shown to increase with Rh substitution, implying a stabilization of LS Co^{3+}. Thermal conductivity was found to be almost exclusively due to lattice vibrations, and the enhanced phonon scattering ability of air voids is thought to be the reason for decreased thermal conductivity compared to previous studies on $YCoO_3$ of less porosity. A maximum ZT of 0.085 was achieved with x = 0.3 and x = 0.7 at 750 K, and is linearly increasing with temperature at that point.

(69g) Patterned Co-Culture of Breast Cancer Cells and Stromal Cells As An In Vitro Breast Cancer Model

Daverey, A., University of Nebraska-Lincoln
Drain, A., University of Nebraska-Lincoln
Crone, K., University of Nebraska-Lincoln
Kidambi, S., Michigan State University
Originally presented on: 11/4/2013 10:18:00 - 10:36:00

The paracrine signaling pathways between breast cancer cells (BCCs) and stromal cells are critical for epithelial-mesenchymal transition (EMT) and metastasis. The transformation of normal cells to malignant cells and malignant cells to metastatic cells may depend on the interaction between the cells and the cell microenvironment. Several cancer studies have focused mainly on the mutated and highly proliferated cancer cell; however, during the last decade, the role of the tumor microenvironment (TME) has been shown to play a significant role in the malignant evolution of neoplasia and cancer progression. Considering the rebuilding of TME, in the present study, we engineered a platform to co-culture breast cancer cells (BCCs) with stromal cells. The use of cell systems or in vitro models that more closely recapitulates the in vivo like signaling in BCCs would be desirable to increase the possibility of translating results of culture models into patient care. It is known that the TME of the mammary gland is composed of ECM and stromal cells in a well ordered architecture that is disrupted during tumor initiation and progression. Therefore, we engineered an in vitro patterned co-culture model having structural similarities of in vivo by recreating the cell-cell interaction in the TME. Recent advances in micro-fabrication techniques have made it possible to engineer surfaces that mimic in vivo mechanical and structural aspects of cellular microenvironment. Using the same approach we have established the patterned co-culture of BCCs and stromal cells. Results showed that co-culture of BCCs with stromal cells affected the proliferation, gene and protein expression as compared to monoculture, thus, modulates the biology of breast cancer cells. Our in vitro patterned co-culture model is capable of controlling cell placement and size of the cells arrays and provides a valuable tool to study paracrine cell-cell interactions.

(71a) I'm a Chemical Engineer. Why Should I Become a PE?

Parrish, W. R., Consultant
Jensen, C. D., Colorado School of Mines
Originally presented on: 11/4/2013 8:30:00 - 8:55:00

Many chemical engineers think Professional Engineering licensure is needed only when they are required to stamp drawings. As a result relatively few practicing chemical engineers consider obtaining Professional Engineer licensure. This presentation briefly outlines the value of becoming licensed from both the personal and business perspective. It presents the most common areas of engineering practice where having a PE can be beneficial or even mandatory by state law.

Licensure is based upon education, examination and experience. The presentation discusses both current and possible education requirements. It discusses the variability of experience requirements between various states. The presentation gives an overview of the Fundamentals Exam (FE) and Principles and Practice of Engineering exam (PE) required for licensure. It briefly discusses the process for making sure the exams are relevant and graded fairly. The FE exam will be making major changes in 2014 and an overview of these changes will be presented.

(71b) Fundamentals of Engineering Examination and Enhancing Student Performance

Peters, R. W., University of Alabama at Birmingham
Kirby, J. T., University of Alabama at Birmingham
Hosch, I.,
Originally presented on: 11/4/2013 8:55:00 - 9:20:00

Graduating students, in the present global economic environment/job market, face increasing difficulty and scrutiny while trying to secure their first engineering positions in the job market. Among the many entry-level requirements, employers are increasing mindful of a student's licensure potential. While favorable for the majority of engineering disciplines, civil and environmental engineering graduates are all but required to secure this certification prior to job placement. In an effort to increase student performance on this crucial examination, University of Alabama at Birmingham (UAB) Civil Engineering developed a FE style review course for graduating seniors in 2009. The student performance on the FE test (before and after implementation of the FE style review course) are highlighted. This paper will explore the ramification of this continuing effort and offer comparisons of pre/post course treatment. Furthermore, lessons learned and a perspective on student motivations will be highlighted.

(71c) The New Computer-Based FE Exam for Chemical Engineers

Mcdowell, D.,

Originally presented on: 11/4/2013 9:20:00 - 9:45:00

Beginning January 1, 2014, NCEES will offer all fundamentals exams in a computer-based format exclusively through approved Pearson VUE testing centers. This transition has been in development for almost three years and has led to many significant changes to include the decision to develop 7 free-standing, discipline-specific FE exams based on curriculum-specific content, a shorter exam length, year-round testing at locations throughout the country and quicker results.

This presentation will provide information on the NCEES content review process that was used to create the new stand-alone computer-based FE Chemical engineering exam and include a detailed comparison of the old and new exam specifications

(71d) Licensure - An Academic Perspective

Wickramasinghe, S. R., University of Arkansas
Spicer, T. O., University of Arkansas
Originally presented on: 11/4/2013 9:45:00 - 10:10:00

Most Chemical Engineering faculty members are not licensed. Why? Is it important for faculty to be licensed? A few departments have in the past required faculty teaching engineering courses to be licensed. Is this beneficial? Many undergraduates take the Fundamentals of Engineering exam but do not take the Professional Engineering exam. Why? Is the FE Exam relevant for assessing ABET outcomes? Is licensure important for faculty who are conducting research? Should we encourage faculty to obtain their PE license? In this presentation these questions will be discussed.

(71e) Use of the Fundamentals of Engineering Exam As An Engineering Education Assessment Tool

Rockstraw, D. A., New Mexico State University

Originally presented on: 11/4/2013 10:10:00 - 10:35:00

The Chemical Engineering program at New Mexico State University began using the Fundamentals of Engineering Exam as an assessment tool in 2000. It was found over the ensuing years that students in the program were not finding much value in taking the exam. Consequently, the sample size was always too small to provide any useful assessment information. The NCEES report "Using the Fundamentals of Engineering (FE) Examination to Assess Academic Programs" by LeFevre, et al. suggested use of a "scaled score" method to treat the FE data for use as an assessment tool. Collected data was treated by this method, but again data provided little value as the error bars on the scaled scores were of extremely large size relative to the scale of the scores. It was noted by the Ch E faculty that analysis by this method assumes the results are for a population rather than a sample, and does not address whether the sample taking the exam reflect the population.

It is assumed that the FE exam represents a national norm capable of providing a useful assessment of the effectiveness of a program. It was thus suggested that the curriculum be modified so that taking the FE exam is a requirement of graduation. In this manner, the true population would be reflected by the data, and a method of treating the data to perform an assessment could be developed. Because students do not have to pass the exam as a degree program requirement, concern was expressed that students would not put forth a valid effort. A protocol was developed to moderate this concern. Beginning in 2007/08, students in the program were required to sit for the exam during the fall semester of the senior year, with the registration fee paid by the department. Students who do not pass the exam must take the FE a second time at their own expense in the spring semester. The requirement to take the exam a second time at their own expense will help to ensure students will put forth their best effort to pass. In this manner, a delay in graduation timeframe is avoided, yet the department collects more useful data.

Data has been collected by this method for a few years, and analysis of the results have permitted identification of topic-specific strengths and weaknesses in the program. Results and the ensuing student support network will be discussed.

(71f) Panel Discussion with Session Speakers

Cramer, J. J., AIChE
Perl, J., Chicago Chem Consultants Corp
Originally presented on: 11/4/2013 10:35:00 - 11:00:00

This will be an open panel discussion between all speakers and the attendees

(76l) Developing An IEEE Elearning Library Course On Energy Efficiency in Industry
Jalali, Y., Lehigh University

Development of three online learning modules on the topic of energy efficiency in industry has been initiated in affiliation with the IEEE Power and Energy Society (PES). These modules will be placed in the IEEE eLearning Library upon review and completion. The first module addresses introductory aspects of energy efficiency. Topics include how to interpret industrial energy losses, analyze energy management programs, and distinguish the potential for energy efficiency in various equipment and processes. The module was developed based on a comprehensive literature review of energy efficiency in industry, focusing on the U.S. industrial sector. Adult learning and multimedia learning theories have been used in order to create a unique learning platform for informing and motivating industrial professionals in regard to energy efficiency opportunities. The course also provides an informative training resource for engineering graduates. In this paper, comprehensive literature review about energy efficiency in industry is presented. The study then reviews the instructional design of the first module.

(76b) Investigating Belonging for STEM Students

Floyd-Smith, T., Tuskegee University
Wilson, D., University of Washington
Jones, D., University of Washington
Plett, M., Seattle Pacific University
Veilleux, N., Simmons College
Bates, R., Minnesota State University at Mankato

The objective of this study is to understand belonging for STEM students and to understand the relationship between belonging and engagement. For the purpose of this study, belonging, which is defined as frequent and local positive interactions with others, is investigated at the class, major and university levels. Surveys, focus groups and class room observations were conducted at five institutions which included a research institution, a teaching institution, a HBCU, a women's college and a private faith-based institution. Engineering and computer science were target majors. Additionally, a contrast STEM major, chemistry, was identified at all schools but the HBCU where animal science was chosen as the contrast major. For this study, the hypothesized correlation between belonging and self reported classroom engagement was confirmed. Also, results, to date, indicate that belonging does not have a distinct pattern by year in school and, overall, gender differences are not statistically significant. Lastly, students reported family as the community to which they feel most connected. The results of this study will be presented in poster format.

(76g) Using Advanced Modelling and Computation Tools in An Undergraduate Programme

Sorensen, E., University College London

The field of engineering is changing rapidly and the increasingly advanced corresponding information and communication technologies (ICT) are having a profound effect on the way that engineering is now being conducted. Mature, commercially available technologies have been penetrating all fields of chemical engineering, in academia as well as in industrial practice. Modelling, simulation and optimisation technologies (MSO) have become a commodity and they are not an exclusive or even distinguishing feature of the process systems engineering (PSE) field any more (Klatt and Marquardt, 2009). This highlights new challenges for engineering education with a strong need to construct modern curricula and course designs that build student knowledge and skills in a cohesive MSO framework. Systems thinking and systems problem solving have to be prioritised rather than the mere application of computational problem solving methods. Educational activities must engage students in an integrated learning environment that intimately links theory and practice, and students need to be challenged, not just in their knowledge acquisition but in the development of professional skills to enable them to operate in a practical engineering environment (Cameron and Lewin, 2009).

This work focuses on the use of gPROMS® (Process Systems Enterprise, 2013) in a 4th year undergraduate module at UCL in London, UK. The course module is a compulsory 4th year MEng level module within the Department of Chemical Engineering at UCL and is taken by 40 or more students every year. The module is worth 1 unit out of 4 units in the final year. The module runs over two terms and consists of two parts; in Term 1, the students focus on further developing their modelling skills and they learn the use of gPROMS. The students then use these modelling and gPROMS skills in Term 2 when they work in groups of six to produce a technical design report of their analysis of the dynamic performance of a chemical plant. All the course material is made available to the students prior to the lecture or tutorial through Moodle and the students also submit their coursework and receive feedback through Moodle (Moodle, 2013).

In this work, the detailed course structure, and the motivations behind, will be presented, and examples will be given of lecture and tutorial material as well as the different modes of assessment. The results from end-of-module student surveys will also be included which show that the students find the module both highly relevant to their studies and interesting, despite the challenges of first-principles modelling.

References

Cameron, I.T. & Lewin, D.R. (2009). Curricular and pedagogical challenges for enhanced graduate attributes in CAPE. Computers & Chemical Engineering, 33(10), 1781–1792.

Klatt, K.-U., & Marquardt, W. (2009). Perspectives for process systems engineering—Personal views from academia and industry. Computers & Chemical Engineering, 33, 536–550.

Moodle, 2013. http://www.moodle.com/.

Process Systems Enterprise, 2013, http://www.psenterprise.com/gproms.html/.

(76m) "Should We Start This Company?": the Theory and Practice of Starting a Business

Jablonski, E. L., Bucknell University

"Should we start this company?": the Theory and Practice of Starting a Business is essentially a student-project centered course in entrepreneurship, generating new business ideas, and technological product or service design and development through business planning. During the course students develop a Business Strategy including Business Plan, Business Model, and Feasibility Evaluation, for both their own business idea and a local "client". Students also have to identify a customer base and do market research, including surveying if appropriate. Stretch goals for student-driven projects include proposals for expanding capabilities of the product or service, optimizing production or delivery, and identification of new market share. The culmination of the course is to present the groups' business ideas to investors and judges (business pitch competition), and use feedback and knowledge gained from these presentations to make a final report including an analysis of the businesses' potential for success. Other activities include (but are not limited to) connections with students in other courses, guest lectures, and interactions with the Small Business Development Center (SBDC) located at Bucknell University.

Students enroll from all disciplines of engineering, from the School of Management, and from across the university. This course was supported in part by the KEEN (Kern Entrepreneurship Education Network) and thus the course outcomes are tied to that program. The outcomes reflect that students completing the course should recognize that successful entrepreneurs are able to do things that successful engineers should also be skilled at: effectively collaborating in a team, thinking critically and creatively about ambiguous (ill-defined, open-ended) problems, being able to construct and effectively communicate a customer appropriate value proposition, persist through and learn from failure, effectively manage projects through appropriate commercialization or final delivery process, and demonstrate voluntary social responsibility.

There are high expectations in this course, somewhat comparable to senior design. Students are expected to spend 8-10 hours per week *outside of class* being productive to meet the deliverables deadlines. There is also a focus on personal accountability; for example, students are explicitly told that not all of their outside class hours will be spent with their group because it is important to learn when time is best spent together and when it is more efficient to delegate tasks to individuals. I also have the explicit expectation that although different members of each group will take on different roles in the projects, all members are expected to be able to answer any reasonable question regarding progress, justification of choices, etc. In this way each member is personally accountable for understanding the larger scope of the projects, so even though it is a team-based course, no one can "coast" in an otherwise strong group. Groups present on their progress on a regular basis in addition to the formal presentations scheduled during the course so that each team member can be actively involved in the discussion of their team's progress.

Another important focus of the course is professionalism. Students are expected to conduct themselves individually and in their groups as they would if the course were a job situation. Unacceptable behavior includes being disrespectful to others in class, inappropriate speech or gestures, being late to class, and not pulling their weight in a group situation. While this may seem like a small detail, it is essential to have this be emphasized in order for the course to function smoothly.

The projects in the course generally fall into three categories, and can represent for-profit or non-profit entities:

a) Development of a novel physical product or process, from concept to commercialization

b) Development of a new service (e.g. website or business or home delivery of service), from concept to launch

c) Other: by approval, may combine elements of various businesses, integrated format (e.g., optimizing/enhancing an existing operation)

The course is organized into five phases: Idea Generation, Initial Project Planning, Intermediate Developments, "The Pitch", and Reflection. During **Idea Generation** each group is given direction in highly energetic and productive forms of ideation (brainstorming) and critical selection. From there, each team selects several (2-3) ideas to pitch to the whole class. Selection of which project to pursue from that point involves some form of basic market research using their classmates as focus group. Groups then move to **Initial Project Planning** where the main deliverable is a seven minute explanation of the project plan and scope, including refinement from initial ideas, and a brief outline of planning. This point is critical because it is treated as an oral defense of the business idea the group has chosen to pursue. The students must provide specific details about the expectations of the final product or service and outcomes of the business planning process. For client-initiated ideas the students must provide a refined explanation of the initially proposed business concept. After initial project planning some ideas may fail and groups may re-form. This is also treated in a professional manner – students on teams whose ideas failed must apply to be hired by another business. To help with this step we have a session on how to best characterize your own skills, so each student is better able to manage their own hiring process. Once new new groups have been formed they delegate tasks to complete the **Intermediate Developments and Preliminary Business Pitch** for their own business idea. The main deliverable for this stage is a ten minute presentation about what your group has learned through literature and market research, culminating in a single product or service that will be pursued. This presentation must include a description of potential customers, applications, and a "reach" market. The presentation is evaluated by their peers and their feedback is used to help formulate the "pitch". **The Pitch** is actually done during the Bucknell Business Pitch Competition. The students are not graded on whether or not they advance to the finals – they are graded on their preparation, professionalism, and self-reported feedback from the judges. The Business Pitch Competition occurs when only 2-3 weeks of class are left, so the final deliverable for the course is for the teams to take everything they have learned from the course, their peers, and the judges, and any assistance they have received from the consultants at the SBDC and complete the final business plan and video pitch. There are also final presentations that would be comparable to a Venture Plan Competition.

One of the methods used to keep the teams moving forward each week is extensive, directed blogging about each phase of the process using prompt questions. These blogs allow everyone on the team and the instructor to track the progress of each business idea.

In order for students to be successful in this course they not only must possess the technical knowledge of how their product or service would be launched, they need to demonstrate an understanding of the business sense that would be needed to 'get it off the ground'. Overall, it is a great complement to the typical engineering senior design course.

(76h) The "N-Whys": Targeting the Highest Levels of Bloom's Taxonomy in Research

Sanders, J. R., Tennessee Technological University
Arce, P. E., Tennessee Technological University
Albritton, S., Tennessee Technological University
Kramer, K., Tennessee Technological University

Research experiences are critical components in the professional development of students, but while such opportunities for discovery are tremendously important, research in general is a difficult process. To emphasize this point, one can look to the words of Albert Einstein who stated, "If we knew what we were doing, we wouldn't call it research." While numerous approaches have been identified to improve student learning in the classroom and to teach research skills, fewer validated methods are available to efficiently train and engage students in the process of research project development in which the students are empowered to be working specifically towards solving societal problems. Traditional approaches ("solutions") used in directing student research involve providing literature to review, providing lab experiences, encouraging students to attend seminars and research group meetings. While all of these are commonly used, we contend that these approaches (though tremendously important) represent lower levels on the updated Bloom's taxonomy (i.e., "Remembering", "Understanding", "Applying", and "Analyzing") and as such are inadequate (alone) to promote the highest order thinking skills ("Evaluating" and "Creating"). In the High Performance Learning Environment (Hi-PeLE^{TM}) which is being pioneered in chemical engineering at TTU, a focus is placed on both the acquisition and transfer of knowledge to real-world problems, and student groups are coached through a process of identifying societal needs and producing prototypes that are responsive to those needs. In such cases, they are not told the need, rather it is up to the students to work collaboratively, canvas the landscape, and identify a need. Thus, within this environment, the highest order thinking skills are needed both in the research setting and in the classroom. The "5 Whys?" is a tool that aids the user in identifying the "root cause of a problem." Using this methodology, the problem solver asks the question "Why?" several times until the "root cause" is determined, at which point a solution can be implemented. It is hypothesized that this approach can be used (once a "need" or a "problem" has been identified) to enhance a student's ability to formulate a research project, an activity believed to fall within the highest levels of Bloom's updated taxonomy. Once the problem/need is identified, the question "Why?" will be asked until a "root cause" is identified. The root cause then forms the basis for creating a measurable research objective to be pursued by the student researcher. In other words, it is expected that a research project can be easily formulated ("created" after this analysis is completed) that seeks to provide a solution to the "root cause." In this session, the authors will share their perspectives and case studies on this approach.

(76k) Engineering Pre-Service STEM Teachers to Introduce Chemical Engineering to K-12 Students in Bordertown: Laredo

Gupta, A., Texas A&M International University

This work commits to the advancement and diversification of STEM education in Texas through a novel approach of creating a one-day workshop on "Chemical Engineering" as a part of the activities for secondary pre-service STEM teachers. The goal of the program is to fulfill the needs of industries in regions of South Texas serviced by Outreach Strategy Team's Pilot Program at Texas A&M International University (TAMIU). Situated on the border with Mexico, TAMIU, a Member of the Texas A&M University System serves predominantly Hispanic population by providing academic services to a high number of economically disadvantaged and first-generation students, who, due to Laredo's isolated geographical location have very limited access to higher education. The majority of the teachers in Laredo received their education from TAMIU, targeting the pre-service teachers at TAMIU will expand our future teachers' knowledge of chemical engineering, and aid in Texas A&M Engineering recruitment program.

The highlights of the one-day workshop on Chemical Engineering are: 1) familiarize the pre-service teachers with the chemical engineering curriculum and the profession, 2) introduce the concepts of basic unit operations, block and process flow diagrams, materials, and energy balances, and 3) engage them on a hands-on laboratory activity to produce their own biodiesel. The laboratory activity is supplemented with a question set that includes calculations appropriate for an introductory-level course, and flow diagrams for the process.

This one-day workshop constitutes three modules. In the first module, basic unit operations are discussed, followed by the generation of block diagrams. The concept of recycling is introduced, and the pre-service teachers modify the streams on their generated block diagrams to "avoid wastage". Activities include proposing a flow diagram of a simple two-step process, identifying the basic components of process flow diagrams and guess the unknown process. The second module, covers the mass based material and energy balances. The concept of conservation of mass is introduced using mass-based chemical reactions, which is further applied to a unit operation problem using the block diagram generated in the first module.

The third laboratory module provides an overview of the scientific and engineering aspects of biodiesel production supplemented with a hands-on-activity on the biodiesel production using soybean, and waste vegetable oil obtained from the TAMIU diner. The recorded mass and volumes from the activity is to use to understand unit conversions and stoichiometry. Flow diagrams of the process are generated followed by simple material balance calculations. At the end of the workshop, the pre-service teachers are surveyed for their knowledge on chemical engineering program, and opinion of the workshop.

[YMA1]You mean the one-day workshop, right? Maybe better to say "workshop"

(76f) The Effect of Undergraduate Research On Student Performance

Himstedt, H., University of Minnesota
Hernandez, P.,
Brown, M.,

Undergraduate research, in some form, is present in every chemical engineering department; however, its extent varies greatly. In most research university chemical engineering departments undergraduate research plays a critical role but often difficult to characterize role. Coinciding with an initiative from the author's department to increase availability of undergraduate research, the net benefit of undergraduate research was sought. Does the time and effort put into training undergraduate researchers benefit the research project, the student, or the PI?

The effect of undergraduate research on engineering skills, student confidence, and performance in coursework were evaluated over multiple semesters using a self-efficacy questionnaire. Engineering skills and student confidence improved rapidly after starting undergraduate research. There was no distinct correlation between participation in research and coursework performance. Additionally, performance increases seemed to plateau after rapidly increasing over the first two semesters of participation in research. Results were used to compare chemical engineering students in this department with other engineering students, all undergraduates at the institution, and undergraduates nationwide.

(76e) Experience in Using the Case-Discussion Method in Chemical Engineering Courses

Roman-Martinez, A., Technical University of Denmark
Alejo Gonzolez, M. G., Universidad AutÃ³noma de San Luis Potosi
Cervantes Nino, S. E., Universidad AutÃ³noma de San Luis Potosi

This work aims to encourage the use of the Case Discussion Method in courses of Chemical Engineering, presenting an example used in the classroom which requires multiple and several applications of principles of chemical sciences and engineering. We aim to emphasize that the discussion of actual cases of Chemical Engineering in the classroom with an appropriate methodology, facilitates the fulfillment of the learning objectives to be obtained with students, such as the development of problem solving skills, communication skills and ethical, social and environmental awareness

We present our experience in using the Case Discussion Method to teach principles of Chemistry and Environmental Sciences in Chemical Engineering courses. Also, we encourage the use of this method of learning in the areas of basic and engineering sciences, as it is not widely used in these fields of knowledge, because, due to the method approach, this has not been used significantly for the study of the subjects of Chemistry, Physics, Mathematics and Engineering. However, this method is very well accepted in other areas such as Management, Law, Medicine and Finance.

We start by presenting the educational intentions that led us to use the case method, and then we present the case used, objectives, content and methodology used in the courses as well as the results of the progress and learning achieved by the students of Chemical Engineering in our Institution.

(76c) Impact of Summer Bridge Programs On Students Sense of Belonging and Retention to STEM Majors

Tomasko, D. L., The Ohio State University
Waller, R.,
Ridgway, J.,
Olesik, S. V., The Ohio State University
Upton, J.,
McGee, M., The Ohio State University
Barclay, L.,
Harkin, K.,

The Ohio State University has run a multidisciplinary STEM bridge program for entering freshman for four years. The program emphasizes math, science, engineering, and communication subjects as well as college success skills. It exposes students to the rigor of college coursework and highlights the difference in study approaches from their high school experiences. Evening professional development programs focus on study skills and resources available across the university. Results from this program indicate significant peer to peer bonding and creation of a support network that reinforces a student's motivation and drive to succeed. Although participants continue to change majors during their first year in school, they have shown increased persistence (retention) to majors in the STEM fields compared to a non-participant control group. Using qualitative indicators drawn from longitudinal surveys of the participants and controls, this poster will discuss possible correlations between program activities and impacts on student attitudes towards school and their choice of major.

(76d) Introduction of a Biomedical Engineering Module At Southwestern Indian Polytechnic Institute Tribal College

Pascal, J. A., University of New Mexico
Vadiee, N., Southwestern Indian Polytechnic Institute

Of the tribal colleges in the United States that offer engineering programs, none include a bioengineering focus or module. Thus, we have developed a 1 - month module to introduce students enrolled in our ENGR 105 course, Introduction to Engineering and Design at Southwestern Indian Polytechnic Institute (SIPI), to the fundamental concepts of biomedical engineering with a specific focus on diseases that have a significant impact on Native American communities (i.e., diabetes, heart disease, cancer, alcoholism). Over the past two trimesters, we have also created a biomedical engineering project for the course in which students seek to further understand the role that biotechnology plays in the diagnosis and treatment of diabetes in Native Americans while raising awareness within their own tribes. So far, the students have researched the causes, spread and increase in diabetes diagnoses in the Navajo Nation, and the history of biotechnology advances in diabetes diagnoses and treatment (e.g., blood glucose monitors, insulin pumps). The students have proposed to understand, from an engineering point of view, how blood glucose monitors interface and plan to improve their operation by constructing their own prototype. Here we will present our module and how it has been implemented in the course and the feedback from current students.

(76j) A New Graduate Level Seminar to Prepare Students for the Next Step in Their Careers

Fleming, K., University of Washington
Matthaei, J.,
Richards, J., University of Washington
Pozzo, D., University of Washington
Pfaendtner, J., University of Washington

A new graduate level seminar to prepare students for the next step in their careers

Kelly Fleming, James Matthaei, Jeff Richards, Danilo Pozzo, Jim Pfaendtner

Department of Chemical Engineering, University of Washington

Seattle, Washington, 98105, USA

ABSTRACT

Many doctoral students and postdoctoral researchers who apply to become faculty candidates when they graduate are not prepared for the rigorous application and selection process. Candidates must give one or more hour long seminars on their past and future research, which many have never done before. Additional challenges come in the form of a rigorous 1-2 day interview, making small talk and meeting with potential colleagues in both social and professional contexts. In 2011, the graduate students in the University of Washington (UW) Chemical Engineering department designed a seminar to help bridge the gap between graduate school or postdoctoral research and applying as a candidate for first permanent positions. The seminar series takes place each summer and is comprised of 7-8 students and postdoctoral researchers from around the country. In an effort to maximally impact the graduate students at UW ChE, all facets of organization, selection and execution of the seminar are executed by the students themselves.

Applications to the seminar are comprised of an abstract, C.V., and a letter of recommendation. A team of four "lead" students are selected each Spring to handle receipt and organization of applications. Thereafter a panel of ~15 current grad students is selected to complete a daylong meeting in the style of an NSF review panel. An explicit goal of this step in the process is to exposure our grads to some of the most accomplished young scholars in the field who are actively seeking permanent positions. Our hope was to help the graduate students self-calibrate and provide a wide basis for comparison in terms of things like publication and presentation records.

During the first year of the panel students were faced with the difficult task of balancing strictly quantitative scoring metrics with other valuable factors such as contributions to diversity or qualitative "intangibles". To facilitate a professional and fair process, the UW ADVANCE program was asked to provide workshop to our students during the 2nd and 3^{rd}years. The workshop helped the students address the issues in a constructive way and more fairly make panel selections of people who excel in a broad range of metrics.

This poster reviews quantitative analysis of surveys given to three populations: 1) the graduate students who attended the seminar, 2) graduate students who served on the panel and 3) the selected speakers who visited UW to give the seminar. All groups gave enormously positive feedback, stating that the seminar helped them prepare for interviews and their future careers. All the students involved wanted to continue the seminar in future years. We also use the poster to review the structure and organization of the new seminar and provide perspective on how the concept could be translated to other departments or institutions.

(76i) A Didactic Proposition: Energy Audit in the Laboratory of Materials I of Facultad De Ciencias Quimicas of Universidad AutÃ³noma De Nuevo Leon
Castillo, M.,

A Didactic Proposition: Energy Audit in the Laboratory of Materials I of Facultad de Ciencias Químicas of Universidad Autónoma de Nuevo León.

Ma. Teresa Castillo-Escobedo, Nancy D. Leal-Sanmiguel, Cynthia Estefanía González-Moreno, Felipe de Jesús Cerino-Córdova, María de la Luz Díaz de León, Jose Ángel Loredo-Medrano.

Universidad Autónoma de Nuevo León, Facultad de Ciencias Químicas Av. Universidad s/n. Cd. Universitaria. San Nicolás de los Garza Nuevo León, MÃ©xico.

Email: mtcastillo2002@yahoo.com.mx, felipejccuanl@yahoo.com.mx, ml.diaz@hotmail.com

One of the principal tasks of the chemical engineer is to develop or modify processes minimizing material, energy consumption and cost. The energy balances course is taught in the 5th semester of the Bachelor's degree in chemical engineering of the Facultad de Ciencias Químicas of Autonoma de Nuevo León Universty (UANL), Mexico. In this course, the students must do final projects where the knowledge acquired are applied in a practical case. This project represents 10% of your semester final rating. One of these projects was performing an energy audit to the laboratory of materials I. Measurements were made such as: indoor temperature, outdoor temperature, time of use of appliances and conditioners air, conditioner air and appliance setting temperature, weather, number of persons, month of year, laboratory location and size, , construction material laboratory, number of devices, energy in the form of heat radiated by appliances, heat convection (flows of air between the interior and the exterior). The energy used in the laboratory of materials I in the form of heat that enters through Windows exposed to the Sun all day, the power consumption of devices used during the period of measurements is 11,732,508.6 KJ. In this project, the students understanding the importance of using the energy and its environmental impact. The students proposed to improve the energy use in the laboratory via the installation of efficient lighting control systems and fluorescent lamps (32W), the insulation of windows, the use of light colors on the walls and ceiling and to avoid dark windows, the application of periodic servicing and maintenance to the equipments.

(76n) Linking Community College Students to Engineering

Palanki, D. S., University of South Alabama

This project presents the results of an NSF grant to support transfer students from community colleges to the University of South Alabama. It will be shown how this project stimulated enrollment, enhanced retention in engineering programs at the University, and increased the technical work force in the Mobile area. The targeted community colleges have large minority populations and many first generation students. In particular, this project addresses financial barriers of minority community college students by providing scholarships each year to students from area community colleges. The students were placed in learning communities and provided with faculty and student mentors and academic support systems. Participants interacted with K-12 students, broadening the impact of the program. Data on the "what worked" and "what did not work" will be presented and a rational methodology for facilitating the transition from community college to the university will be discussed. In particular, we will show how performance in basic mathematics and science courses community colleges correlated with performance in engineering courses in the univeristy. Furthermore, academic and non-academic issues that the students faced when they transitioned to the university, will be discussed.

(76a) Engineering Camp At Bucknell University: Motivating Pre-College Students

Jablonski, E. L., Bucknell University

Engineering Camp: A college experience for pre-college students

Each year Bucknell University hosts a week-long on-campus residential experience for roughly 100 post 7th -11th grade students focused on introducing the disciplines of engineering. Engineering camp introduces students to engineering disciplines through introductory seminars, demonstrations, laboratory experiments, and design challenges. The program improves students' awareness of the breadth of engineering and emphasizes the benefit of developing skills in STEM. The camp is offered in grade-based parallel sessions geared to the audience, and campers can return in subsequent summers. Importantly, the camp provides a college-like experience that is integral to building academic capital and a sense of belonging for students, particularly those from under-resourced schools.

During Camp students live in dormitories, eat in the university dining hall, go to class each morning, have laboratory sessions each afternoon, enjoy an active "social life" (under close supervision), and get a full college experience. The main goal of the camp is to introduce students to engineering disciplines in a way that motivates and prepares them for undergraduate majors in STEM fields. Camp features highly active classroom and laboratory sessions that introduce these technical topics not through lecture or 'cook-book' laboratory, but through active, collaborative, and problem-based learning approaches shown to be not only more effective at fostering student understanding than traditional approaches, but also superior for retaining the most diverse audience of learners. The camp particularly reaches out to students in urban and rural under-resourced schools who might not otherwise be exposed to topics in engineering and technology before graduation, and therefore would not have prepared themselves for or even considered an undergraduate major in the STEM fields.

In addition to classes in several engineering disciplines (chemical, biomedical, mechanical, electrical, civil, environmental), there are special topic discussions on everything from sustainability and green engineering to biomaterials and nanotechnology. Among the goals of the camp are for students to become familiar with a variety of engineering disciplines and to practice engineering design. Several of the assignments during the week have include elements of design and innovation, including "Redesign Your Town Green", where students are asked to reduce the carbon footprint of residents of their towns by thinking about making communities more pedestrian-friendly or starting public programs that would encourage recycling or composting. Students are asked to propose new designs for common packaging as a way to reduce waste, to think about logical improvements to devices with which they are familiar, or to reverse-engineer processes or products. Throughout the camp, students make presentations and write reports; at the end of the camp, students give professional presentations to their families, friends, and fellow campers. All together, these activities significantly boost students' confidence, interest, and identity in science, technology, and engineering.

During the camp, students spend more than 35 hours in classes and laboratories; an additional 28 hours of extracurricular activities are also scheduled. The significant laboratory component of the camp provides an opportunity to participate in intellectually stimulating hands-on experiments. Technical sessions indicated include faculty-lead sessions on structural engineering, biomechanics, food science, materials science, nanotechnology, and others as described above.

Engineering Camp at Bucknell was established in July 2008 with prior NSF support. In its first year, this program welcomed 26 8th–10th grade students aged 13-16 (10 girls, 16 boys); all but one were from local area schools, and many were from disadvantaged rural schools- schools in areas with a high percentage of low-income families and/or having limited course offerings in upper-level mathematics and science in the high school. Since its inception, camp has grown significantly to incorporate three tracks (post-7-9th graders, 10th graders, and 11th graders) and a significantly expanded number of campers from a significantly expanded geographic reach of over 28 states: 2009 – 51 students, 2010 – 52 students, 2011 – 83 students, 2012 – 97 students, 2013 expected enrollment of 124 with another 40 on a growing wait-list. The camp introduces students to numerous engineering disciplines and faculty members from each department in the College of Engineering at Bucknell are involved as instructors, 16 faculty members in all for the 2012 session. For the 2011, 2012 and 2013 sessions, there were more applicants for slots than there were resources to meet them. The program is expected to grow to 200 students in 2014.

Campers come from a variety of educational and economic backgrounds. Currently, Camp recruits some students from partner schools in target rural and urban areas. An additional goal for Engineering Camp is to remove the mystique about going to college and majoring in a technical field. Many of the campers, particularly those from under-resourced urban and rural areas, would be first-generation college students, without the family academic capital that could help them prepare for university processes. Camp provides a safe environment to become comfortable with residential living learning environments. Both class and laboratory content are challenging but tailored to be interesting to students aged 13-16. By engaging these students in activities that closely mimic those of typical undergraduates in an engineering or technical major, their apprehension about college in general and difficult curricula in particular are all but eliminated.

The real problem with the current pre-college initiatives is that we are unable to reach all those students and schools that are most in need of this type of enrichment. The current tuition of Engineering Camp is too great a price for the most disadvantaged students. Parents are dissuaded from inquiring about programs for their children if they presume they will be unable to afford to allow their children to participate. These are the students we need to be targeting to make the greatest impact. These programs also need to target traditionally underrepresented groups, including women and minorities, by reaching out to schools through classroom visits to talk about the Engineering Programs. Students who feel there are barriers to their participation in programs of this type must be identified and encouraged, and financial support must be provided if real changes in undergraduate enrollments in STEM fields are to be impacted.

Many pre-college students, particularly those from disadvantaged school districts, are not exposed to engineering or related technological topics prior to their undergraduate education. It is hoped that an experience like Engineering Camp will motivate interest in pursuing a technical degree in the sciences and engineering.

(76p) The Return to School Decision

Long, R. Jr., NMSU

The decision to return to school is a big one. The cost and effort are heavy. And the expectations are uncertain and the outcome is undecided. So the actual return to school is often put off or not done at all. Oftentimes the individual comes to the conclusion that a "mistake" has been made. If a faculty member can capture the individual at this time this can be helpful to the person and encourage the indivdual. The teacher can suggest tools and methods that a younger person may not know about and thus encourage a career change or a more productive direction.

(85a) An Example Of Success Achieved With The Collaboration Mexico - Canada: A Scientific Project Sponsored By Conacyt - Mexico and Its IMPACT On Students

Serrano, B., Universidad Autonoma de Zacatecas.
de Lasa, H. I., University of Western Ontario
Moreira, J., University of Western Ontario
Vazquez, J. L. Sr.,
Hernandez, M. A.,
Guevara, S.,
Garcia Bermudez, S.,
Originally presented on: 11/4/2013 8:30:00 - 8:50:00

In November 2012, the National Science and Technology Council of Mexico (CONACYT) selected as a case of success the completed project 83144, Basic Science Call, 2007 – 01. The project's title is "Photocatalytic Reactors with Different Scale and Novel Photocatalysts for the Water Decontamination", being Dr. Benito Serrano Rosales the technical responsible. In addition, this project was presented in a conference in Can Cun, Mexico, with title "National Conference of Research of Basic Science, Successful Cases, CONACYT, Calls 2006 and 2007".

The general objective of the proposal was to synthesize, characterize and test novel catalysts doped or impregnated with metals, in the photodegradation of organic pollutants in water by using solar and UV light. Reactors of different sizes, 7, 35 and 300 liters, are used to establish reaction mechanisms, reaction rates, irradiation fields, and reactor efficiencies.

This research project was regarded as a success case because six papers were published in high ranking journals (three in preparation), two books as invited editors with one of them being re edited in China) and one book chapter, and because the following reasons:

Technical reasons:

The following original objectives planned in the protocol were achieved and published in the papers, books and presented in conferences and preceding: 1) Synthesize, characterize and test several photo catalysts, with different model compounds. 2) Propose a Unified Kinetic Model to describe the photo degradation of phenol, using six photo catalysts, in a broad range of concentrations and perform kinetic modeling. 3) Simulate the irradiation field of the Photo CREC Reactor using Monte Carlo. 4) Calculate the reactor efficiencies during the experiment span based on the hydroxyl radicals and intermediates. 5) Explain the effect of iron and other metals impregnated on the catalyst surface, during the degradation of phenol. 6) Use several metals when doping the catalyst.8) Utilize different reactor sizes in all experiments (7, 35 and 300 Liters). These objectives were achieved and published in papers, book chapters and presented in various conferences.

Academic Reasons: There was an impressive formation of human resources, thesis, summer research stays of students both in Mexico and Canada.

The following thesis were completed: Three doctoral, two master in science and eight undergraduate and the following thesis are in preparation: Four doctoral, three master in science and seven undergraduate. The work was divulgated in 23 conferences, Mexican and international.

There was mobility of students, researchers, faculties. The students sent to perform research stays: 1) In CREC – UWO - Canada, two graduate, six undergraduate. 2) In Universite Laval, Quebec City, four graduate, one undergraduate. Also, several students from Queretaro, Cohauila, performed summer research stays in the laboratory of UAZ.

National and international relationships with other universities.

The relationships with Mexican, Canadian and American universities were reinforced and increased. The following universities and researchers were involved:

1.- Universite Laval, Quebec Canada, Dr. Serge Kaliaguine. Involved students: Sarai Guevara Castillo (UAZ), Alfonso Talavera López (UAM-I), Omar Valdez Martinez (UAZ), Juan Carlos Paniagua Rodriguez (UMSNH).

2.- The University of Western Ontario, London Ontario Canada, Dr. Hugo de Lasa. This university was the main collaborator, and seven Mexican students have performed summer research stays, three Mexican students completed doctoral studies, and three students are pursuing the doctorate. Due this academic and scientific relationship we could publish papers, books, obtain a grant from IDRC-Canada to launch the Institute of Advanced Studies Canada – Mexico (IASCM) and to organize conferences in Mexico and Canada.

3.- Universidad Autonoma Metropolitana – Iztapalapa, Dr. Gustavo Fuentes and Dra. Gretchen Lapidus, the following students were involved: Alfonso Talavera Lopez and Oscar Joaquin Solis Marcial.

4.- Universidad Michoacana de San Nicolas de Hidalgo, Dr. Rafael Maya Yescas, student: Juan Carlos Paniagua Rodriguez.

5.- Universidad Autonoma del Estado de Mexico, Dra Reyna Natividad and Dr. Armando Ramirez Serrano, students Eduardo Martin del Campo and Oscar Alvarado.

6.- Universidad Autonoma de San Luis Potosi, Dr. Edgar Moctezuma Velazquez, student Alfonso Pinedo Escobar.

7.- Universidad Nacional Autonoma de Mexico, Dr. Camilo Arancibia Bulnes and Dr. Rosa Ramirez Zamora, students Sayra Lissette Orozco Cerros, Sergio Cuevas and Mario Alberto Hernandez Mazatan.

8.- University of Tennesee Campus Chatanooga, Dr. Jim Henry, students Jose Alberto Gonzalez Guerrero, Adan de Jesus Zacarias Balderas.

It is possible to see that the technical aspects were fulfilled and exceeded; this was a proposal of basic science, with the main goals of generating scientific knowledge. Also, a very important point was the extension of the Heterogeneous Photocatalysis for water to the purification of air and hydrogen production through water dissociation. Based on the results and development of this project, these topics were proposed and initiated as new research lines, taking advantage of the experience and existing infrastructure. Also, several additional projects were written and approved due this project.

It is considered that this proposal had its most important impact in the development and formation of the involved students, inasmuch as undergraduate students were interested in scientific research and were motivated to pursue graduate studies. The graduated students were encouraged to continue with the doctoral program, they met professors and researchers from other universities, institutions, countries, and this eventually will serve to them to get jobs.

The success of this proposal was based on the enthusiastic participation of researchers, students and collaborators, but was facilitated by the excellent academic technical and scientific collaboration between UAZ and UWO – Canada, enhanced and promoted by the IASCM, created in 2005, an institute whose objectives are (among others): 1) Develop collaborative research in areas of strategic importance for both Canada and Mexico. This would consist of developing, testing and implementing technologies for the protection of the environment; the use of clean and renewable energy resources; as well as research on related socio-cultural, health, legal and migration issues. 2) Contribute to the training of Canadian and Mexican Master and PhD graduate researchers and university professors, in their respective fields.

This success case is an example of the importance of the international collaboration and its multiplicative effect on the formation of human resources, which leads to obtain more productivity in technical, scientific and academic areas.

(85b) Dynamic Design of a Cryogenic Air Separation Unit

Schmidt, S.,
Clayton, R.,

Originally presented on: 11/4/2013 8:50:00 - 9:10:00

Abstract

We present the dynamic design of an air separation unit (ASU), which is used to separate oxygen, nitrogen, and argon products from air at very low temperatures. This unit can be used as an isolated plant to produce the products to be sold or as a unit joined to a larger plant, also called a "piggy back" plant, to produce products to be used in an additional process [1]. These gases are commonly used in refineries and oil recovery efforts. This process is convenient because the raw material is available at no cost and in limitless amounts. The primary cost in the process is compression. In order to minimize costs, heat integration is used throughout the process. This particular ASU will produce 1500 metric tons of 99.5% oxygen, 5000 metric tons of 99.5% nitrogen, and 58 metric tons of 95% crude argon each day to a customer. Heat integration will eliminate all costs from heating and cooling in the process. The compressor capital cost will be $16.5 MM[1]. The venture guidance appraisal for this process is $118.5 MM. When running at full capacity, the plant will sell $113.9 MM worth of products. With a total annual cost for equipment and utilities of $39.0 MM, this plant will yield a yearly profit of $73.4MM. An integrated control structure has been designed to maintain product purity and reject atmospheric (temperature and pressure) disturbances, as well as flowrate disturbances. An overview of both a feed control and on demand scheme is included in the following pages, along with disturbance testing for both schemes.

(85c) Navigating the Natural Gas Boom: Policy Option for Ensuring a Safe, Clean, and Economical Natural Gas Revolution

Crothers, A., RWTH Aachen University
Winkler, P., Air Products and Chemicals, Inc. (Retired)
Originally presented on: 11/4/2013 9:10:00 - 9:30:00

The recent revolution in natural gas has revealed incredible economic opportunity for this nation. There are copious natural gas reserves in the US, particularly in shale repositories. Capturing these reserves will yield massive economic growth in direct energy production, transportation, and downstream product synthesis industries. Despite these benefits, there are continual controversies on how best to utilize these newfound resources and how to mitigate the potentially severe environmental implications. It is important that policy makers decided on how best to incentivize safe shale gas mining, regulate global trade, and promote responsible utilization of these limited resources. Obtaining natural gas from shale repositories requires hydraulic fracturing which has controversial public health and environmental implications. Moreover, while gas is cleaner than coal and oil, it does not provide a completely nonpolluting and sustainable solution our scarce energy supplies problem. Providing regulatory certainty to these challenges of natural gas would provide a strong foundation for growth.

(85d) A Wavelength Shifting Polymer to Enhance the Growth and Production of Cyanobacteria

Nelson, T.,
Salem, D. R., South Dakota School of Mines and Technology
Winter, R. M., South Dakota School of Mines and Technology
Originally presented on: 11/4/2013 9:30:00 - 9:50:00

Cyanobacteria consume carbon dioxide while producing oxygen and desired genetically engineered products. When grown in photobioreactors, cyanobacteria can provide both oxygen and fuel during long term space travel and extraterrestrial colonization. The overall goal of the research team is to optimize the production of linalool, a terpene alcohol, by the cyanobacteria genus anabaena grown in a photobioreactor. The goal of this particular sector of the research is to modify the light delivered to the anabaena in order to stimulate growth rate and production of fuel. Polymer films, poly(methyl methacrylate), were prepared with fluorescent molecules which absorb photons of non-essential wavelengths and shift them to wavelengths more beneficial to cyanobacteria photosynthesis. These films were prepared by using a solvent casting method and tested for fluorescence emissions. Ultraviolet light was converted by Coumarin 460 to match the 440 nm absorption peak of anabaena (chlorophyll a). Green light (approx. 532 nm) was red-shifted to higher wavelengths by inexpensive Rhodamine fluorescent dyes. These energy shifts are expected to increase the cyanobacteria growth and alcohol production rate by increasing the available specific energy required by these processes. A current study is being done to observe growth rate, by optical density, and fuel production, by HPLC, of the anabaena when stimulated with particular LED lighting arrays characterized by a spectroradiometer. Latter tests will use a solar spectrum of light that is modified by the wavelength shifting polymer. Results of this study will indicate the quintessential wavelengths for maximum fuel production. Future work includes up-conversion of infrared light using triplet-triplet annihilation and increasing fluorescent efficiencies by surface plasmon resonance from metallic nanoparticles and films. This wavelength shifting technology could be used in other applications that enhanced photosynthesis would benefit, such as other photo-reactors, crop production, and development of LEDs used for grow lights.

(85e) Exploring Multiphase Viscous Drop Impact With a Bulk Fluid

Logia, A. N., Stanford University
Walker, T. W., South Dakota School of Mines and Technology
Fuller, G. G., Stanford University
Originally presented on: 11/4/2013 9:50:00 - 10:10:00

Exploring Multiphase Viscous Drop Impact with a Bulk Fluid

Miscible fluid systems are all around us from the mixture of water and dish soap in the kitchen to the dissolution of oxygen into blood. Building our intuition about how miscible and immiscible fluid systems differ can significantly further our understanding of fluid mechanics.

We have developed an experiment to explore how a droplet impacts a bulk fluid interface. A drop of dense, viscous fluid is ejected from a syringe, free falling into a bath of water. A high-speed video is captured by triggering a laser photo diode system. These liquid-liquid drop impact systems can be inherently complex due to the time dependent interfacial tension between the two phases. Their interactions can produce a variety of unique shapes and morphologies including toroids, liquid columns, and captured bubbles. We aim to understand the conditions under which these various regimes form and the physics governing their formation.

One particularly interesting regime is the "free-surface pendant drop." Under certain conditions, drops will partially wet the bath surface, residing on top of the bath despite the drops greater density. Interfacial tension between the drop and the bath reduces, as the drop dissolves into the bath, and the drop is driven down into the bath once gravity overcomes the interfacial tension. Between the surface of the bath and the bulk of the downward driven drop, a liquid column is formed. The liquid columns formed by these "free-surface pendant drops," as well as many other regimes, could have a variety of practical applications in industry.

(85f) Designing Copper Zinc Oxide as a P-Type Transparent Conducting Oxide for Photovoltaic Applications

Sun, D.,

Originally presented on: 11/4/2013 10:10:00 - 10:30:00

Transparent conducting oxides (TCOs) are electrically conductive materials that are used in thin-film opto-electronic applications, such as solar cells, but major improvements in performance and cost are necessary for TCOs to be used on a commercial scale. We present Copper-doped Zinc Oxide as a low-cost TCO candidate. We determined how the electrical transport properties and optical properties vary with composition by performing density functional theory calculations, including gradient-corrected, hybrid, and Green's functions. With increasing Cu concentration (x), the band gap of Copper-doped Zinc Oxide ($Zn_{1-x}Cu_xO$) decreases due to the down shift of the valence band. However, the optimal value of x is determined by the balance between the increase in electrical conductivity and the decrease in optical absorption. We also compare our results to ternary oxides formed from zinc and copper, and discuss strategies for designing materials with desirable properties.

(138a) NSF Overview

Wellek, R. M., The National Science Foundation
Lighty, J. S., University of Utah
Originally presented on: 11/4/2013 12:30:00 - 12:55:00

Overview of NSF Programs.

The National Science Foundation (NSF) is an independent federal agency created by Congress in 1950 "to promote the progress of science; to advance the national health, prosperity, and welfare; to secure the national defense..." With an annual budget of about $7.0 billion (FY 2012), we are the funding source for approximately 20 percent of all federally supported basic research conducted by America's colleges and universities. In many fields such as mathematics, computer science and the social sciences, NSF is the major source of federal backing.

We fulfill our mission chiefly by issuing limited-term grants -- currently about 11,000 new awards per year, with an average duration of three years -- to fund specific research proposals that have been judged the most promising by a rigorous and objective merit-review system. Most of these awards go to individuals or small groups of investigators. Others provide funding for research centers, instruments and facilities that allow scientists, engineers and students to work at the outermost frontiers of knowledge.

(138b) Highlights of CBET Cluster On Biomedical Engineering and Engineering Healthcare

Srienc, F., National Science Foundation

Originally presented on: 11/4/2013 12:55:00 - 13:15:00

Biomedical Engineering Cluster

The National Science Foundation (NSF) is an independent federal agency created by Congress in 1950 "to promote the progress of science; to advance the national health, prosperity, and welfare; to secure the national defense…" With an annual budget of about $7.0 billion (FY 2012), we are the funding source for approximately 20 percent of all federally supported basic research conducted by America's colleges and universities. In many fields such as mathematics, computer science and the social sciences, NSF is the major source of federal backing.

We fulfill our mission chiefly by issuing limited-term grants -- currently about 11,000 new awards per year, with an average duration of three years -- to fund specific research proposals that have been judged the most promising by a rigorous and objective merit-review system. Most of these awards go to individuals or small groups of investigators. Others provide funding for research centers, instruments and facilities that allow scientists, engineers and students to work at the outermost frontiers of knowledge.

(138c) Highlights of CBET Cluster On Transport and Thermal Fluids Phenomena

Wellek, R. M., The National Science Foundation
Papavassiliou, D. V., The University of Oklahoma
Originally presented on: 11/4/2013 13:15:00 - 13:35:00

Transport and Thermal Fluids

The National Science Foundation (NSF) is an independent federal agency created by Congress in 1950 "to promote the progress of science; to advance the national health, prosperity, and welfare; to secure the national defense…" With an annual budget of about $7.0 billion (FY 2012), we are the funding source for approximately 20 percent of all federally supported basic research conducted by America's colleges and universities. In many fields such as mathematics, computer science and the social sciences, NSF is the major source of federal backing.

We fulfill our mission chiefly by issuing limited-term grants -- currently about 11,000 new awards per year, with an average duration of three years -- to fund specific research proposals that have been judged the most promising by a rigorous and objective merit-review system. Most of these awards go to individuals or small groups of investigators. Others provide funding for research centers, instruments and facilities that allow scientists, engineers and students to work at the outermost frontiers of knowledge.

(138d) Highlights On CBET Cluster On Chemical, Biochemical and Biotechnology Systems

Achenie, L., National Science Foundation
Antos, G., National Science Foundation
Originally presented on: 11/4/2013 13:35:00 - 13:55:00

Chemical, Biochemical and Biotechnology

The National Science Foundation (NSF) is an independent federal agency created by Congress in 1950 "to promote the progress of science; to advance the national health, prosperity, and welfare; to secure the national defense…" With an annual budget of about $7.0 billion (FY 2012), we are the funding source for approximately 20 percent of all federally supported basic research conducted by America's colleges and universities. In many fields such as mathematics, computer science and the social sciences, NSF is the major source of federal backing.

We fulfill our mission chiefly by issuing limited-term grants -- currently about 11,000 new awards per year, with an average duration of three years -- to fund specific research proposals that have been judged the most promising by a rigorous and objective merit-review system. Most of these awards go to individuals or small groups of investigators. Others provide funding for research centers, instruments and facilities that allow scientists, engineers and students to work at the outermost frontiers of knowledge.

(138e) Highlights of CBET Cluster On Environmental Engineering & Sustainability

Wellek, R. M., The National Science Foundation
Gupta, R. B., The National Science Foundation
Originally presented on: 11/4/2013 13:55:00 - 14:15:00

The National Science Foundation (NSF) is an independent federal agency created by Congress in 1950 "to promote the progress of science; to advance the national health, prosperity, and welfare; to secure the national defense..." With an annual budget of about $7.0 billion (FY 2012), we are the funding source for approximately 20 percent of all federally supported basic research conducted by America's colleges and universities. In many fields such as mathematics, computer science and the social sciences, NSF is the major source of federal backing.

We fulfill our mission chiefly by issuing limited-term grants -- currently about 11,000 new awards per year, with an average duration of three years -- to fund specific research proposals that have been judged the most promising by a rigorous and objective merit-review system. Most of these awards go to individuals or small groups of investigators. Others provide funding for research centers, instruments and facilities that allow scientists, engineers and students to work at the outermost frontiers of knowledge.

(138f) Interactive Question and Answer Session With NSF Program Directors

Wellek, R. M., The National Science Foundation
Lighty, J., National Science Foundation

Originally presented on: 11/4/2013 14:15:00 - 14:45:00

The National Science Foundation (NSF) is an independent federal agency created by Congress in 1950 "to promote the progress of science; to advance the national health, prosperity, and welfare; to secure the national defense…" With an annual budget of about $7.0 billion (FY 2012), we are the funding source for approximately 20 percent of all federally supported basic research conducted by America's colleges and universities. In many fields such as mathematics, computer science and the social sciences, NSF is the major source of federal backing.

We fulfill our mission chiefly by issuing limited-term grants -- currently about 11,000 new awards per year, with an average duration of three years -- to fund specific research proposals that have been judged the most promising by a rigorous and objective merit-review system. Most of these awards go to individuals or small groups of investigators. Others provide funding for research centers, instruments and facilities that allow scientists, engineers and students to work at the outermost frontiers of knowledge.

(150a) Race To 2015: The Completion Of The Millennium Development Water, Sanitation, and Hygiene Goal

DeLago, C., SIT, AIChe (American Institute of Chemical Engineers) Chapter
Originally presented on: 11/4/2013 12:30:00 - 12:50:00

Today, one in every five people do not have access to clean water, especially those in third-world countries. Many rivers and lakes have now been contaminated by effluent factory waste causing them to have poisonous amounts of ammonium, phosphate, and arsenic. This, in addition to natural causes such as droughts, over population, and natural disasters have turned the need for water into a fast-growing problem that could soon turn into a global epidemic if proper precautions are not taken. With the turn of the new millennia in 2000, many of the world's leading countries knew something had to be done to aid developing nations through these water crises. As a result, in 2005, the United Nations put into play the Millenium Development Goals, with the seventh goal being to reduce the amount of people without access to clean water and proper sanitation by 50% before 2015. A portion of this goal was prematurely achieved in 2010 with 10 million people worldwide gained access to a sustainable water source.

I would like my project this summer to focus on this initiative by focusing on what the American government agencies and non-profit organizations have done thus far to aid the United Nations in completing this goal and conduct further research into what still needs to be implemented to fulfill the objective. Similarly, I would like to look into what other countries and organizations around the world have done to help meet this goal and conclusively postulate as to whether or not the Initiative will be completed in time. Lastly, I would like to focus on what the United States, the rest of the world, and The U.N. should do in the future to make sure that the Water and Santitaion Crisis remains an issue of the upmost importance.

(150b) Dynamic Design of a Ethylamine Production Plant

Becht, K.,

Wiedorn, A.,

Originally presented on: 11/4/2013 12:50:00 - 13:10:00

The ethylamine production facility documented in this process design produces on average 98 tons/day of triethylamine and 186 tons/day of diethylamine. Fresh ammonia and fresh ethanol are fed into the process with the presence of hydrogen to produce mono, di-, and tri- ethylamine with a byproduct of water. The products are produced in a series of three reactions. Ammonia and ethanol react to produce mono-ethylamine. Diethylamine is produced from monoethylamine and ethanol and triethylamine by diethylamine and ethanol. The unreacted ethanol and ammonia, the monoethylamine and a portion of diethylamine and triethalmine are recycled back into the reactor. This production facility produces a return on investment (ROI) of 42% and will make a profit after four years. The steady state plant design was created from optimizing cost and efficiency and consists of a compressor train, cooled plug flow reactor and a series of six distillation columns. The dynamic simulation of the process focuses on maintaining the product purities and controlling the temperatures due to safety concerns at high process temperatures. A HAZOP review was performed on the reactor temperatures since this is a critical safety concern. The safety of the plant and its operators is also discussed. Issues of flammability, auto ignition temperatures, and construction materials are addressed.

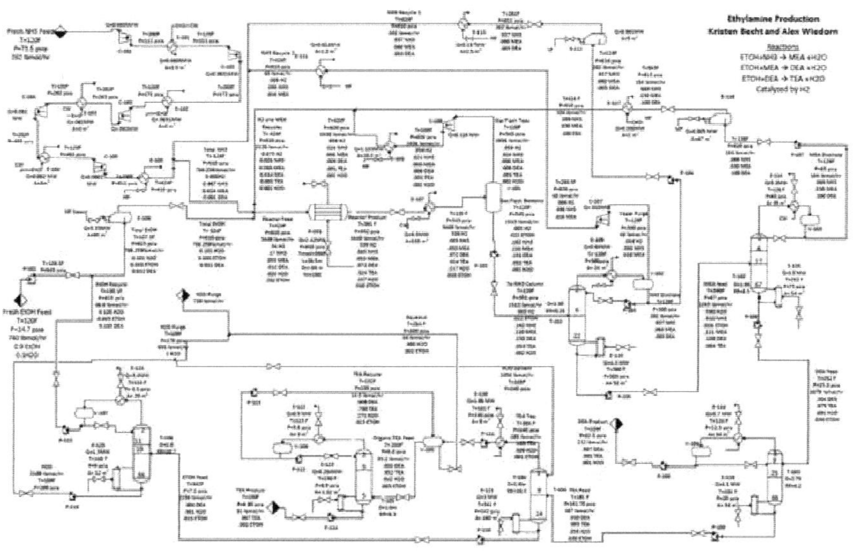

Figure 1. Optimized steady state design of the ethylamine production plant.

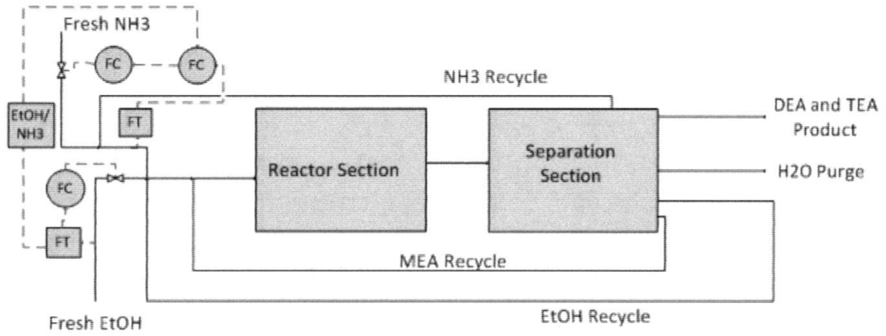

Figure 2. Control scheme for plant wide control (PWC).

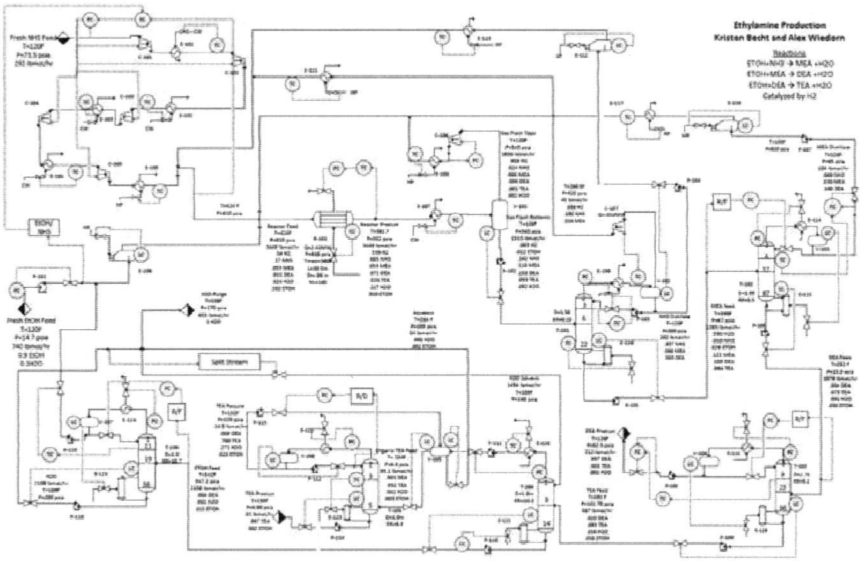

Figure 3. Dynamic design of the ethylamine production plant.

(150c) Acetic Acid and Methanol Recovery Unit

Berard, V. F., Lehigh University
DiNardo, A., Lehigh University
Fortney, S., Lehigh University
Originally presented on: 11/4/2013 13:10:00 - 13:30:00

The goal of this unit is to utilize a stream from the saponification section of a Polyvinyl Alcohol plant situated on the Gulf Coast of the United States. In addition to methanol to be recovered, the feed stream contains large quantities of methyl acetate. Therefore, methyl acetate is hydrolyzed to produce additional methanol and acetic acid for sale. The total amount of methanol is purified to 99.9 weight percent and acetic acid purified to 99 weight percent to meet prescribed specifications. Biodegradable sodium acetate also enters in the feed stream and is removed through a purge stream to be dumped. A second impurity, acetaldehyde, is removed from the plant and incinerated to eliminate the carcinogenic hazard. The optimized steady-state design includes five distillation columns and one reactor and three recycle streams. Unconverted methyl acetate is returned to the front-end with only trace amounts exiting the unit unconverted. An azeotrope between methyl acetate and methanol is broken using extractive distillation techniques, resulting in a second recycle stream for the water solvent. The final recycle arises from the excess water fed to the reactor to increase methyl acetate conversion. At full capacity, the plant recovers 177 million pounds of methanol per year and produces 101 million pounds of acetic acid for sale. Further, a plant wide control scheme is implemented to operate the plant. The control structure is designed to withstand composition disturbances of +/- 10%, a 10 % increase in feed rate, and a 33% decrease in feed rate. An economic analysis shows that the plant will begin to make a profit and consistently make a profit of at least $50 million for the following 10 years with a return on investment of 233%.

(150d) A Re-Examination of US Water Policies That Prevent Sustainable Water Consumption and Reuse

Kusanto, N., Oregon State University
Originally presented on: 11/4/2013 13:30:00 - 13:50:00

Water management and allocation has been integral in the success and development of the United States of America. While ground water has been historically recognized as a personal property right, surface waters required greater regulation. Thus, the first water policies were developed and reflected man's desire to harness the wild rivers for human development. Acts such as the 1920 Federal Power Act and 1922 Colorado River Compact illustrate governmental hopes of controlling water ways for power generation and agricultural allocation among states. The Colorado River Compact, in particular, attempted to fairly divide a shared water resource among users across a number of states based on projected need. As US expansion and development continued, the federal government sought to further centralize control and regulation of waterways through the 1936 Flood Control Act and the 1948 Water Pollution Act. Early water policy failed to protect water supplies, allowing US development to severely deplete and pollute water resources. As such, in the last quarter of the 20th century, focus was placed on protecting drinking water, 1974 Safe Drinking Water Act, and to support research which investigates how the country can safely reclaim and reuse waste water, Title XVI in 1992. Looking forward into the 21thcentury, water policies should reflect a need for sustainable water consumption and increase of water reuse and reclamation, employing the discoveries made in science.

The present need and desire to encourage wise water use and water recycle must be driven by an understanding of the factors which have historically prevented municipalities and industry from water recycling. The current system for water distribution taxes users a nominal fee to subsidize the cost of water transportation. The water, itself, is free. As such, the abundance of free water coupled with lack of public knowledge on water management issues have contributed to excessive consumption of fresh water which has depleted water resources around the US. For example, the average depth in the 174,000 square mile Ogallala aquifer, one of the world's largest, has dropped from 240ft to approximately 80ft. Conservative estimates predict that full depletion of the aquifer will occur in the next 25 years, requiring a recharge time of 6,000 years to reach initial levels. Clearly, current time horizons dictating water use disregard long term water conservation. If the rate of water withdrawal continues to exceed the existing rate of aquifer recharge, the US will soon run out of a necessary resource which is at the heart of the US agricultural belt. Because the irrigated plains produce 20% of American grain and corn, failure to control water use will result in a collapse of the international agricultural market. Concerns for the protection of public health and a decline in public trust in government agencies are at the forefront of a list of challenges which prevent local governments from enacting infrastructure changes to improve water recycle. On a grander scale, the federal government is only able to act in a manner consistent with the U.S. Constitution. Thus, state laws may vary and can come into conflict if aquifers or surface water resources are shared. Recognizing how to address conflicts between states, municipalities, and the federal government is important to finding a solution. Industry must follow the regulations imposed by the three actors above. Strict water reuse regulations lead to cost-benefit analyses which prevent companies from conserving water and choosing to discharge their waste. Likewise, incentives are not currently in place to reward companies who are able to reduce consumption within limited time frames.

A simple mass balance on water usage in the US will illustrate that action must be taken to help conserve water resources. The national government should invest more heavily on alternative future analyses to determine the effects water use has on biophysical and human systems. Increasing scientific literacy in the public sector will also play a key role in gathering support from municipalities. Moreover, incentives, much like solar energy tax breaks, must be implemented to encourage better use of water. Infrastructure costs for water must be externalized to consumers resulting in increasing rates for water use, which will encourage sustainable water consumption. It is suggested that excessive water use be penalized on a tiered system to penalize waste and encourage sustainability. Existing policies must also be reexamined and redefined to reflect research discoveries. The relaxation of water reuse regulation coupled with more strict regulations on water discharge may encourage companies to conserve water. This paper will provide possible solutions for the aforementioned difficulties regarding water reuse.

(150e) Toluenediamine Production Through Hydrogenation in Methanol Solvent

Masel, R.,
Smith, D., Lehigh University
Originally presented on: 11/4/2013 13:50:00 - 14:10:00

A feasibility analysis for a new toluenediamine (TDA) production plant was conducted. The production process selected for investigation was the hydrogenation of dinitrotoluene (DNT) in liquid methanol solvent over a nickel catalyst. The product TDA and reactant DNT have much higher boiling temperatures than the solvent methanol and byproduct water, and the boiling points of the reactants and products are too close to allow for sufficient separation. Therefore, the reactor must be sized sufficiently to achieve nearly complete conversion, as the reactant cannot be separated and recycled. Oligomerization of the DNT and TDA occurs at higher temperatures, generating high boiling tars that deactivate the catalyst and contaminate the product streams. Additionally, the thermal instability of DNT, as well as TDA, tars, and other nitrogen containing byproducts necessitate a temperature limit of approximately 350 F in any streams that could potentially contain the least stable component DNT. The composition of water in the product must be no greater than 0.1 weight percent. Two potential alternatives in distillation of the solvent and water from the products were apparent as a result of both a temperature and composition specification in the bottoms product. The first was a single column at high vacuum requiring refrigeration in the condenser, and the second was two separators with a column at atmospheric pressure using cooling water in the condenser before the vacuum column achieving the desired composition. The total annualized cost (TAC) for the single column was $\$6.361/10^6$, while the TAC for the double column system was less at $\$4.696/10^6$ due to reduced refrigeration costs. Health and safety hazards were considered in a process safety analysis that outlined concerns for implementing the process. A product price of $1.00 per pound of TDA was calculated to yield sufficient return on investment (ROI) of approximately 51.9% with a payback period of 1.9 years. Production of toluene diisocyanate (TDI) from toluenediamine was considered as the primary use, with an estimated price in 2016 of $1.459 per pound of TDI, validating the assigned price of intermediate TDA. The net present value of the project considering five years of operation after startup is $56.2 MM, or $129.3 MM over an operation lifetime of 14 years.

(150f) Experimentation: The Foundation of Career Growth

Worstell, J. H., Worstell and Worstell, Consultants
Orsak, J.,
Arredondo, J., University of Houston
Originally presented on: 11/4/2013 14:10:00 - 14:30:00

Every chemical engineer will design experiments and evaluate experimental data sometime during their career. Thus, it is important for chemical engineering students to have a meaningful laboratory experience early in their scholastic career. At the University of Houston, chemical engineering students begin evaluating experimental data during their Junior year. This situation raises challenges for the laboratory staff. First, designing a laboratory course is difficult when you consider how little time Junior chemical engineering students have available for a laboratory course. During that limited time, students must assemble their experimental apparatus, perform the experiment, then dismantle and clean their apparatus. In some cases, they have as little as ninety minutes for their laboratory period. Second, to be meaningful, such an experience needs to be reflective of the courses they are taking and have taken.

At the University of Houston, we have refashioned the Junior laboratory to meet these parameters. In this presentation, we outline how we devised, designed, tested, and implemented a set of experiments that Junior chemical engineering students can do quickly and efficiently. We also discuss how we integrated the laboratory and course work so that the student could directly apply classroom learning to evaluating experimental data.

(150g) Using Molecular Simulation to Improve the Efficiency of Biomass Processing

Jarin, Z.,

Originally presented on: 11/4/2013 14:30:00 - 14:50:00

Interest in biomass as a feedstock for chemical processes has been increasing over the last several years. In order to compete economically, conversion of biomass to small molecules for production of fuels and value-added chemicals must become more efficient. However, a major hurdle to chemical processing is the robust structure of biomass and subsequent resistance to breakdown. In the past ten years, experiments have shown ionic liquids, a special group of solvents, have the rare ability to dissolve biomass. While there are countless ionic liquids known to the research community, these experiments are not guided by molecular scale design principles, hindering more promising research. This project focuses on providing fundamental knowledge to explain the ionic liquid-mediated biomass dissolution by using molecular dynamics (MD) coupled with enhanced sampling techniques. Enhanced sampling allows the efficient exploration of the free energy landscape of a system by modifying MD algorithms. Specifically, an enhanced sampling technique called metadynamics will be used to explore glucose's conformational free energy surface, which can be used to better understand the mechanism of dissolution in ionic liquids. Insight from these simulations can then be applied to larger sugar oligomers that constitute cellulosic biomass.

(183a) Session Honoring Phil Wankat

Litster, J., Purdue University
Harris, M. T., Purdue University
Noble, R. D., University of Colorado
Mazotti, M., Institute of Process Engineering - ETH Swiss Federal Institute of Technology
Pease, L. F. III, University of Utah
Anderson, T., University of Massachusetts - Amherst
Slater, S., Rowan University
Silverstein, D. L., University of Kentucky
Oreovicz, F.,
Farmus, C.,
Felder, R. M., North Carolina State University
Originally presented on: 11/4/2013 15:15:00 - 17:45:00

This session honors Professor Phil Wankat for a career of over 40 years as a pioneer in Separation Science and Engineering Education. Phil Wankat holds positions as the Clifton L. Lovell Distinguished Professor of Chemical Engineering and the Director of Undergraduate Degree Programs in the School of Engineering Education at Purdue University. He has a BS from Purdue, PhD from Princeton and MSEd in Counseling from Purdue. Phil has made tremendous impact in his career through his writing: two important text books in Separation Processes, as well as *Teaching Engineering* and *The Effective, Efficient Professor: Teaching, Scholarship and Service.* He was also instrumental in establishing the Interdisciplinary Engineering Program at Purdue. Through his research, teaching and mentoring, Phil has made a huge impact on more than one generation of students and faculty colleagues. Speakers at the session are:

Marco Mazotti (ETH, Zurich) "*Three-Column Intermittent SMB Chromatography for Three- Fraction Separations*"

Leonard Pease (Utah) "*Influence of Social networks on Engineering Education*"

Tim Anderson (Massachusetts) "*How Chemical Engineering Students Are Different and the Effective, Efficient Professor*"

Rich Noble (Colorado) "*Thanks to Phil*"

Stewart Slater (Rowan) "*TBA*"

David Silverstein (Kentucky) "*Following in Selected Footsteps of Phil*"

Frank Oreovicz (Purdue) "*Reflections on an Enquiring Mind*"

Cristina Farmus (Purdue) "*Phil's Legacy at Purdue*"

Rich Felder (North Carolina State) *"OK, So There Are Some Superhuman Professors"*

Phil Wankat "*The Last Word*"

(195a) Proposal Writing Tutorial

Good, D. T. A., University of Maryland Baltimore County
Rorer, G., National Science Foundation
Originally presented on: 11/4/2013 15:15:00 - 16:15:00

The National Science Foundation (NSF) is an independent federal agency created by Congress in 1950 "to promote the progress of science; to advance the national health, prosperity, and welfare; to secure the national defense…" With an annual budget of about $7.0 billion (FY 2012), we are the funding source for approximately 20 percent of all federally supported basic research conducted by America's colleges and universities. In many fields such as mathematics, computer science and the social sciences, NSF is the major source of federal backing.

We fulfill our mission chiefly by issuing limited-term grants -- currently about 11,000 new awards per year, with an average duration of three years -- to fund specific research proposals that have been judged the most promising by a rigorous and objective merit-review system. Most of these awards go to individuals or small groups of investigators. Others provide funding for research centers, instruments and facilities that allow scientists, engineers and students to work at the outermost frontiers of knowledge.

(195b) Interactive Breakout Panels

Wellek, R. M., The National Science Foundation
Lighty, J., National Science Foundation
Originally presented on: 11/4/2013 16:15:00 - 17:15:00

The National Science Foundation (NSF) is an independent federal agency created by Congress in 1950 "to promote the progress of science; to advance the national health, prosperity, and welfare; to secure the national defense…" With an annual budget of about $7.0 billion (FY 2012), we are the funding source for approximately 20 percent of all federally supported basic research conducted by America's colleges and universities. In many fields such as mathematics, computer science and the social sciences, NSF is the major source of federal backing.

We fulfill our mission chiefly by issuing limited-term grants -- currently about 11,000 new awards per year, with an average duration of three years -- to fund specific research proposals that have been judged the most promising by a rigorous and objective merit-review system. Most of these awards go to individuals or small groups of investigators. Others provide funding for research centers, instruments and facilities that allow scientists, engineers and students to work at the outermost frontiers of knowledge.

(209a) Building a College-Wide Safety Culture: Opportunities and Challenges

Ziemer, K. S., Northeastern University
Sacco, A. Jr., Northeastern University

Originally presented on: 11/4/2013 15:15:00 - 15:45:00

Texas Tech University's Whitacre College of Engineering (WCOE) is building a safety culture among their students, faculty, and staff by approaching both the curriculum and the culture, with the work environment culture believed to be the most important in affecting change of behavior in both classrooms and laboratories. WCOE has set a goal of ZERO safety incidents while setting the standard for University Laboratory Safety and Safety Awareness of all faculty, students, and staff. Key strategies include incorporating alumni and industrial support (both in spirit and financial), increasing awareness of the university as a workplace and the opportunities and responsibilities to model safe behaviors and practices, approaching safety as a matter of pride for the school, and replacing many "punish bad safety practice if you see it" systems with self-assessment ownership and rewards for exemplary safety practice. The program will be officially kicked-off in September of 2013.

The opportunities and challenges encountered in developing the WCOE program as well as the initial assessment of progress will be discussed. In addition, a dean's perspective on the unique leadership challenges will be shared. It is hoped that the experiences discussed will convince others that the ABET requirements for Process Safety in the Chemical Engineering Curriculum provide opportunities beyond correcting a program criterion deficiency. By thinking broader than ABET, chemical engineers and chemical engineering departments have the opportunity to lead their academic institutions toward producing higher quality engineering graduates at all levels, and increasing financial stability for academic institutions.

(209b) Teaching Process Safety Systems: Some Practical Applications

Klein, J. A., DuPont
Vaughen, B. K., BakerRisk
Originally presented on: 11/4/2013 15:45:00 - 16:15:00

Engineering education must provide students a practical understanding of ‘conservation of life’ and its equivalence in importance with other fundamental engineering laws of conservation. This requires that students develop a real world view of effective process safety programs, which necessarily include safety culture and leadership, process safety systems, and operational discipline. In particular, process safety systems and methodologies must be introduced and used to identify and assess process hazards and to evaluate and manage the associated process risks through appropriate design, multiple layers of protection, and rigorous operation and maintenance of process equipment. This paper describes how teaching process safety systems with a systemic “life cycle” approach can help ensure that students understand and can apply important process safety concepts and methodologies in support of ‘conservation of life’ and other design and operational goals.

(209c) Teaching Safety in Undergraduate Chemical Engineering Laboratory and Chemical Engineering Design Courses

Dunn, R. F., Polymer and Chemical Technologies, LLC
Debelak, K. A., Vanderbilt University
Originally presented on: 11/4/2013 16:15:00 - 16:45:00

Process and product safety are of increasing importance in the design and operation of chemical engineering processes. In general, a number of chemical process and product failures have been the result of design deficiencies such as the improper selection of materials of construction for the intended end use, the improper processing and/or fabrication of these materials into the end product and the improper maintenance of the chemical process and/or product, in addition to other factors. Furthermore, failure analysis of several chemical plant accidents has also resulted in the identification of numerous process design deficiencies. Scientific methodologies and tools used for identifying the potential mode of failures for polymer products and chemical processes including, but not limited to, what-if analysis, failure mode and effects analysis (FMEA) and hazard and operability analysis (HAZOP). Proper instruction and application of these methodologies, in addition to the use of case studies, are essential elements of the undergraduate chemical engineering curriculum. Specifically, efforts are in progress to incorporate safety methodologies and a safety culture into the chemical engineering undergraduate laboratory courses and senior design courses. Experience and lessons learned in applying and reinforcing safety in these courses in the undergraduate curriculum will be presented.

(209d) Start the Revolution ~ Ensuring There Is the Top of the Food Chain: Using Process Safety As The Capstone

Howat, C. S. III, C. S. Howat & Associates

Originally presented on: 11/4/2013 16:45:00 - 17:15:00

"Look what's coming down the street! Start the revolution." Synthesize, evaluate, modify, keep it safe and everyone goes home. How many heard that beyond lip-service in their education? How many practiced it? The safety revolution is underway and the past isn't good enough for our undergraduates. It wasn't good enough for us. Join! Look at our history, modify the future.

Design scenarios have one set of general and specific specifications and one desired capacity and quality. Even in mulit-use, high value added batch processing with multiple products, one set of specifications lead to one set of capacities, conditions and qualities. Certainly this is difficult at the undergraduate level. Most schools didn't have capstone courses so ABET responded in the 80's mandating it. But, this is the 21st century. Our license to operate is tenuous. We can't continue as we did.

How many ways can this solution (design or operation) go wrong? The top of the food chain looks down and asks that question. Education is dedicated to developing a fundamentals foundation; and high-level integrative-, critical-, synthesis- and creative-thinking skills. Curricula are pressured by new-area emergence and industry desires. Design, the cornerstone experience to skill development, no longer can look to specific industries or applications no matter what we were taught. The capstone course must become more inclusive, applicable, and responsive. Risk-focused safety courses apply to any industry and develop high-level thinking skills and integrate fundamentals.

This paper publishes a manifesto, once practiced, describing capstone process safety. The paper shows how this f fits into the curriculum, generally, and conceptual process design hierarchy, specifically. Then, it turns the tables overthrowing that discussion into preparing students for industrial practice, asking the question: How many ways can this go wrong credibly? This paper focuses on the development of the capstone course using process safety as the medium. Topics include the objectives, content, evaluation, ABET and, most importantly, our product – students ready to practice. . Specific recommendations are given using actual examples of what can be taught. Listeners are encouraged to integrate their industrial problems, no matter the area of practice, into this course. The target of the paper is the professor, instructor or lecturer who is interested in teaching widely applicable skills and integrating fundamentals. More to the point, the target is the student who wants to be the best prepared to practice in *any* industry.

(209e) Experience Using Inexpensive Water Overflow Experiment to Demonstrate SIS Concepts

Richmond, P. C., Lamar University
Benson, T. J., Lamar University
Rasel, M. A. K., Lamar University
Originally presented on: 11/4/2013 17:15:00 - 17:45:00

This paper describes our experience using an inexpensive hands-on water overflow experiment to improve undergraduate chemical engineering students' understanding of Safety Instrumented Systems (SIS). The experiment was designed to demonstrate how adding a SIS improves the probability of failure on demand for the process activity of filling a tank, a very serious issue for tank farms used to store petroleum products. This lab experience was coupled with classroom instruction on how SISs are used to protect chemical process operating personnel and equipment.

Two Arduino microcontroller boards and a pair of relays were used to create the control systems and pressure sensors were used to measure the level. A submersible pump and acrylic tank were used to represent the process. One microcontroller was used as the basic control system (BCS) and it was programmed to fill and then empty a water tank with a predetermined failure probability of overfilling the tank. The SIS device was created using the other Arduino board and a different tank level instrument separate from the basic control system. The SIS device was programmed with a second independent failure probability. The failure probabilities were implemented in the Arduino codes with pseudo random numbers.

The students used a Layer of Protection Analysis (LOPA) to calculate the SIS probability of failure required for a given BCS failure probability and a desired mitigated failure rate. The experiment was then used to physically demonstrate how the SIS reduced the probability that the water filling operation would result in water overflowing from the tank. We tested their understanding before and after they used the experiment to see if it improved their understanding of SISs.

(210a) Rapid Skin Permeabilization By The Simultaneous Application Of Dual-Frequency Ultrasound

Maa, R.,

Originally presented on: 11/4/2013 15:15:00 - 15:35:00

Transdermal drug delivery is an attractive delivery route because of the ease of access of the skin. However, the skin acts as an effective barrier. Therefore, large molecules must be delivered with needles, which are painful and can increase the transmission of infectious diseases. Low frequency ultrasound has previously been studied for its ability to painlessly increase the permeability of the skin. Despite this work, ultrasound has not yet attained wide clinical use due to limitations on the resulting permeability. We hypothesized that skin permeability can be enhanced by coupling a high-frequency ultrasound signal to the existing low-frequency treatment. Aluminum foil pitting experiments, along with in vitro tests performed with porcine skin, tested the simultaneous application of two distinct frequencies ranging from 20 kHz to 3 MHz to examine their effect on the formation of localized transport regions (LTRs) within the skin. These experiments demonstrated that skin permeability and resulting LTR size both increased as a result of dual-frequency treatment compared to low-frequency treatment alone. Glucose (180 Da) and inulin (5000 Da) transdermal flux experiments further confirmed these results by demonstrating enhanced delivery through and into the dual-frequency-treated skin. This methodology can be a viable option for transdermal drug delivery.

(210b) N/a

Zeidman, J., N/a

Originally presented on: 11/4/2013 15:35:00 - 15:55:00

The controlled production of artificial antibodies for therapeutic treatments is a rapidly growing and profitable field. One primary method of antibody production is transient transfection in mammalian cells most often using polyethyleneimine (PEI) as the transfection agent and Human Embryonic Kidney (HEK) or Chinese Hamster Ovary (CHO) stable cell lines. Despite its importance, transfection is not well understood and leads to variable results based on many factors including medium used, transfection and incubation times, viable cell density (VCD) at inoculation, PEI concentration, and DNA concentration. The quantity and quality of antibody produced after transfection are of the upmost importance. In this experiment, a specific transfection protocol was optimized for antibody titer in HEK cells. Design of Experiments software was utilized to plan experiments to maximize antibody titer by varying the parameters of inoculation VCD, PEI concentration, and transfection time. Experiments were carried out in a six-chamber HexaScreen® mini-bioreactor with a control and scale-up performed with a 2 liter roller bottle system. Using the HexaScreen, 15 different trials were performed over the given parameters to conclude that the optimal combination was an inoculation VCD of 3 million cells/mL, a PEI concentration of 35 mg/L, and no transfection time, which would yield an antibody concentration over 40 mg/L. The roller bottle scale-up demonstrated inconsistency and scale-up problems that would need to be addressed with future experiments. While further experiments are necessary, this project demonstrated the feasibility of this system for better understanding and optimizing transient transfection and some of the relations between antibody titer and transfection parameters.

(210c) A Novel Formulation For Non-Swelling Peg Hydrogels For Application In Spinal Cord Injury (SCI)

Fu, M.,

Originally presented on: 11/4/2013 15:55:00 - 16:15:00

The notion of using hydrogels, which are 3-dimensional cross-linked polymer networks, for the delivery of biologically relevant molecules has been around for decades. However, with regard to administering hydrogels to anatomically sensitive areas like the spinal cord, there still exist 3 main challenges: the mechanical stress conferred by the swelling of the hydrogel, the rate of degradation of the gel network to control drug release, and the capacity of the hydrogel to maintain the efficacy of the drug over time. In this project, all three areas are addressed. Testing of various novel formulations for PEG hydrogels has yielded a network that exhibits temperature-sensitive syneresis, which allows for complete control of swelling. In addition, specific additional modifications of the chemistry to alter the gel matrix's susceptibility to hydrolysis has allowed for tailoring of the degradation of the hydrogel. As a result, precise tailoring of therapeutic drug release rate from the gel is attainable. Finally, the efficacy of incorporating stabilizing excipients into the gel network as a method of preserving the enzymatic activity of therapeutic proteins is being investigated and has shown tremendous preliminary potential. In particular, the incorporation of trehalose has been shown to confer stabilizing properties to the hydrogel as a whole with respect to the enzyme Chondroitinase ABC, which has therapeutic applications in spinal cord injuries.

(210d) The Quantification Of Mitochondrial Location and Morphology Using Matlab

Bussie, B., AUBURN UNIVERSITY
Kerscher, P.,
Lipke, E. A., Auburn University
Originally presented on: 11/4/2013 16:15:00 - 16:35:00

The analysis and quantification of biological images is essential for researchers in many disciplines. For proper statistical analysis and results justification, hundreds of cells frequently have to be analyzed by manual counting due to the limitations of image processing tools. In order to better understand the function of human cells and tissues and to be able to use that knowledge to both prevent and treat disease, we need to be able to quantify visual information contained in biological images accurately in a time-efficient, reproducible, and unbiased manner.

A MATLAB program was created to analyze fluorescently labeled organelles within cells. Specifically, the program was used to locate and quantify mitochondria and nuclei of different cell types, including human induced pluripotent stem cells (hiPSC), human embryonic kidney cells (HEK-293), and a human breast cancer cell line (MCF-7). The program extracts information about the number of mitochondria within the cell, as well as location of the mitochondria relative to the nucleus. Information about the morphology of the mitochondria, such as shape and connectivity was also determined. Compared with manual interpretation, it was found that the MATLAB program was quicker, more consistent, and bias-free. When applied to differentiating stem cells, this program can reveal how the organization and morphology of mitochondria changes with cell maturation and differentiation.

(210e) Analysis Of CELL Proliferation Within A 3D Tissue-Engineered Cancer MODEL

Clary, J.,
Pradhan, S., Auburn University
Lipke, E. A., Auburn University
Originally presented on: 11/4/2013 16:35:00 - 16:55:00

Breast cancer is one of the most common forms of cancer throughout the world. Researchers are attempting to develop an artificial platform for growing cancer cells outside the human body for the study of the disease. The current study aims to create an *in vitro* environment that accurately mimics tumorigenic behavior. To accomplish this, human breast cancer MCF7 cells were encapsulated in polymer hydrogel microspheres composed of poly(ethylene glycol) and fibrinogen (PEG-Fb) using an oil-in-water emulsion technique and maintained in culture long-term. The microspheres were analyzed for size changes using ImageJ. SEM imaging was conducted on the microspheres to visualize the distribution of cells within the polymer matrix. Additionally, the MCF7 cells were stained with Hoechst 33342 dye after enzymatic-dissociation of the hydrogel microspheres for measurement of cell proliferation using the total DNA assay. The results obtained from the analysis indicated that total DNA content increased and the cells proliferated within the polymer hydrogel system over time. Thus, an artificial 3D platform for long-term culturing of cancer cells has been established. Further biochemical assays on the developed system are necessary to make it more closely mimic native breast cancer conditions so that cancer therapies may be reliably studied *in vitro*.

(210f) Membrane Compatibility With Switchable Polarity Draw Solutions For Use In Forward Osmosis Applications

Coscia, B. J., University of Connecticut
Reimund, K. K., University of Connecticut
Wilson, A. D., Idaho National Laboratory
McCutcheon, J. R., University of Connecticut
Originally presented on: 11/4/2013 16:55:00 - 17:15:00

Water scarcity is problem being faced worldwide and present in every continent. Close to one-fifth of the world's population has difficulty acquiring safe water, and the problem is worsening as populations continue to grow in poorer countries. As the availability of unimpaired freshwater sources dwindles, cew water sources, such as the oceans and saline groundwaters, must be tapped. However, desalination technologies are very expensive due to a high energy requirement. Forward osmosis (FO) is a process which may be able to replace or become integrated with existing desalination technologies like reverse osmosis. FO relies on an osmotic agent, or draw solution, to drive water from a saline water source by osmosis. The water dilutes the draw solution and a secondary separation process where the solute is removed and recycled. This secondary separation step is key to FO process economics. Recently, a classification of draw solutes, known as switchable polarity solvents (SPSs) have emerged. These solvents become highly miscible with water upon sparging with CO_2 and phase separate when stripped with an inert gas. One such SPS is N,N-dimethylcyclohexylamine. This SPS has shown promise in early testing at Idaho National Laboratories. However, because it is a solvent, membranes, which are polymeric, may not be compatible with the solution. . This study examines the longevity of conventional osmotic membranes after exposure to relevant SPS solutions. Membrane performance metrics, such as permeance and salt rejection as well as osmotic flux, are presented.

(224a) ABET Accreditation: Updates and Insights

Lewis, R. S., Brigham Young University
Ludlow, D. K., Missouri University of Science and Technology
Originally presented on: 11/5/2013 8:30:00 - 8:55:00

This session, hosted by the Education and Accreditation Committee, will provide a brief review regarding the ABET accreditation process. In addition to briefly reviewing ABET General Criteria and the accreditation visit, discussion will also focus on aspects associated with the Chemical Engineering Program Criteria. Invited speakers only for this session, though the other ABET-related session (Best Practices) will accept abstracts.

This session will be chaired by Randy Lewis of Brigham Young University and Doug Ludlow of the Missouri University of Science and Technology. Topics for the session include:

- preparing your program before the ABET visit

- evaluator preparation prior to an ABET visit

- setting up an ABET visit schedule

- post visit interactions

- accreditation issues

- evaluator expectations

- accreditation resources

This will be an interactive session, with short talks by invited speakers (to be determined after the June 2013 ASEE meeting) and time for discussion with the audience. Confirmed speakers include: Ron Danner, Tom Hanley, and Gary Patterson.

If you are a new ABET evaluator, or if your program is about to undergo an ABET visit, or if you are interested in engineering accreditation, this session is for you!

(245a) Implementation of Tower Gardens On An Urban University Campus

Peters, D. R. W., University of Alabama at Birmingham
Barbour, N.,
Gaston, S.,
Burns, M.,
Originally presented on: 11/5/2013 8:30:00 - 8:55:00

As society faces the problems of food deserts, diseases of civilization due to malnourishment, and countless environmental harm such as pollution or droughts, creating sustainable communities may serve as a solution to cope with these problems. Implementing green spaces in a form of vertical gardens will have positive environmental effects and offer improved access and quality of fresh produce in the urban as well as rural areas. Growing produce in tower gardens and consuming more unprocessed foods, can potentially reduce rates of undernourishment and many illnesses of longevity.

The Tower Garden® utilizes aeroponics to recycle 100% of its nutrients and water. Furthermore, the system reduces the need for use of pesticides, insecticides, or herbicides. Vertical gardens also use less than 10% of both water and land that are used in traditional farming. They have an undeniable benefit of high yield fresh, nutritious, produce.

Several Tower Gardens are being installed on building rooftops on the University of Alabama at Birmingham (UAB) campus. Raised soil bedded roof gardens are already in-place on several buildings. Knowledge is being gained as to what is most successful in a roof garden in terms of soils and plant materials. Also under study is the growth of plant materials under varying degrees of water and nutrients, and differing locations with respect to sunlight. The use of the Tower Garden will provide the researchers with another system to test and compare with the success of the raised soil bedded gardens already in place. The rooftop gardens promote a sustainable environment as they enhance cooling on rooftops and provide increase insulation for the host building rooftop. This serves to reduce energy use and lowers the building operating costs. The Tower Gardens add two additional benefits to the positive nature of vegetated roofs: (1) the plants grow in an air/water/nutrient mix, with most of the water returned to the atmosphere via evaporation, and (2) the main plant sources are edible produce. For this research project, the Tower Gardens make use of building collected rainwater and equipment condensate as the water source, thereby reducing the storm water runoff from rooftops.

All the benefits of vertical farming contribute to a more sustainable planet.

(245b) New Developments in Implementing Green Engineering in the Chemical Engineering Curriculum

Shonnard, D. R., Michigan Technological University
Allen, D. T., The University of Texas at Austin
Nguyen, N., U.S. Environmental Protection Agency
Austin, S., Chemical Engineering Branch, US EPA
Originally presented on: 11/5/2013 8:55:00 - 9:20:00

The goal of this presentation is to outline updated green engineering and sustainable engineering content for undergraduate and graduate chemical engineering education. The work of chemical engineers is central to the development of sustainable chemical and energy production systems and to meet global human needs. Incorporating environmental issues and sustainability into the design of engineered products, processes, and systems generally means evaluating environmental, economic, and social performance. This presentation will summarize two approaches to incorporate green and sustainable engineering into undergraduate chemical engineering education; **i.** a stand-alone elective course, and **ii.** modules for required core chemical engineering courses. The modular approach, one or more chapters in a forthcoming 2nd edition of "Green Engineering: Environmentally-Conscious Design of Chemical Processes", Pearson-Prentice Hall, can serve as stand-alone supplementary materials for core chemical engineering courses. For example, in a thermodynamics course, environmental equilibrium partitioning and fate properties of chemicals could be a supplemental topic and these properties can be predicted with group contribution or other structure-activity relationships. In a mass balance course, these environmental partitioning and fate properties can be used in conservation of mass problems involving accidental releases of pollutants to various environmental media; air, water, soils, and sediments. These and other examples of incorporating green engineering concepts, methods, and tools in core chemical engineering courses will be presented.

(245c) Incorporating Process Integration Methodologies Into Senior Capstone Design Courses

Dunn, R. F., Polymer and Chemical Technologies, LLC
Debelak, K. A., Vanderbilt University
Originally presented on: 11/5/2013 9:20:00 - 9:45:00

Significant efforts have been devoted to the development of process integration design methodologies over the past several decades. Some of the main developments in process systems analysis using process integration design tools are the identification of heat exchange networks, mass exchange networks and water allocation networks. These systems-based tools are widely applicable and are now essential elements of undergraduate process design instruction. Furthermore, numerous opportunities exist to incorporate these tools in the analysis of senior capstone design projects. In addition, senior design projects based primarily on process integration design methodologies have recently been developed. Experience and lessons learned in applying process integration in senior design projects in the undergraduate curriculum will be presented.

(245d) Assembling Engineering for Sustainable Development Programs

Jensen, C. D., Colorado School of Mines

Originally presented on: 11/5/2013 9:45:00 - 10:10:00

This talk will present four years of action research aimed at developing an Engineering for Sustainable Development (ESD) degree program at a Higher Education Institution known for expertise in mineral extraction, geosciences, materials, and science-technology-engineering-mathematics topics. The action research presented will highlight six stages of research that when incorporated into the education process illustrate the breadth and depth that ESD programs can take. The action research presented continues to evolve and recent outcomes (e.g. a collaborative international summer curriculum) will be presented. This research is a testament that traditional engineering programs can be progressive and make use of holistic concepts and not only reductionist education philosophy or principles.

(280a) Insights Into a Productive Academic Career: A Research Perspective

Grover, M., Georgia Institute of Technology

Originally presented on: 11/5/2013 8:30:00 - 8:55:00

This session is the annual Young Faculty Forum (YFF) for the Education Division of AICHE. In this session, invited speakers will discuss issues of interest to the advancement and promotion of young faculty and those applying for faculty positions. This session includes two presentations on developing a successful academic career along with a few presentations from directors of several divisions within the Chemical, Bioengineering, Environmental, and Transport Systems (ENG / CBET) Division of the National Science Foundation (NSF).

The presentations from faculty members include one from a research perspective and one from an educational perspective. Both of these faculty speakers are established faculty members from US universities. There will be time at the end of the talks for the NSF speakers to answer any additional audience questions in a break out session.

This particular presentation is from a faculty member with insight into developing a successful research program.

(280b) Insights Into a Successful Academic Career: An Educational Perspective

Felder, R. M., North Carolina State University

Originally presented on: 11/5/2013 8:55:00 - 9:20:00

This session is the annual Young Faculty Forum (YFF) for the Education Division of AICHE. In this session, invited speakers will discuss issues of interest to the advancement and promotion of young faculty and those applying for faculty positions. This session includes two presentations on developing a successful academic career along with a few presentations from directors of several divisions within the Chemical, Bioengineering, Environmental, and Transport Systems (ENG / CBET) Division of the National Science Foundation (NSF).

The presentations from faculty members include one from a research perspective and one from an educational perspective. Both of these faculty speakers are established faculty members from US universities. There will be time at the end of the talks for the NSF speakers to answer any additional audience questions in a break out session.

This particular presentation is from a faculty member with insight into developing the educational aspects of professorship.

(280c) Presentations By NSF CBET Program Managers

National Science Foundation, C., National Science Foundation

Originally presented on: 11/5/2013 9:20:00 - 9:45:00

This session is the annual Young Faculty Forum (YFF) for the Education Division of AICHE. In this session, invited speakers will discuss issues of interest to the advancement and promotion of young faculty and those applying for faculty positions. This session includes two presentations on developing a successful academic career along with a few presentations from directors of several divisions within the Chemical, Bioengineering, Environmental, and Transport Systems (ENG / CBET) Division of the National Science Foundation (NSF).

The presentations from faculty members include one from a research perspective and one from an educational perspective. Both of these faculty speakers are established faculty members from US universities. There will be time at the end of the talks for the NSF speakers to answer any additional audience questions in a break out session.

This particular presentation is from a several division leaders in the CBET Division of the National Science Foundation (Speakers # 3-5)

(280d) Break-Out Session With NSF Program Managers
National Science Foundation, C., National Science Foundation
Originally presented on: 11/5/2013 9:45:00 - 10:10:00

This session is the annual Young Faculty Forum (YFF) for the Education Division of AICHE. In this session, invited speakers will discuss issues of interest to the advancement and promotion of young faculty and those applying for faculty positions. This session includes two presentations on developing a successful academic career along with a few presentations from directors of several divisions within the Chemical, Bioengineering, Environmental, and Transport Systems (ENG / CBET) Division of the National Science Foundation (NSF).

The presentations from faculty members include one from a research perspective and one from an educational perspective. Both of these faculty speakers are established faculty members from US universities. There will be time at the end of the talks for the NSF speakers to answer any additional audience questions in a break out session.

This particular presentation is a break out session for audience questions and answers with the NSF CBET Program Managers. (Speakers #3-5)

(282a) Experiences Obtaining Program Assessment Data From Capstone Design Projects

Schlup, J. R., Kansas State University

Originally presented on: 11/5/2013 12:30:00 - 12:55:00

Capstone design courses and related project work should present excellent opportunities for program assessment.Â This is well recognized by many programs as evidenced by the use of capstone design projects in program assessment plans.Â The question then becomes how to acquire the data desired without the process becoming too burdensome.Â During this presentation, a description of the current capstone design course sequence at Kansas State University will be described along with an explanation of how these courses are employed in program assessment. Â The current process evaluates the design projects while simultaneously providing data for assessing student outcomes.Â The rubric utilized in grading capstone design reports will be presented as well as how aspects of these data are subsequently incorporated into the assessment of specific student outcomes.

Obviously programs vary in their approach to assessment and in how their design courses are integrated into their respective assessment activities.Â During the presentation, alternative approaches to obtaining assessment data from design courses will be outlined.Â Example rubrics that have been employed at other institutions will be presented supporting their approach to assessing student outcomes.Â The examples considered will be from diverse institutions and programs to provide a broader perspective and hopefully to stimulate creative approaches meeting the needs of different situations.

The intended outcomes for this presentation are that participants will gain (1) an introduction to several alternate rubrics for design project evaluation and/or student outcome assessment through design projects, (2) a exposure to one example in which a rubric for evaluating engineering design projects has been employed simultaneously for both project evaluation and program assessment, and (3) insight into integration of design project evaluation and program assessment.

(282b) Consistent Professional Skills Development Across Longitudinal Core Courses: Placement Outcomes and Demonstration of ABET Criteria

Blowers, P., The University of Arizona
Sorooshian, A., University of Arizona
Originally presented on: 11/5/2013 12:55:00 - 13:20:00

Students often assume that a strong GPA is enough for generating good offers when they graduate, but employers also desire well rounded candidates who are more than just a GPA. The University of Arizona faculty have implemented a series of longitudinal course assignments that instruct students in other skills that aid in landing early internships and in generating awards throughout their time as students. These assignments are blended with developmental advising across all four, or more, years of student participation in the program. This work describes those activities, their placement, and the successes that result from them.

In the sophomore year, students in the Material and Energy Balance series participate in two rounds of resume development skill building exercises. After an hour long introduction, students turn in one version with their name and another that is anonymous. In the next class meeting, the anonymous ones are used to ask, "Who? What? When? Where? And Why" questions that elicit higher levels and descriptive bullet points and more well formatted unique resumes. Students then turn in an improved resumes and receive feedback from the faculty on areas where they can improve the written text, along with developmental advising about elements of their professional development that are missing or are weak. These areas include the obvious scholarship through GPAs, but also include encouragement for students to engage in leadership, research, and internship opportunities. This feedback occurs immediately prior to the UA fall career fair.

In the sophomore year, the lecture period prior to each of the four midterms is used to practice oral interview skills in class. In each of the four rounds, students are given between 1 and 3 typical resume prompts and then asked to write a detailed verbal answer that is then submitted and evaluated. In the first round in class, students partner with those near them, generally well-known peers, and then answer the prompts orally and are evaluated by their peer on their ability to sell skills explicitly, be specific, be positive, quantify, make eye contact, and use natural non-verbal communication skills. In round two, students are paired randomly and answer one prompt they had submitted in written form and one new prompt. In round three, students are again randomized in pairs and now answer two prompts they have not prepared for in written form, but that get at the underlying skills. In round four, students are randomized and have to answer typical off-the-wall questions interviews ask and are given the reasoning behind why companies use these and how to handle them.

In the junior year, students again are required to turn in a resume and it is more critically evaluated. Again, developmental advising is used to guide students into leveraging their prior experiences into higher levels of engagement, or to get them started if they have not taken initiative yet. Clear communication describes the typical student trajectory of those who have not begun activities. Written and verbal oral practice in interviewing questions again takes place in the lecture prior to each midterm so students can continue to assimilate the material on the exam without trying to learn new material at the same time. The resume development takes place prior to the spring career fair, while the oral interview questions are spaced throughout interview season for summer internships.

In the senior year, students again submit resumes and receive final advice prior to the fall career fair. In some years, if timing allows, students are randomly picked to verbally answer a new prompt in front of the class and then the class gives constructive criticism and positive feedback. The randomization is done by drawing cards or rolling dice in the first 3 minutes of class and then the student giving their answer in front of the class. This activity was anticipated to be a high stress situation, but students have responded very favorably and comment that their real interviews are much less stressful because they were only with one or two interviewers and not 40-60.

Early placement into NSF REU programs and internships is strong, while graduating senior placement is even stronger. It is typical that 35% of our graduate go into PhD and MS programs with full funding. Approximately 10-15% of each class signs full time offers in early October for May start dates. 15-20% of students are still looking for employment opportunities in the last month before graduation, with only a few students not having offers prior to graduation. Student feedback during senior exit interviews shows the students strongly value the constant and positive support in developing areas that are often neglected in engineering core courses.

(282c) A Regimen to Zero Findings

Rockstraw, D. A., New Mexico State University
Mitchell, M. C., New Mexico State University
Originally presented on: 11/5/2013 13:20:00 - 13:45:00

The Chemical Engineering Department at New Mexico State University completed an ABET review in October 2012 with zero findings for the second consecutive time. Strengths of the programs identified in the Draft Statement of Accreditation are summarized as (1) the alumni, (2) the students, and (3) the faculty. This discussion will highlight the continuous improvement and management of change processes employed in the program, as well as the constituency relationships and the distribution of responsibilities that have led to a strong environment of cooperative learning, peer-assessment, and self-assessment. Program assessment has become an inherent feature and an accepted routine within the Chemical Engineering program that is regularly discussed and clearly communicated among the constituency. During the 2012 review, the NMSU Chemical Engineering program identified a method to streamline the collection of assessment data concerning the meeting of course objectives. Since the fall 2012 review, the department has designed and constructed a relational database that permits the pre-planning of each assessment measurement, as well as provides a highly efficient means of collecting and reporting the assessment data.

(282d) Developing Quantifiable Assessments in Advance of An ABET Visit: University of Connecticut

Mustain, W. E., University of Connecticut
Burkey, D. D., University of Connecticut
Originally presented on: 11/5/2013 13:45:00 - 14:10:00

In advance of UConn's planned ABET visit in the fall of 2013, the university held a mock review in the spring of 2012. Data collection was carried out as it had been previously, and a draft report compiled such that the mock evaluation team could provide feedback in advance of the actual report. While the chemical engineering program received a six-year accreditation during the 2006-2007 academic year, the mock evaluators pointed out that most of the assessment being carried out was subjective in nature with an over-reliance on survey data and self-reporting of performance by students and faculty.

In response to this criticism, the departmental ABET committee examined the best practices as seen in various presentations of the AIChE Education Division, as well as the ASEE Chemical Engineering Division, and developed a new, quantitative system for performing assessment and providing a formal method for continuous improvement at the course and program level. At the program level, the ABET a-k criteria are mapped to various courses, such that all of the a-k outcomes are covered by the curriculum. This mapping is reviewed on a yearly basis at the faculty retreat. At the course level, faculty are now mentored on how to write student outcomes for their class that can be quantitatively assessed by mapping them to various problems and assignments that demonstrate competence in these areas. Additionally, the outcomes are mapped to the ABET a-k criteria that were identified for that class at the retreat. Thus, we are in the process of shifting to an outcomes-based curriculum, in which all assignments can be mapped to a desired outcome for the course, and the student attainment of the outcomes can be ascertained through the grading of these assignments. Faculty were also mentored in rubric generation, such that assignments can be fairly and uniformly evaluated should the instructor change over time. Graduate teaching assistants were employed to help document and implement the system, such as to reduce the faculty burden on data collection once the initial selection of problems and mapping were completed.

Thus, after the initial investment in developing assessable student outcomes and performing the initial mapping of outcomes to problems and a-k criteria, a quantitative assessment of student performance was obtained that not only addressed the previous criticism of a largely subjective assessment of student learning, but was also much more useful in identifying specific problem areas within individual courses that the faculty could address in future offerings, directly addressing the concept of continuous improvement. Under discussion for future implementation are focus groups in the core areas (transport, thermodynamics, and reaction engineering) to evaluate the courses in those areas and find synergies, identify and deploy best practices, mentor new faculty, and provide a further mechanism for continuous improvement.

(282e) ABET Impressions: Experiences During the First International Accreditation Process

Medellin-Rodriguez, F. J.,
Alejo Gonzolez, M. G., Universidad Autonoma de San Luis Potosi
Cervantes Nino, S. E., Universidad Autonoma de San Luis Potosi
Montante Montelongo, A. D.,
Roman-Martinez, A., Technical University of Denmark
Originally presented on: 11/5/2013 14:10:00 - 14:35:00

In this presentation we aim to share our experiences and lessons learned during the first ABET accreditation process for the educational program in Chemical Engineering of our institution in Mexico.

In our country, one of the main goals to achieve in undergraduate programs is to formalize an international evaluation culture. The ABET accreditation process has become the milestone of a long-term process that can involve all educational actors in the dynamics of continuous review and self-analysis in order to improve educational processes in chemical engineering programs. Therefore, the international accreditation goes beyond meeting a number of requirements in order that the educational program is accepted.

The international accrediting agency ABET has clear, precise and open criteria, which allow us to pose the reality of the institution in terms of what has been achieved so far and what is to be achieved in the medium and long term. Therefore, the referents to accreditation, not only are the external criteria of accrediting agency, but the historical, academic and cultural referents of our institution pursuing the international accreditation. From this, the necessary adjustments to improve the program arise.

In this sense, the impact of international accreditation institution can be evidenced in two dimensions, one internal, which is more immediate to identify and the other external, which can be identified and assessed in the medium and long term. Both dimensions are not mutually exclusive, but rather, are nurtured in a continuous process, through ongoing evaluation mechanisms that have arisen during the process of international accreditation.

During this presentation, we will expose first the reasons that led us to participate in the international accreditation process, then, we will explain our identification process of the so-called "outcomes", to continue with the presentation of our assessment process, as well as our experience with ABET evaluators, to conclude with the results of the assessment and action plans for continuous improvement in the short, medium and long term.

(336a) John Prausnitz, Humanist and Epistolarian

Debenedetti, P. G., Princeton University

Originally presented on: 11/5/2013 12:35:00 - 12:55:00

John Prausnitz's towering influence on our profession has helped shape modern Chemical Engineering. Undoubtedly this influence can be traced in large part to his creation of the discipline known as molecular thermodynamics, which has enabled engineers to make accurate predictions of important physical properties through a judicious blend of intuition and rigor. Remarkably, John has also shaped the careers of many practitioners of Chemical Engineering by reminding us of the profoundly human dimensions of our craft and discipline. He has done this through his scholarly writings, his formal speeches, and, most notably, his personal correspondence with countless colleagues, disciples and former students. Drawing on examples from his papers and letters, I will pay tribute to John Prausnitz the eloquent humanist and virtuoso epistolarian, whose insatiable intellectual curiosity has enabled him, in the course of a remarkable career spanning more than half a century, to enlighten, inspire, and educate an entire profession.

(336b) Berkeley in the Early 60's

Eckert, C. A., Georgia Institute of Technology

Originally presented on: 11/5/2013 12:55:00 - 13:15:00

First of all I was a maverick; I had done two degrees at MIT, and everyone who was able to do so remained there for their PhD. My advisor and department head, Ed Gilliland told me I was crazy to leave MIT and that Berkeley was "no good." But Tom Sherwood told me about a hot young faculty member there -- John Prausnitz. I had no intention of doing thermodynamics, but his energy, enthusiasm, and creativity seduced me. Thermodynamics was dull rote stuff that would get properties from other properties. But Molecular Thermodynamics could get something from nothing, using "enlightened empiricism."

Really what appealed to me most is that it seemed that every day John came into the lab with five new ideas. He was so prolific that we, his students, formulated a defense, and that was to have more new ideas. I don't believe that at the time we realized that that was his plan all along, but it worked and we opened lots of new doors. Another technique he used was to put his students into competition with each other. He would tell me that another student was really doing better than I was, so I had better work harder. Eventually I compared notes with the other student, who said he was being told that I was surpassing him. So we knew what was going on, but it it elevated us both.

John's great skill was motivating students and colleagues to achieve important results. For example, he was the antithesis of an experimentalist, but he knew what experiments could do, and he knew to whom to send us to work out the details. For example, we were using IR at the time, ages before the advent of Fourier transform. There was no quantitative chromatograph commercially available, so we built one and used it successfully.

Another case was computers. Fortran had just come out and it seemed that a whole class of new problems had become tractable. But John would not even look at computer output – we had to make graphs. The department head, Don Hanson, came up with a plot to get John computer-active, and outlined the possibility of setting up computer programs for calculating separations. Several of us worked on it, and the upshot was that we came up with what was really the first monograph on applications to separations, full of computer programs. John coauthored the book and made very substantial contributions to the text, but still never looked at the output.

One of the outstanding applications of computers was the opportunity to use novel excess Gibbs energy expressions based on entropy more than enthalpy. These really quite inaccessible with slide rules, but facile with Fortran. At one point George Scatchard from MIT came to Berkeley and gave a talk of such an equation, developed with his student Wilson. It had not yet been published (and would not be published until some years later), but John's students copied the form and started using it. Soon they had demonstrated clearly its superiority to the enthalpic VLE models such as Margules and van Laar, and they published the results (with attribution) before Wilson actually published what we now know as Wilson's Equation. This led to improved entropic equations still used, such as NRTL and UNIQUAC, but still without John becoming computer-active.

John did always want to learn. I remember when he decided to learn statistical mechanics. He assigned three of his students to go learn about it and teach him. One took the course in Chemistry, one in Physics, and I took the Mechanical Engineering course from the youngest (and youngest-looking) professor at Berkeley, Prof Tien. He was a wonderful teacher, and later became the first nonwhite Chancellor of Berkeley. But then I digress; the three of us started a class for John and the rest of the group, and he went on to use the results most powerfully.

One of the most striking things I learned about John early on was his ability to communicate to a group. Other seminar speakers would inundate the audience in equations attempting to be paragons of elegance. John was different – I always emerged from his presentations saying "My goodness that was easy. I wish I had thought of that." He did not try to appear elegant; rather he solved important problems simply.

I went to Berkeley thinking I might want to be a university professor, and by his example and guidance, John intensified that attitude. He has stayed in touch with me, and with his other academic progeny, for now nearly half a century, always challenging and elevating us.

(336c) Fluctuation Solution Theory and Messages Heard From A Master: Reprise

O'Connell, J. P., University of Virginia

Originally presented on: 11/5/2013 13:15:00 - 13:35:00

The objective of physical properties modeling is to judiciously use fundamentals, data, approximation, computation, and generalization to optimize descriptions of chemical systems, especially for process design. John Prausnitz masterfully devised, demonstrated, applied, and taught - to generations of students and professionals - some extraordinarily successful strategies for comprehending and treating an incredible richness of important systems, substances, and conditions. Further, he has shown their value in many spheres of intellectual work and human endeavor.

Some time ago, I presented a talk that shared my appreciation of John's influence on my 3 decades of research on extending and applying fluctuation solution theory, the molecular theory of solutions due to Kirkwood and Buff. I phrased a set of "messages" that I could imagine hearing from him about how to proceed at moments of decision along the way. In the last 15 years, our work has made further progress that was aided by more messages I heard from within me because of John's impact.

The talk will reprise the original theme and format, describe what we have achieved in recent years, and articulate additional messages.

(336d) Prausnitizian Phase Equilibria in Soft-Contact-Lens Copolymer Hydrogels

Radke, C. J., University of California Berkeley
Prausnitz, J., University of California at Berkeley
Originally presented on: 11/5/2013 13:35:00 - 13:55:00

Prausnitizian Phase Equilibria in Soft-Contact-Lens Copolymer Hydrogels

C.J. Radke[1,2] and J.M. Prausnitz[1]

[1] Chemical and Biomolecular Engineering Department, University of California, Berkeley, CA, 94720-1462

[2] Vision Science Group, University of California, Berkeley, CA, 94720-2002

85th Birthday Celebration of J. M. Prausnitz

Thermodynamics and Transport Properties

AIChE San Francisco, 2013

Nearly 50 % of first-time SCL wearers drop out because of discomfort. Wear comfort increases with increasing lens water content making water saturation a critical variable. John pioneered the language to tackle such problems and continues to apply them toward solution of practical problems. We report studies with undergraduate students on the swelling of 2-hydroxyethyl methacrylate (HEMA)/methacrylic acid (MAA) hydrogels over the complete range of MAA content. Gel water content strongly increases near the pK of MAA. At low pH, water content is insensitive to the MAA/HEMA ratio, whereas at high pH, small amounts of MAA increase water saturation dramatically, but less so as MAA copolymer fraction increases. Application of multipolymer Flory-Huggins-Rehner theory proves inadequate as do Donnan electrostatics. We propose specific complexation of water with charged acid groups of MAA along with charge-dependent ion-binding of salt counterions. John's appetitive for mentoring students through insightful research remains alive and well at Berkeley.

(336e) The Chemical Engineers' Role in Energy and Economic Security
Trainham, J.,
Originally presented on: 11/5/2013 13:55:00 - 14:15:00

The Chemical Engineers' Role in Energy and Economic Security

James A. Trainham

RTI International and NC State University

Transportation—not power generation—is the real energy challenge.

Energy Independence for the U.S. is really North American Energy Independence because Canada, Mexico and the U.S. are interconnected, linked by geography, geology, infrastructure and economics. Together we can detox North America from OPEC's monopoly.

Until now no substitutes for oil existed. Unconventional natural gas (UG) reserves are plentiful and increasing. The opportunity is for UG to diversify the fuel choices. UG will change the geopolitics of the world as UG recovery and use technology, applied globally, substantially reduce oil dependencies.

Chemical engineers must lead this migration to UG, reducing the investment intensity of UG based fuels in the short-term and in the longer-term reduce the carbon footprint of UG usage..

These circumstances and the leadership legacy of Professor Prausnitz demand that we chemical engineers develop the fundamental understanding to make the new technologies economic reality and eliminate transition barriers.

(336f) How to Transform a Field of Study and Define a Discipline

de Pablo, J. J., University of Wisconsin-Madison

Originally presented on: 11/5/2013 14:30:00 - 14:50:00

In a career spanning over five decades John Prausnitz transformed chemical engineering. Through his pioneering research efforts and visionary studies, the field of molecular thermodynamics gradually evolved into a discipline where some of the most fundamental experimental observations and theoretical constructs found an outlet in engineering practice. Indeed, one of John's lasting imprints on chemical engineering and on his students' psyche has been that, to the extent possible, research should lead to solution of important problems. That evolution continues today, and builds on John Prausnitz's text books and research monographs, his nearly 700 papers, many of them seminal, and those of hundreds of students and collaborators around the world that have been influenced by his ideas. In this presentation I will describe several recent, exciting developments in chemical engineering that can be traced back to John Prausnitz's original contributions and those of his co-workers, and I will attempt to anticipate how those developments might shape engineering for years to come.

(336g) Nature-Inspired Chemical Engineering â€“ Pathways to Innovation and Sustainability

Coppens, M. O., Rensselaer Polytechnic Institute

Originally presented on: 11/5/2013 14:50:00 - 15:10:00

Some of the most urgent challenges facing us today are related to energy, lack of fresh water, and how to translate discoveries at the nanoscale to macroscopically relevant scales for practical applications, as in scalable manufacturing.

My presentation will give a brief overview of our nature-inspired chemical engineering (NICE) approach, which addresses these challenges by a rigorous approach that takes guidance from Nature. In particular, we seek to discover fundamental mechanisms (rather than imitation out of context) behind desirable traits in natural systems – like scalability, robustness and efficiency – and use this insight to guide the design and synthesis of engineered systems that have the same properties.

Particularly promising for innovation are three fundamental and ubiquitous natural mechanisms, which we see as most timely for exploration: hierarchical transport networks (such as in lungs and trees), force balancing (e.g., steric vs. electrostatic effects), and dynamic self-organisation (as in regular pattern formation induced by perturbations, and emergence of robust properties as a result of collective phenomena behind much simpler entities).

The NICE methodology allows us to discover innovative solutions to technological problems. In my presentation, I will illustrate this thematic approach via a few examples, e.g., how the structure of trees and leaves guides the design of fractal fluid distributors and hierarchically structured catalysts; how chaperones present us with ideas on how to stabilize enzymes for therapeutics; and how bacterial communities teach us lessons on how to design adaptive, robust materials.

Taking ideas from biology to architecture, and cognizant of the context in product and process design as a boundary condition to engineering practical solutions, nature-inspired chemical engineering brings the humanities back to engineering.

(336h) A Variational Field Theory for Ion/Charge Correlations in Electrolyte Systems

Lue, L., University of Strathclyde
Originally presented on: 11/5/2013 15:10:00 - 15:30:00

Many biological systems and industrially important formulations consist of solutions containing highly charged colloidal particles. For these solutions, electrostatic and dispersion forces typically play a major role in determining their structure and thermodynamic properties. Much of our understanding of the role of electrostatic interactions in these systems is based on Poisson-Boltzmann theory which, although quite successful, has many well documented limitations, such as its failure for systems with multivalent ions where ion-ion correlations are significant. In this work, a variational field theory approach is developed, which removes many of these limitations.

The key physical motivation behind the theory is to treat the short and long wavelength fluctuations in the system within different approximation schemes. At short range, the systems are strongly coupled, and at long range the systems are weakly coupled. To describe both these regimes within a single theory, we split the interaction into a short and long range contribution. The long range behavior is often well approximated by a mean field theory, or including first order fluctuation corrections, while the short range behavior can be captured by a virial expansion, or other liquid state methods suitable to describe particles with short range pair interactions.

This approach is able to properly account for the many-body effects of the electrostatic interactions, even for strongly charged particles, and it also naturally handles the coupling between electrostatic, induction, and dispersion interactions. For weakly coupled systems, this theory approaches the Poisson-Boltzmann theory, while for strongly coupled systems, the theory resembles the strong coupling expansion. The theory also performs well for intermediate couplings. In addition, the accuracy of the theory can, in principle, be systematically improved.

The theory can be applied to a wide variety of problems, such as systems with different geometries and conditions (e.g., dielectric interfaces) and particles with different shapes and charge distributions. We examine its application to determine the effective interaction between charged, low dielectric particles in electrolyte solutions. This fairly simple system exhibits a range of interesting phenomena, such as like-charge attraction and electrostatic depletion, that can not be captured with mean field approximations. The resulting theory is shown to gives accurate predictions in comparison to Monte Carlo simulation results.

(336i) Becoming a Prausnitzian

Wu, J., Delft University of Technology

Originally presented on: 11/5/2013 15:30:00 - 15:50:00

Becoming a Prausnitzian

Jianzhong Wu

Department of Chemical and Environmental Engineering and Department of Mathematics

University of California, Riverside, California 92521

On several occasions after a technical presentation someone would come to tell me in private that my talk sounds very Prausnitzian. I take it as a great compliment and enjoys a sense of pride for being affiliated with the big name in chemical engineering. I once shared this experience with the master. While he was apparently pleased with the comment, he told me that he was not quite sure whether it means anything because a Berkeley professor always has a strong influence on his/her graduate students. I worked with John Prausnitz first as a visiting fellow then as a graduate student and also as a postdoctoral researcher. His influence on my personal and professional developments has been enormous. I will share some of that experience in this symposium to celebrate his Life Time Accomplishment Award from the American Society of Engineering Education (ASEE).

(336j) Prausnitz Talks About Prausnitz

Prausnitz, M. R., Georgia Institute of Technology
Originally presented on: 11/5/2013 15:50:00 - 16:00:00

Prausnitz Talks About Prausnitz

Mark R. Prausnitz, Chemical and Biomolecular Engineering, Biomedical Engineering, Georgia Institute of Technology, Atlanta, GA

This talk will highlight a series of professional accomplishments in the career of John Prausnitz and describe their impact and their context from the perspective of a family member using anecdotes and photos.

(366a) Determinants of Calcium Reabsorption in the Rat Kidney

Edwards, A.,

Originally presented on: 11/5/2013 15:20:00 - 15:38:00

We have developed a mathematical model of calcium (Ca^{2+}) transport along key segments of the rat nephron to investigate the factors that promote hypercalciuria (i.e., abnormally high levels of calcium in the urine), one of the leading causes of kidney stone formation. The reabsorption of Ca^{2+} occurs passively in the thick ascending limb, and its urinary excretion is finely adjusted downstream, in the distal convoluted (DCT) and connecting (CNT) tubules. Our model represents the detailed epithelial transport of Ca^{2+} and other solutes in these segments; it accounts for the specific apical and basolateral transporters of each cell type. Solute concentrations, volume, pH, and electric potential in the tubular lumen and the epithelial cells are determined, as a function of position along the tubule, by solving steady-state conservation equations coupled to flux expressions. Model simulations reproduce experimentally observed variations in cellular Ca^{2+} uptake as a function of extracellular pH and Na^+ concentrations. The model predicts complex interactions between the reabsorption of Ca^{2+} and that of other solutes in the nephron; in particular, urinary Ca^{2+} excretion appears to be strongly modulated by the transport of NaCl. Our results also suggest that the DCT and CNT can adapt to upstream variations in Ca^{2+} transport, but not always sufficiently to prevent hypercalciuria.

(366b) Effects of Electrostatic and Intermolecular Interactions On Ultrafiltration of Pegylated Proteins

Zydney, A. L., The Pennsylvania State University
Ruanjaikaen, K., Pennsylvania State University
Originally presented on: 11/5/2013 15:38:00 - 15:56:00

It is now well-established that protein transmission through ultrafiltration membranes is strongly effected by electrostatic interactions, with the magnitude of these interactions very effectively described using the partitioning model developed by Bill Deen and co-workers. This presentation will examine the extension of this theoretical framework to describe the ultrafiltration behavior of pegylated proteins produced by covalent attachment of a polyethylene glycol chain to a native protein. Experimental data were obtained for the transmission of pegylated α-lactalbumin through both unmodified and negatively charged composite regenerated cellulose membranes over a range of solution ionic strength and PEG concentrations. The attachment of the PEG altered the protein charge, the effective size, and the electrostatic potential surrounding the protein. The pegylated α-lactalbumin was strongly retained by the negatively charged membrane due to the significant electrostatic interactions in this system. The transmission of the pegylated protein also increased with increasing PEG concentration due to the increase in the solute partition coefficient arising from unfavorable intermolecular interactions in the bulk solution. A simple model was developed to account for the effects of the electrostatic and intermolecular interactions on the ultrafiltration of the pegylated proteins based on an extension of the classical Smith and Deen analysis. The pegylated protein was treated as an effective hard sphere with the surface charge shifted to the outer edge of the PEG by accounting for the alteration in the electrostatic potential associated with the reduced electrolyte concentration within the PEG layer. The experimental results were in good agreement with model calculations, providing a framework for analyzing the behavior of pegylated proteins in batch ultrafiltration and diafiltration processes.

(366c) Problems in Dermal Absorption As Vehicles for Teaching Diffusion and Mass Transfer

Nitsche, J. M., University at Buffalo, The State University at New York

Originally presented on: 11/5/2013 15:56:00 - 16:14:00

Quantitative understanding of molecular diffusion through the skin is crucial to the effective development of topically applied and transdermally delivered drugs, as well as the assessment of risks associated with chemical exposures. Theoretical models of dermal absorption[1] can strongly support advancement in these areas by providing mechanistic insights into solute flux pathways, reasonable predictions of transdermal transport rates and subsurface concentration levels, and a quantitative framework to correlate experimental data and guide measurements. The strong barrier properties of skin (physiologically advantageous) are precisely what make it hard to effect transdermal delivery. So difficult is the challenge that some authors ask whether breaching the barrier is effectively holding up a crucially important avenue of pharmaceutical development, i.e., "is delivery technology keeping up?"[2] Assessments for contact allergy and other elements of dermal risk must go forward in the face of increasing prohibitions on animal testing.[3] Fundamental predictive understanding of molecular diffusion through the multiphase, multiscale architecture of skin lies at the core of meeting these challenges, and it represents an area where chemical engineers can make important contributions to the pharmaceutical enterprise.

More or less idealized presentations of dermal absorption problems present good opportunities to teach a variety of aspects of mass transfer in a context that makes their practical significance very tangible. Among them are transient diffusion through effectively infinite- and finite-thickness membranes, hindered diffusion through media with fibrous and other types of microstructure, homogenization/coarse-graining of transport equations, partitioning equilibria at phase boundaries, electrostatic effects on transport, and techniques for solving differential equations. These areas are ones in which Prof. William Deen has made seminal advances in research, both on fundamental problems and applications to renal physiology, and has empowered several academic generations of grateful students as a classroom teacher and academic mentor.

This talk describes the use of dermal absorption problems as teaching vehicles in undergraduate and graduate transport courses at the University at Buffalo. It presents a series of examples of increasing complexity, emphasizing mechanistic understanding and the value of analytical methods. Important lessons learned from Prof. Deen are identified for each.

1. Mitragotri S, Anissimov YG, Bunge AL, Frasch HF, Guy RH, Hadgraft J, Kasting GB, Lane ME, Roberts MS 2011. Mathematical models of skin permeability: an overview. *Int J Pharm* **418**(1):115-129.

2. Cross SE, Roberts MS 2004. Physical enhancement of transdermal drug application: is delivery technology keeping up with pharmaceutical development? *Curr Drug Deliv* **1**(1):81–92.

3. Basketter D, Pease C, Kasting G, Kimber I, Casati S, Cronin M, Diembeck W, Gerberick F, Hadgraft J, Hartung T, Marty JP, Nikolaidis E, Patlewicz G, Roberts D, Roggen E, Rovida C, van de Sandt J 2007. Skin sensitisation and epidermal disposition: The relevance of epidermal disposition for sensitisation hazard identification and risk assessment - The report and recommendations of ECVAM Workshop 59. *Altern Lab Anim* **35**(1):137-154.

(366d) Modeling of Mass Transport in Semiconductor Nanocrystals and Single-Layer Epitaxial Islands On Crystalline Substrates

Maroudas, D., University of Massachusetts Amherst
Originally presented on: 11/5/2013 16:14:00 - 16:32:00

Species transport underlies a number of important assembly processes in electronic materials used in the fabrication of microelectronic, optoelectronic, and photovoltaic devices. In this presentation, we focus on the analysis of such species transport and its impact on the synthesis of semiconductor quantum dots with optimal electronic function and on the external-field-enabled nanopatterning of crystalline substrate surfaces.

First, we address the optimization of the function of quantum dots (QDs) of ternary compound semiconductors, such as $In_xGa_{1-x}As$, $ZnSe_{1-x}S_x$, and $ZnSe_{1-x}Te_x$, by precise tuning of their electronic band gap through control of the QD composition (x). We report results on compositional distributions in ternary QDs and how they affect the QDs' electronic band gap. We follow a hierarchical modeling approach that combines first-principles density functional theory (DFT) calculations and classical Monte Carlo simulations with a continuum model of species transport in spherical nanocrystals. In certain cases, the model predicts the formation of a concentration boundary layer near the nanocrystal surface, i.e., the formation of core/shell-like structures with shell regions rich in the surface segregating species. A systematic parametric analysis generates a database of transport properties, which can be used to design post-growth thermal annealing processes that enable the development of thermodynamically stable QDs with optimal electronic properties grown through simple one-step colloidal synthesis techniques.

Next, we analyze the current driven morphological evolution of single-layer epitaxial islands on crystalline solid substrates. We develop and validate a continuum model for the island current-driven dynamics. Mass transport is dominated by adatom diffusion on the island edge; electromigration, misfit-strain-driven diffusion, and edge diffusional anisotropy are accounted for by the model. Simulations based on the model show that the dependence of the stable steady island migration speed v_m on the inverse of the island size is not linear for larger-than-critical island sizes. In this nonlinear regime, we report morphological transitions, Hopf bifurcations, and instabilities for various surface crystallographic orientations and island misfit strains. Proper rescaling of v_m gives a universal linear relationship for its dependence on island size. Our findings set the stage for designing numerical simulations and experiments of the current-driven evolution of epitaxial island populations and for exploring the collective island dynamical phenomena that govern pattern formation on the crystalline surface.

(366e) Two reaction-transport problems for Bill Deen

Shvartsman, S. Y., Princeton University

Originally presented on: 11/5/2013 16:32:00 - 16:50:00

Bill Deen's work as an educator and scientist provides a great example for anyone interested in engineering analysis of complex systems. His remarkable ability to identify important problems, from transport in kidneys to Nitric Oxide dynamics in cells, and provide solutions with clarity fitting for teaching in graduate and undergraduate courses is a hard act to match. The elegance of his research can be approached only asymptotically. I will present two examples from our recent work in this direction. Both problems are related to mechanisms of cell communication and lead to reaction-diffusion problems that are difficult to solve at first glance. In the first of these problems, which comes from ligand binding to cell surface receptors, one must solve reaction-diffusion PDEs with fine-grained boundary conditions. In the second problem, related to tissue patterning by chemical gradients, one is interested in low-dimensional approximation of concentration dynamics. In both of these cases, we could derive simple analytical results that relate quantities of interest to parameters of the original problems, very much in the style of Chapter 3 of Bill's classical textbook.

(366f) The Effect of Solute-Solute Interactions On Diffusion in Gels and Micelle Solutions

Phillips, R. J., University of California, Davis

Originally presented on: 11/5/2013 16:50:00 - 17:08:00

Many controlled release systems in food and medical applications, as well as bioseparation methods, occur under conditions where solute-solute interactions have a significant impact on rates of diffusion. To improve the current understanding of these processes, a rigorous application of the principles of fluid mechanics and mass transfer is essential. In polymer gels such as agarose, the presence of gel fibers screens hydrodynamic interactions between solutes, while enhancing non-hydrodynamic interactions because of volume exclusion. The result is in increase in the effect of concentration on rates of diffusion, relative to what is present in gel-free colloidal solutions. As a second system, we also consider the diffusion of hydrophobic solutes in micelle solutions. In such solutions, multicomponent effects are not negligible: a gradient in surfactant concentration can induce or even eliminate fluxes in the hydrophobic solute.

(366g) Intracellular Time and Length Scales in Receptor-Mediated Cell Signaling

Lazzara, M. J., University of Pennsylvania

Originally presented on: 11/5/2013 17:08:00 - 17:26:00

In response to upstream stimuli including the activation of receptor tyrosine kinases, protein complexes are formed that are involved in the process of cellular signal transduction. In many cases, the linkages between proteins in these complexes are formed by reversible binding interactions between phosphorylated tyrosines and cognate Src homology 2 (SH2) domains. Given that phosphotyrosines on receptors and intracellular signaling proteins are under constant regulation by protein tyrosine phosphatases (PTPs), it seems likely that PTPs could exert substantial control over the persistence of signaling complexes in time and space, but this has not been previously explored. Here, through implementation of a combined experimental and computational approach, we demonstrate the validity of this hypothesis. Specifically, we show that linear, chain-like protein complexes nucleated by the activation of the epidermal growth factor receptor (EGFR) and held together by phosphotyrosine-SH2 interactions are highly transient species that fall apart and reform many times during the overall time scale of signal transduction, with the linkages in the chain regulated by PTPs. We use our model to quantitatively identify the time and length scales over which signaling complexes can remain in complex in response to EGFR activation, and identify the activity of PTPs as the most important determinant of these time and length scales through sensitivity analysis. This new view of protein complex formation and regulation by PTPs has important implications for our fundamental understanding of signal transduction dynamics under normal conditions and for our understanding of how signaling may be perturbed in diseases characterized by oncogenic kinase mutations or perturbations to PTP expression or activity.

(418a) Department Heads Forum

Dickinson, R., University of Florida
Lalvani, S., Miami University
Originally presented on: 11/6/2013 8:30:00 - 8:52:00

This session is the annual meeting of chemical engineering department heads and chairs. Department chairs will share their experiences on important topics relevant to leading chemical engineering departments. This year's topic will include trends in chemical engineering education including online education, creating a culture of safety, and managing a large-enrollment masters program. In addition, a representative from NSF will discuss funding opportunities, and annual reports will be provided by CACHE and on the chemical engineering faculty academic salary survey.

AGENDA

"Creating and Managing a Large-Enrollment Masters Program in Chemical Engineering", Richard B. Dickinson, Professor & Chair, Chemical Engineering, University of Florida. (8:30-8:50)

Open round table discussion: Enrollment trends and the Future of ChemE education. Shashi B. Lalvani, Professor and Chair, Chemical and Paper Engineering, Miami University (8:50-9:20)

"Creating a Culture of Safety", Al Sacco, Jr., Dean, Whitacre College of Engineering, Texas Tech University (9:20-9:40)

Break (9:40-9:50)

ABET Report - Jeffrey J. Siirola, Department of Chemical Engineering, Carnegie Mellon University (9:50-10:10)

NSF Funding Report. Pramod Khargonekar, Assistant Director for the Directorate of Engineering, National Science Foundation (10:10-10:30)

Chemical Engineering Faculty Salary Survey –Geoffrey L. Price, Professor and Chairman, Department of Chemical Engineering, University of Tulsa (10:30-10:45)

CACHE Report - Joseph T. Golab, PhD, INEOS Technologies (10:45-11:00)

(428a) Learning By Doing: Students, Experimentation, and Process Development

Worstell, J. H., Worstell and Worstell, Consultants
Arredondo, J., University of Houston
Originally presented on: 11/6/2013 8:30:00 - 8:51:00

Every chemical engineer will design experiments and evaluate experimental data at some point during their career. Thus, it is important for chemical engineering students to have a meaningful experience in their Senior laboratory course. Unfortunately, during a one semester course, there is hardly time to design, build, and test an experiment. At the University of Houston, we encourage chemical engineering students to take an Independent Study course during their rising Senior summer. For that course, they are given an experiment to research, design, build, punch, and commission. If successful, the experiment is incorporated into the Junior or Senior laboratory sessions. Not only do these summer students learn experimentation from the grass roots upward, but they get a feel for process development, since what they do during the summer is a miniature process development project.

This presentation discusses the positive and negative aspect of this concept, a concept that provides evergreen experiments for the Junior and Senior laboratory sessions, detailed experience at experimental design, and experience at process development.

(428b) Use of Pre-Laboratory Orientation and Safety Videos for the Landolt Iodine Clock Reaction Experiment in Undergraduate Reaction Engineering

Wenzel, J. E., Kettering University
Beck, B.,
Rodgerson, T.,
Seeley, S.,
Originally presented on: 11/6/2013 8:51:00 - 9:12:00

In order to better prepare students for the chemical engineering laboratories at Kettering University, pre-laboratory videos were prepared for several experiments for the Unit Operations, Mass Transfer and Separations, and Reaction Engineering Laboratory Courses. These videos are provided to students on Blackboard in three segments: an introduction of the background and principles of the experiment, a safety orientation, and a demonstration of the experimental procedure. Each segment is between two and ten minutes in length. The videos supplement the written laboratory manual and broaden the scope of a "flipped classroom" to include a "flipped lab." This reduces pre-laboratory orientation time since Chemical Engineering laboratories at Kettering University are allotted two hours and provides students the opportunity to view the demonstration multiple times and enter the lab environment feeling confident of the expected procedure. Prior to performing the experiment, the students are orally quizzed. For this paper the use of laboratory orientation videos are discussed for an experiment involving the Landolt Iodine Clock Reaction, an introductory experiment regressing activation energy, frequency factor, and reaction order for the Chemical Engineering Reaction Engineering Laboratory course. Students were surveyed prior to and following the laboratory and these results and course outcomes will be discussed.

(428c) Initiated Chemical Vapor Deposition (iCVD) As a New Take On a Kinetics Lab

Burkey, D. D., University of Connecticut
Anastasio, D., University of Connecticut
Suresh, A., University of Connecticut
Originally presented on: 11/6/2013 9:12:00 - 9:33:00

In this work, we describe the introduction of an initiated chemical vapor deposition reactor into the senior capstone laboratory at the University of Connecticut. Initiated chemical vapor deposition is a relatively new technique in which an initiator and a monomer species are volatilized and fed into a reactor at low pressure, where the initiator is activated (usually thermally or photolytically) and begins a polymerization reaction with the monomer. The thin films so produced are typically uniform and conformal, and can be broadly functionalized by varying the identity of the monomer species. These thin polymeric films find broad use in the semiconductor, biomedical, and specialty coating industries.

UConn purchased a turn-key iCVD reactor from GVD Corporation (Cambridge, MA), though with the proper expertise and facilities, it is very possible to fabricate a home-grown version of this system. The experiment was deployed for the first time in the fall of 2011 as a three-day experiment, and then expanded to a six-day experiment in the spring of 2012. Students are charged with exploring the reaction mechanism of one of a series of alkyl acrylate monomers, as well as determining kinetic parameters for their selected monomer. This experiment requires them to combine knowledge of equilibrium thermodynamics, mass transfer, polymer chemistry, and reaction kinetics to successfully complete the lab. In the fall of 2012, we asked the student groups working on the lab to each study a different monomer. They were able, with very little guidance, to reproduce the kinetic data produced via iCVD published in literature and make comparisons to their own data. Groups uniformly enjoyed the experiment and felt it represented a good addition to the lab as an example of modern chemical engineering technology.

(428d) Automated Process Control Laboratory Experience: Simultaneous Temperature and Level Control in a Continuously Stirred Tank Reactor System

Levinson, J. A., Lafayette College
Tennyson, G. T., Lafayette College
Maase, E. L., University of Massachusetts - Lowell
Originally presented on: 11/6/2013 9:33:00 - 9:54:00

A process control laboratory experience has been developed using a continuously stirred tank reactor system that permits simultaneous level and temperature control using water as the process medium. This work was originally completed as a senior honors thesis project, and the resulting system has been successfully incorporated into the process control block of a junior-level unit operations laboratory course. Use of the apparatus over multiple laboratory sessions provides students with a hands-on experience that illustrates the concepts of system characterization (e.g., calibration, determining operating ranges, understanding electromechanical component specifications, etc.), open-loop process response measurements, and closed-loop response and PID tuning. A controller program to interface with the valves and sensors and to perform data logging was constructed in LabView, employing a graphical user interface. As part of the experience, all sensors and valves of the system are to be characterized and calibrated. The two system processes—i.e., temperature and height—are run in an open loop manner, with data collection providing a means to determine appropriate process models. Process parameters from these models permit the estimation of controller tuning values (i.e., controller gain and time constants) through formula- or software-driven means (e.g., Control Station Loop-Pro). SISO configurations are then employed to test various control settings for tuning purposes, with the use of hand-tuning techniques to refine these values. The system can ultimately be operated in a MIMO configuration without and with decoupling gains; the typical oscillatory behavior without decoupling is demonstrated due to the inherent process interactions.

(428e) Teaching Process Design Through Alginate Bead Synthesis

Branch, K.,
Butterfield, A., University of Utah

Originally presented on: 11/6/2013 9:54:00 - 10:15:00

The creation and analysis of alginate beads has been found to be effective in teaching concepts of mass transfer [1]. We have designed and implemented a freshman-level teaching module in which students create an automated process for the production of alginate beads, introducing them to the principles of process design, conservation of mass, pump characteristics, and fluid dynamics.

Alginate beads are created by dripping alginic acid into a calcium chloride solution, transforming the outer shell of the droplet into a hydrogel. The resulting beads may be separated from the calcium chloride and loaded with a dye, creating a model drug-delivery system. By placing the loaded beads into water, mass transfer of the dye out of the bead may be characterized. Our teaching module uses this process to create an open-ended design project in which the students are able to be create a wide variety of processes of their own design. We explain the basic chemistry to the students and provide students with the chemicals, pumps, tubing, containers, a dye, and mesh. The students may also request additional equipment.

The design goals of the project are to create an automated system which produces uniform, spherical alginate beads with a specified size and release rate. The student's process must be able to convert 5 ml of alginic acid solution into beads continuously without interference, separate out the resulting beads, and recycle the calcium chloride solution. After their beads are created, students may quickly determine size and eccentricity distributions using MATLAB's image processing toolbox and a webcam. Then the rate of diffusion may be determined by tracking the amount of dye that diffuses out of the beads over time using a spectrophotometer. The students may go back and alter several parameters in order to create more uniform beads that are closer to the design goals. Some of the parameters that may be adjusted include the concentrations and flow rates of the alginate and calcium chloride solutions, the drip height and tip diameter, the residence time of the beads in the calcium chloride solution, and the flow rate of air around the alginic acid droplets.

Due to the number of goals and parameters that can be adjusted, this project may be adapted to various levels of difficulty. We have used this project for teaching freshman and senior students in the chemical engineering program, and for junior high and high school students as part of the Science Olympiad. The real-world applications of alginate beads make it attractive as a teaching module. According to a student survey in our chemical engineering freshman design laboratory, this was the most-enjoyed project out of seven, and it was also ranked the highest in the students' assessment of which teaching modules were most educational.

[1] Farrell, Stephanie, and Jennifer Vernengo. "A Controlled Drug-Delivery Experiment Using Alginate Beads." *Chemical Engineering Education* 46.2 (2012): 97-109. Print.

(428f) Photobioreactor Design & Algae Biodiesel Production Teaching Module

Butterfield, A., University of Utah

Originally presented on: 11/6/2013 10:15:00 - 10:36:00

The area of photobioreactor (PBR) design for the purpose of producing algae biofuel holds many opportunities for teaching core chemical engineering concepts. Concepts of reactor design and microbial growth are of obvious importance. Additionally, the necessity to keep algae suspended in a PBR introduces concepts of fluid dynamics. Light penetration and dissolved gas concentrations provide additional design consideration which may be related to beer's law and thermodynamics. Solvent extraction and two phase reactions may also be introduced during the processing of the algae into biodiesel.

Lab modules based on algae biofuels are also highly desirable among our student body. Biofuels are of significant interest to a large portion of students due to their potential as a carbon neutral source of energy. Furthermore, because the technological hurdles in scaling-up production of algae crops have yet to be overcome, this area is also an interesting example to students of the type of unsolved problems faced by practicing engineers.

To capitalize on this interest and educational potential, we have developed a PBR/biodiesel teaching module that has found wide use in both our curriculum and outreach efforts. In this open-ended design module, students are given access to a wide variety of containers, air and liquid pumps, spargers, tubing, and lighting options. We have amassed a materials list of inexpensive, yet effective components which allow this project to be accomplished with little investment. Each team begins with the same concentration of algae and nutrients from a common source. After being given some instruction and preliminary data, students are asked to design a reactor to grow algae as quickly and efficiently as possible, within several other design constraints. Over a two week period students track algae growth using optical density measurements. While the algae are growing, students convert a variety of biological oils into biodiesel, compare their products using a range of analytical techniques, and complete a project on scaling-up their process. Once the module is completed the students should be able to use their data to identify phases of batch microbial growth, and calculate doubling times and maximum growth rates within their reactor.

In student surveys, this teaching module has been ranked well above average in both enjoyment and educational effectiveness by students. The teaching module or some variant has found use from our freshman design laboratory to our senior capstone lab, proving itself to be versatile in its sophistication. Furthermore, PBR design projects have been an important outreach tool for our department, used in several summer camps and in a two-week introduction to chemical engineering course taught by our outreach team for a local high school.

(428g) Development of Model Based Visualization Tool for Demonstration of Basic Separation Processes in Unit Operation Laboratory Courses

Elsaid, K., Texas A&M University at Qatar
Stijepovic, M. Z., Texas A&M University at Qatar
Originally presented on: 11/6/2013 10:36:00 - 10:57:00

Unit Operation Laboratory courses are considered a crucial part of chemical engineering syllabus. In these courses the student are able to apply the theory and basic practices they have learned in different chemical engineering core courses such as fluid mechanics, heat transfer, thermodynamics, mass transfer and separation processes, chemical reaction engineering.....etc. Chemical engineering departments are equipped with up to date facilities to support students in gaining knowledge related to the different physical phenomena taking place during the different unit operation experiments with more focus on complicated operations related to separation processes, reaction engineering. Membrane separation is one of the newly developed separation process where separation of components takes place through semi-permeable membrane allowing certain component(s) or phase to pass through and rendering the other from passing through. In addition it can be applied in wide range from microfiltration up to reverse osmosis RO, and on wide scale from simple home tap water filtration up to large scale RO desalination plants.

In this paper we describe the development of visualization tool to demonstrate membrane separation process with practical application to RO membrane for desalination purposes. Due to the complex nature of different phenomena taking place during separation processes over membrane surface, it is very helpful to have such visualization tool that helps to have an improved understanding and insight vision of nature and behavior during such membrane separation process. The model to be utilized by the visualization facility will be based on solution diffusion and hydrodynamic models which are the common models used to describe membrane separation. In this paper water desalination using reverse osmosis RO membranes, one of the most common and extensively used desalination process worldwide, will be modeled and visualized in a 3-D environment utilizing the Immersive Visualization Facility CAVE at Texas A&M University at Qatar. The developed visualization tool will help the students by giving insight vision to what happening on micro-scale on membrane surface and within the membrane itself. In addition the tool can be used to examine wider range of operating conditions that might be behind the capabilities of lab facilities. In addition the tool can be used to check the performance of the laboratory unit, given that the used model has been initially validated, or even to substitute for the lab unit when being not available.

(510a) Science The World: The Development of Structured, Research-Inspired Discussions, Demonstrations, and Experiments For K-12 Curricula

Landherr, L. J., National Institute of Standards and Technology
Originally presented on: 11/6/2013 12:33:00 - 12:50:00

Past research has proposed that students learn better when they are provided with a greater sense of structure in their education; students build associations between their previously acquired knowledge and their current classes and can thus branch out towards understanding new concepts if their curricula have been well connected. Our research group attempts to address these needs by designing and developing new curricula for the entire K-12 spectrum by incorporating current engineering research into STEM discussions, demonstrations, and experiments. Connecting directly with K-12 educators nationwide, series of structured curricula have been designed and implemented for student groups in grades K-2, 3-5, 6-8, and 9-12, allowing for better connection between concepts to help improve students' understanding and to encourage more young scientists. Curricula examples include friction and lubrication experiments with comparison to drag-reducing polymers and household materials, chemical synthesis and physical elasticity based on polymer networks, groundwater contamination involving remediation treatment and filtration systems, and drug delivery through diffusion rates and concentration. This presentation will describe some of our current efforts with assessment of their effectiveness.

(510b) Adhesion Forces At Work: Daughters and Parents Working Together to Learn About Interfaces

Sorg, V., Cornell University
Rivera, C., Cornell University
Dimiduk, K., Cornell University
Daniel, S., Cornell University
Originally presented on: 11/6/2013 12:50:00 - 13:07:00

Over the last four years, the graduate students in the Chemical and Biomolecular Engineering Department at Cornell have organized a popular outreach program for 10th grade girls and their parents from rural areas to further their interest in STEM fields and provide an early start on college planning. The day-long program includes six hands-on lab modules, college admission information sessions, and a talk highlighting STEM career opportunities. The lab modules focus on teaching college-level concepts that the students are unfamiliar with, such as DNA extraction, packing fractions, and surface tension. A unique aspect of this program that was developed this year was a new parent and daughter laboratory module based on discovering and exploring interfacial phenomena. In this module, the daughters and their parents work together to conduct hands-on experiments that are designed and led by graduate students. During the session, we used i<clicker technology to collect data in real time from the parents and the students to assess learning outcomes. Additionally, we collected feedback forms to determine opportunities for improvement. In this presentation, we will provide an overview of the entire program, present some of the experiments developed for the parent and daughter lab session, and discuss detailed feedback from students and parents about the session.

(510c) Biomaterials Design Projects and Hands-On Activities for High School Students

Adolph, E. J., Vanderbilt University
Prieto, E. M., Vanderbilt University
Guelcher, S. A., Vanderbilt University
Originally presented on: 11/6/2013 13:07:00 - 13:24:00

High school students are not often exposed to the field of biomaterials. We have developed and implemented several interactive biomaterials activities for high school students. The activities performed at two events (Vanderbilt Summer Academy and Biomaterials Day) will be described during this presentation. 11th and 12th grade students at the Vanderbilt Summer Academy completed a biomaterials design project requiring them to create polyurethane scaffolds that could be used to deliver a drug and provide mechanical support for a given tissue. Students participated in hands-on activities that involved making polyurethane scaffolds, measuring release kinetics and determining release mechanism of dye from various scaffolds, and performing tensile tests to create stress-strain curves and determine mechanical properties of the scaffolds. Students worked on the design project for a total of ~20 h over the course of three days. High school students attending Biomaterials Day participated in a networking lunch, poster session, and hands-on activities. The activities lasted half an hour each and included extracting your own DNA, measuring the strength and Young's modulus of biomaterial scaffolds, and making your own lab-on-a-chip.

(510d) Fabrication of Jell-O Milli-Fluidic Chips for Hands-On Education of Hemodynamics and Blood Cell Adhesion

Mitchell, M., Cornell University
Syracuse, D.,
Archer, S. D., Cornell University
Castellanos, C., Cornell University
King, M. R., Cornell University
Originally presented on: 11/6/2013 13:24:00 - 13:41:00

Fabrication of Jell-O Milli-fluidic Chips for Hands-On Education of Hemodynamics and Blood Cell Adhesion

Michael J. Mitchell[1], David Syracuse[2], Carlos A. Castellanos[1], Shivaun D. Archer[1], Michael R. King[1]

[1]Department of Biomedical Engineering, Cornell University, Ithaca, New York USA

[2]TST BOCES High School, Ithaca, New York USA

Introduction: Advances in microfluidics have allowed for the development of devices that can recreate the complex microenvironment of the bloodstream. In particular, our lab has recently developed microscale flow devices that mimic postcapillary venules to study the basic mechanisms of cancer metastasis (1) and inflammation (2,3). Such devices have also been adapted to kill cancer cells in the bloodstream (4), and isolate circulating tumor cells from patient blood (5). As the clinical benefits of microfluidics are becoming apparent, it is essential to educate young students on basic fluidic principles and their potential impact on human disease diagnosis and treatment. A simple fabrication technique was demonstrated using Jell-O to teach high school students the essential components of blood flow, and its subsequent effects on cell adhesion in inflammation and cancer.

Materials and Methods: Model milli-fluidic devices were fabricated using a soft lithography technique consisting of foam plates, coffee stirrers, single- and double-sided tape, and Jell-O. Briefly, coffee stirrers were cut, connected, and arranged in various formations and adhered to foam plates using double-sided tape. Single-sided tape was placed on top of the coffee stirrers to create a smooth surface. A Jell-O liquid was prepared as per manufacturer's instructions, poured onto molds, and left to cure inside a refrigerator for 2 days. Jell-O chips were then peeled from molds and placed onto aluminum plates. Inlets and outlets were created within the chips, and liquid was perfused through the device using disposable transfer pipets. For blood cell adhesion experiments, chips were peeled and placed onto glass slides, and incubated with a solution of soluble E-selectin protein for 2 hours. Human neutrophils were perfused through E-selectin coated Jell-O chips, and assessed for cell adhesion using brightfield microscopy.

Results and Discussion: High school students successfully fabricated model milli-fluidic devices consisting of varying fluidic networks (Figure 1). Comparisons of Jell-O chip versus polydimethylsiloxane (PDMS) microfluidic chip fabrication in biomedical research were highlighted throughout the fabrication process. The concept of pressure-driven flow was demonstrated using Jell-O chips in the absence of an outlet, as fluid could not flow due to air present in the channel, which could not escape due to the absence of an outlet. High school students calculated flow rate by measuring the flow of a set volume of fluid through the device over time (Figure 2). The concept of laminar flow was demonstrated to students via perfusion of both a light and a dark-colored fluid through a Y-shaped Jell-O chip, as fluids perfused through channels simultaneously did not mix. Jell-O chips were also used to demonstrate cell adhesion, as E-selectin coated devices successfully induced human neutrophil adhesion. Graduate students, so to prevent contamination and exposure of hazardous materials to high school students, demonstrated all cell adhesion studies.

Conclusions: We have shown that simple, inexpensive Jell-O microfluidic devices can successfully demonstrate the basic principles of blood flow in a hands-on high school laboratory setting. The fabrication and use of devices can benefit students by introducing them to current biomedical technologies and applications, and encourage the pursuit of engineering and science degrees at the university level.

Acknowledgements: This study was supported by a National Science Foundation GK-12 grant DGE:0841291.

Figure 1: Schematic to fabricate Jell-O chips using soft lithography. (A) Top view of a negative mold consisting of coffee stirrers and a foam plate. (B) Side view of coffee stirrers adhered to foam plate using tape. (C) Liquid Jell-O is poured onto the mold. (D) Chip is solidified over using refrigeration and then peeled from plate. (E) Chip is adhered to aluminum plate or foil for use.

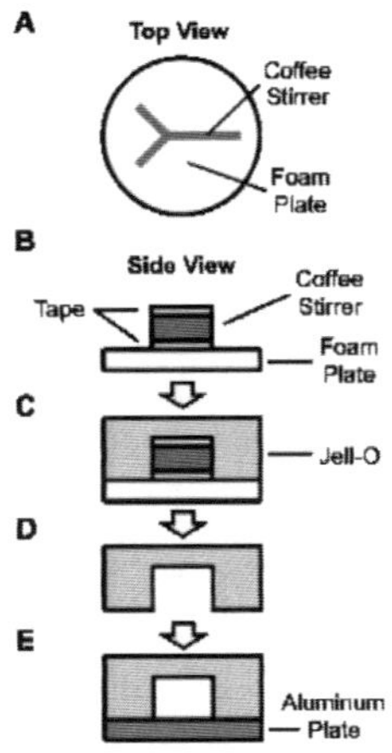

Figure 2: Perfusion of fluid through a Jell-O chip.

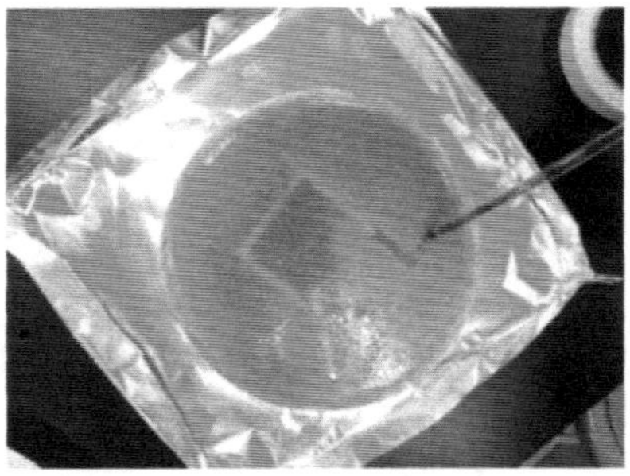

References:

1. Yin X, Rana K, Ponmudi V, King MR. Knockdown of fucosyltransferase III disrupts the adhesion of circulating cancer cells to E-selectin without affecting hematopoietic cell adhesion. Carbohydrate Research. 2010;345:2334–42.

2. Mitchell MJ, King MR. Shear-Induced Resistance to Neutrophil Activation via the Formyl Peptide Receptor. Biophysical Journal. 2012 Apr 18;102:1804–14.

3. Cao TM, Takatani T, King MR. Effect of extracellular pH on selectin adhesion: theory and experiment. Biophysical Journal. 2013 Jan 22;104(2):292–9.

4. Mitchell MJ, Chen CS, Ponmudi V, Hughes AD, King MR. E-selectin liposomal and nanotube-targeted delivery of doxorubicin to circulating tumor cells. Journal of Controlled Release. 2012 Jun;160:609–17.

5. Hughes AD, Mattison J, Western LT, Powderly JD, Greene BT, King MR. Microtube Device for Selectin-Mediated Capture of Viable Circulating Tumor Cells from Blood. Clinical Chemistry. 2012 Feb 16;58:846–53.

(510e) Simple Introductory Experiments Involving Pharmaceutical Engineering Concepts

Slater, C. S., Rowan University
Savelski, M. J., Rowan University
Struck Jannini, A., Rowan University
Krause, D., Rowan University
Originally presented on: 11/6/2013 13:49:00 - 14:06:00

Laboratory experiments exposing undergraduate students to topics in pharmaceutical engineering have been developed. These experiments were created to be used in the lower level engineering curriculum and in outreach activities by introducing students to pharmaceutical terminology through basic science or engineering concepts. The pharmaceutical concepts included: drug delivery, drug formulations, quality control, pharmacology, active ingredient function, and pharmaceutical processing. Several example labs are described have included the following. A reverse engineering lab consisted of the analysis of a common asthma and COPD (chronic obstructive pulmonary disease) drug delivery device, the Advair Diskus®. Through this lab students investigate how an active ingredient in particulate form can be delivered and repeatedly administered over a month dosage period. Student can also compare this to the function of other orally inhaled asthma medicines. Simple statistical analysis of solid dose drug tablets is an experiment quite useful for outreach activities in demonstrating quality control. Antacid and an Alka-Seltzer® effervescence reaction experiments tie together nicely basic chemistry principles and stoichiometric relationships. Fundamental mass transfer experiments use novel strip film dosage forms to illustrate diffusion of novel drug delivery methods. This work is part of the educational outreach efforts of the NSF ERC for Structured Organic Particulate Systems.

(510f) Labs and Experiments At a Distance

Henry, J., University of Tennessee at Chattanooga
Serrano, B., Universidad Autonoma de Zacatecas.
Originally presented on: 11/6/2013 14:06:00 - 14:23:00

This paper presents the benefits to student's learning that can accompany the ability to access remote laboratory experiments. The presenters have developed experimental systems that can be operated remotely. We use the phrase "action at a distance" for this capability. Studnets anywhere with network connection can conduct chemical engineering laboratory experiments from anywhere at anytime, day or night, irrespective of time zone. This allows students at any location or university to use specialized or unique or expensive equipment without regard to their university's capabilities. This paper specifically touches on remote distillation, remote multi-tank water level control system and remote thermal conduction and convection experiments.

(510g) Development of a Multi-University Graduate Certificate Program in Bioenergy and Sustainable Technologies

Schlup, J. R., Kansas State University
Bellmer, D., Oklahoma State University
Rezac, M. E., Kansas State University
Carrier, D. J., University of Arkansas
Humburg, D.,
Originally presented on: 11/6/2013 14:23:00 - 14:40:00

Widespread interest in green and sustainable industries has resulted from a general recognition of the need for systems that meet societal needs without long-term degradation of the environment. With the growing interest in bioenergy and sustainable technologies, there is a need for individuals with the knowledge and skills necessary to develop and sustain biobased enterprises.Â

In 2006, industrial practitioners with a focus on biobased materials and biofuels processing were surveyed regarding the experiential and educational needs of their current and future employees. The overwhelming message from the survey was that the employee pool lacked knowledge and experience with biobased processes.Â The survey findings demonstrated that there exists an urgent need for specialized training that is interdisciplinary in its approach and which is accessible for on-campus students as well as industry personnel who would like to pursue distance education while working full-time.

Since that time, four Universities have collaborated to develop a new graduate certificate program that prepares participants for careers in the emerging biofuels and biobased products industries by enhancing their knowledge in renewable resource development. Â In order to optimize resources and to utilize expertise at the partner institutions, the program is being delivered via distance education through each of the partner institutions, making it accessible anywhere in the world. The multidisciplinary graduate program involves several different areas of study that directly support biobased product and bioenergy development, including chemical engineering, biological and agricultural engineering, plant sciences, agronomy, and agricultural economics.

This presentation will describe the development of the academic program along with the rationale behind selection of a graduate certificate format.Â The curriculum and its implementation in a distance learning environment will be defined.Â The Agricultural Interactive Distance Education Alliance (AG*IDEA), which is an academic board within the Great Plains Interactive Distance Education Alliance (Great Plains IDEA), provides an administrative structure for the program.Â The relationships between the member institutions and with AG*IDEA will be explained.

(513a) Educational Needs in Particle Technology
Jacob, K., The Dow Chemical Company
Originally presented on: 11/6/2013 12:30:00 - 12:45:00

Bulk solids are ubiquitous in our lives today - take a look at the room around you, nearly everything has a significant particle technology component. Consequently, it is no surprise that a new undergrad entering industry has approximately an 80% chance of encountering or dealing with particles in their first project. Yet he or she is significantly ill-equipped to handle the demands of solids processing/particle technology. Merrow, both in the mid-80s and in 2001, has highlighted the significant challenges engineers face when dealing with solids due to a lack of education in the area and the resulting consequences: considerable downtime, lost production, start-up difficulties, etc. Industry has responded with post graduate training for their engineers or with the creation of industrial engineering science laboratories dedicated to particle technology that are on equal footing with more traditional chemical engineering disciplines such as reaction engineering, separations, fluid mechanics, heat transfer,etc. It is very important, if not imperitive, that solids processing/particle technology be incorporated into the chemical engineering curriculum. We will examine the challenges the chemical engineering community faces to meet this need.

(513b) Teaching Mixing to Undergraduate Chemical Engineers

Grenville, R. K., Philadelphia Mixing Solutions Ltd.

Originally presented on: 11/6/2013 12:45:00 - 13:00:00

Chemical engineers take raw materials and mix them so that they react to form valuable products and by-products.Â The products and by-products are then separated.Â While these steps are being performed the components may be heated or cooled as necessary.Â Of these four basic operations, three are covered in the undergraduate chemical engineering curriculum.

Smith (Trans. IChemE, 1990) estimated that the US chemical industry loses approximately $ 1 billion a year due to problems associated with poor understanding of the role of mixing on the processes being operated.Â Given the importance of mixing as a discipline, it is surprising that there are very few universities that offer a formal course on the subject.

After graduation engineers can take continuing education courses or webinars on mixing organized by, among others, the Center for Professional Advancement, the American Institute of Chemical Engineers and the British Hydromechanics Research Group.Â Some larger companies that employ mixing specialists offer in-house seminars on mixing to their employees.

If a formal course on mixing cannot be included in the curriculum due to, for example time constraints, a single lecture could be included in other courses, such as fluid mechanics, reaction engineering and transfer processes, that would show how the concepts being taught can be extended to take the role of mixing into account.Â These modules could be prepared as webinars aimed at undergraduate students rather than those offering continuing education for engineers in industry.

(513c) Addressing Educational Gaps for Process Safety in Both Academia and Industry

Vaughen, B. K., BakerRisk

Originally presented on: 11/6/2013 13:00:00 - 13:15:00

Members of AIChE's Safety and Health Division (S&H) have been working for decades with AIChE's Center for Chemical Process Safety (CCPS) to develop process safety-related continuing education materials for practicing engineers. To help address the needs of chemical engineering educators, members of the S&H Division help bridge the gap between industry and academia through the Safety and Chemical Engineering Education (SAChE) committee. One of the SAChE committee's goals is to help instructors who don't have extensive process safety experience prepare for and teach chemical process safety examples and lessons in their classes. Extensive process safety guideline books, webinars and "safety certificates" (available through SAChE) have been developed and are currently in use today. Integrating process safety system principles, such as inherently safer design and process hazards analyses, into an already overloaded chemical engineering curricula has been one of the SAChE committee's challenges, and is a "hot topic" today, especially now that the Accreditation Board of Engineering and Technology (ABET) added a "process safety" expectation into its audit protocol in the Fall of 2012.

(513d) Engineering Risk Rountable

Matonis, D., Flucomp Enterprises, Inc.

Originally presented on: 11/6/2013 13:15:00 - 13:30:00

The inherent complexity in manufacturing operations through construction have put more pressure on companies to focus on risk management beyond regulations and standards compliance. However, many individual facilities use a multitude of systems and methods and descriptions to define and assess engineering project and construction risks along with different models to score the severity and likelihood of risk and to control it. This creates isolated pockets of risk knowledge, and impedes the ability to accurately share lessons learned and best practices effectively within one company let alone across an industry. The cost of capital, resources and personnel will continue to rise while regulatory requirements and sustainability initiatives become more stringent, integrated, and expensive. All of these factors create a strong need for strategic operational risk and information management uniformity and education.

(513e) Training Chemical Engineers for a Career in Process Development

Hollar, W., SAUDI BASIC INDUSTRIES CORPORATION

Originally presented on: 11/6/2013 13:30:00 - 13:45:00

Training undergraduate chemical engineers for careers in process development represents a unique challenge. Traditional chemical engineering unit operations remain the basis for much of the chemical industry both in the United States and globally. The expansion of chemical engineering into non-traditional areas such as biochemical processes, materials and interdisciplinary areas such as nanotechnology has placed additional driving forces on chemical engineering departments to provide basic training in these areas. This can potentially come at the cost of reducing the depth of training in areas such as process control, process design and unit operations. This comes at a time when the need for young chemical engineers to develop competitive, sustainable processes in energy production and delivery, globally competitive commodity chemicals, and unique high value specialty chemicals continues to rise.

Careers in process development require the conflicting requirements of both depth and breadth in the field of chemical engineering for optimal performance. While much of the training occurs on the job, the education at the university represents the solid base upon which breadth and depth is built. Some opportunities for increasing the level and effectiveness of the interaction between industry and academia are proposed:

Invited lectures by industry chemical engineers in relevant class topics

Heavier participation of industry in classes such as senior design courses

Alternatives for in-depth training in specific chemical engineering topics

Increased funding and opportunities for student research and internships in industry

(513f) Responding to the Needs of Industry: Perspectives From the Education Division

Justice, F.,

Originally presented on: 11/6/2013 13:45:00 - 14:00:00

The membership of the Education Division, by definition have a keen interest is building into the undergraduate curricula those elements that will make one of our products ,"new Chemical Engineering practitioners", valuable to the industries that employ them. Like any business, the 'feedback' from the marketplace is one of the truest inputs which when accurately measured can point the way to continuous improvement activities that assure the graduates are both well-grounded in fundamentals and able to quickly come up to speed in emerging areas.

Getting and verifying this market place feedback and communicating it to Education professionals is crucial to evolutionary improvements of the curricula and teaching methods.

(513g) Responding to the Needs of Industry: Nuclear Engineering Division
Olson, J., Process Engineering Design Manager
Originally presented on: 11/6/2013 14:00:00 - 14:15:00

The nuclear industry needs engineers to maintain and design the future facilities for both power generation and waste cleanup. Chemical engineers, due to their unique background and training, are highly sought-after resources in the nuclear industry. Bechtel categorizes engineers as Process/Chemical Engineers, or Nuclear Engineers. Both work on nuclear industry jobs, but their focus is different.

Process/chemical engineers apply the principles of chemistry to solve problems involving the production or use of chemicals and bio-chemicals. They design equipment and processes for large-scale chemical manufacturing/cleanup, plan and test methods of manufacturing products and treating byproducts, and supervise production. In BSII, Process/Chemical Engineers are responsible for developing Process Flow Diagrams to include process calculations and producing System Design Descriptions. They prepare equipment process designs and are responsible engineers in the purchase of process equipment. Often, they are responsible for engineering studies up to and including pilot scale testing.

Nuclear engineers research and develop the processes, instruments, and systems used to derive benefits from nuclear energy and radiation. They design, develop, monitor, and operate nuclear plants to generate power. They may work on the nuclear fuel cycle—the production, handling, and use of nuclear fuel and the safe disposal of waste produced by the generation of nuclear energy—or on the development of fusion energy. Some specialize in the development of nuclear power sources for naval vessels or spacecraft; others find industrial and medical uses for radioactive materials, as in equipment used to diagnose and treat medical problems. In BSII, Nuclear Engineers develop the calculations associated with nuclear facilities and provide direct support to other engineering disciplines for nuclear related matters.

(513h) Responding to the Needs of Industry: Sustainable Energy Forum

Knox, P., Eco Chem Strategies

Originally presented on: 11/6/2013 14:15:00 - 14:30:00

This is an abstract for the Needs of Industry session, with perspective from the Sustainable Energy Forum. It is an invited talk. This paper will discuss the skills level of graduating B.S. chemical engineers as it relates to the needs of sustainable energy. A panel discussion will follow all of the invited talks.

(513i) Panel Discussion: Responding To Industry's Needs

Rau, N., Sackett Solutions & Innovation
Visco, D. P. Jr., The University of Akron
Originally presented on: 11/6/2013 14:30:00 - 15:00:00

This is a panel discussion featuring all of the authors who have presented in this session, moderated by the session chair and the session co-chair. Audience members will have an opportunity to ask questions of the presenters. Nine different speakers will present and represent the leadership, past or present, from their division or forum.

(574a) Electives in the Undergraduate Chemical Engineering Curriculum

Anderson, T. J., University of Florida
Lepek, D., The Cooper Union
Piergiovanni, P. R., Lafayette College
Vigeant, M., Bucknell University
Davis, B. J., The Cooper Union
Originally presented on: 11/6/2013 15:15:00 - 16:30:00

This invited session focuses on electives in the undergraduate chemical engineering curriculum. In this session, we will address these four key questions on ChemE education: How have electives developed over the past 25 years? How do electives spread to new institutions? How can we bring chemical engineering to new audiences? And finally, what is the current state of elective offerings in chemical engineering education?

Tim Anderson, University of Florida, Distinguished Professor, Director of the Florida Energy Systems Consortium

The Value Proposition of Elective Engineering Courses

Elective courses intend to add a valuable learning experiences that exposes students to a subject beyond the scope of the basic curriculum. Obvious value is added through expanded knowledge, preparation for a specific area of chemical engineering, and practice in applying chemical engineering science. Less obvious are the opportunities to interact with students in other disciplines and appreciate the generality of chemical engineering sciences. The promised value proposition is exemplified by the experiences in teaching materials-based elective course.

Daniel Lepek, The Cooper Union; with Charles Coronella, University of Nevada Reno; Kimberly Henthorn, Rose-Hulman Institute of Technology; Christine Hrenya, University of Colorado Boulder; Timothy Raymond, Bucknell University; Martin Rhodes, Monash University

Particle Technology: Best Practices for Teaching this Elective from Multiple Institutions (see attached abstract).

Polly Piergiovanni, Lafayette College

“Indigo: A World of Blues” – a Chemical Engineering Elective for a Liberal Arts Audience

This talk will describe a course aimed at a mixed audience of engineers and liberal arts students in an exploration of the use of indigo dye across different eras and cultures. Encompassing chemical reaction, mass transfer, and process design and its interrelation with environmental, social, and historical impact, the topic is ideal for learning chemical engineering in context.

Margot Vigeant, Bucknell University and Benjamin Davis, The Cooper Union

Results of the Survey on Undergraduate Electives

We will share the first look at the 2013 Chemical Engineering Education Survey on electives. Survey results will indicate the typical number of electives offered and taken at US institutions, the disciplinary breadth available, and the extent to which student choices lead to concentrations or other distinctions.

(574b) Particle Technology: Perspectives and Best Practices From Multiple Institutions

Lepek, D., The Cooper Union
Coronella, C., University of Nevada, Reno
Henthorn, K. H., Rose-Hulman Institute of Technology
Hrenya, C. M., University of Colorado at Boulder
Raymond, T., Bucknell University
Rhodes, M., Monash University
Originally presented on: 11/6/2013 16:30:00 - 16:55:00

Although the study of particle technology is not commonly required in most undergraduate chemical engineering curricula, it has major impact on traditional technologies within the petrochemical and pharmaceutical industries, as well as newer technologies involving nanostructured materials and renewable energy systems. In this presentation, perspectives and best practices on teaching this course as an elective will be provided by instructors from multiple institutions.

(574c) Introduction of a Small Business Client Project in Upper Level Materials Science and Engineering Elective

Wakabayashi, K., Bucknell University

Originally presented on: 11/6/2013 16:55:00 - 17:10:00

An introductory, survey course in materials science and engineering is a popular offering as a requirement or a technical elective at many chemical engineering institutions. The course content and learning objectives of such lower level materials courses are common and well-established [1-3]. On the other hand, developing and teaching an upper level, comprehensive elective course in materials science and engineering can be challenging, partially because of varying expectations from chemical engineering students with basic materials science knowledge. Depth vs. breadth, science vs. engineering, application to the real world, and student engagement are some of the key components that are addressed in the author's version of such a course.

The present paper shares a new component in Bucknell University's Advanced Materials Science and Engineering elective, in which a local entrepreneur in need of an engineering consulting service at Bucknell Small Business Development Center (SBDC) was identified as the client for the class, and the class conducted a 14 week-long product redesigning project with him. The technical content followed the systematic materials selection methodology of Ashby [4], while the project flow and logistics were modeled after a typical client-based engineering capstone design course [5]. Major observations from this new course activity include enhanced communication, critical thinking, and time and project management skills. The project structure, results, and feedback from the students and client are presented.

[1] W. D. Callister and D.G. Rethwisch, Fundamentals of Materials Science and Engineering: An Integrated Approach, 3rd ed., John Wiley and Sons, 2008.

[2] J. F. Shackelford, Introduction to Materials Science for Engineers, 7th ed., Pearson Prentice Hall, 2009.

[3] D. R. Askeland, P. P. Fulay, and W. J. Wright, the Science and Engineering of Materials, 6th ed., Cengage Learning, 2011.

[4] M. F. Ashby and K. Johnson, Materials and Design: The Art and Science of Material Selection in Product Design, 2nd ed., Butterworth-Heinemann, 2009.

[5] A. J. Dutson, R. H. Todd, S. P. Magleby, and C. D. Sorenten, Journal of Engineering Education 86, 17 (1997).

(574d) Nanotechnology in Chemical Applications

Hill, P. J., Mississippi State University

Originally presented on: 11/6/2013 17:10:00 - 17:25:00

Given the emergence of nanotechnology globally over the last two decades, and its potential to influence both the national and global economies, there is a need to train engineers in nanotechnology. Nanotechnology in Chemical Applications is a new 3 credit elective course that aims to introduce upperclassmen to nanotechnology concepts from colloid and interfacial science. The approach is to present fundamental concepts from colloid and interfacial chemistry, and then link these concepts to nanotechnology. The presentation includes an overview of the course structure, topics covered (e.g. surface forces, colloid characterization, self-assembly and stability, and applications to nanotechnology), and resources used. The course emphasizes differences in material properties between the nanoscale and the bulk scale. Portions of this presentation, but not the entire presentation, have been presented previously as a poster[1].

1. Hill, P. J., Myers, O. J., Koshka, Y., Thibaudeau, G., and Henington, C., A Multidisciplinary Undergraduate Nanotechnology Education Program. *2013 ASEE Annual Conference & Exposition.*

(574e) Characterizing PM2.5 in a Mountain Valley: A Community Engagement Project

Silcox, G. D., University of Utah
Kelly, K.,
Kotchenruther, R.,
Kuprov, R.,
Originally presented on: 11/6/2013 17:25:00 - 17:40:00

The University of Utah, Lowell Bennion Service Center encourages instructors to collaborate with community partners in their courses. Such courses appear on a student's transcript and are given a Community Engaged Learning designation. Chemical Engineering 5305/6305, Air Pollution Control Engineering, has collaborated with the Utah Division of Air Quality on two such projects related to PM2.5 in the Salt Lake Valley.

In the first project, PM2.5 concentrations were measured in residential and nonresidential areas of Salt Lake City, Utah during persistent, stable, wintertime conditions. Concentrations generally decreased with increasing elevation; decreases of up to 30 percent were measured as elevation increased from 1300 to 1600 m. During a particular 8-day episode, concentrations of PM2.5 increased linearly with time by about 6 micrograms per cubic meter per day. During the mix-out of stable conditions, higher elevation sites were sometimes exposed to clean air as much as 24 hours before lower elevation sites.

In the second project, positive matrix factorization (PMF) was used to identify sources of PM2.5 in the Salt Lake Valley. The main contributor is secondary ammonium nitrate. The analysis also showed that ammonium chloride is an important source of secondary aerosol (10 – 15 % of total PM2.5) during stable wintertime conditions. Comparison of the source attribution results to county-level emissions inventories show generally good agreement except for wood smoke and dust which contribute more to PM2.5 than suggested by the inventories.

Some of the students are genuinely interested in the projects and enjoy working with and collecting data. Comments from the course evaluations include, "PM 2.5 project was an effective learning experience", "the group project let me get hands-on experience with a testing device and forced me to work in a group", and "another great aspect of the course was using the EPA PMF program because it was interesting to study real data and find potential sources of different emissions".

(581a) The Chemical Engineering App - Increasing Accessibility of Course Materials, Reference Data and Problem Solving Tools

Bara, J. E., University of Alabama
McLemore, J. P., University of Alabama
Originally presented on: 11/6/2013 15:15:00 - 15:40:00

The Chemical Engineering App for iPad and iPhone were developed to facilitate access to course materials, reference data and calculations for students, educators and professionals in chemical engineering and other STEM disciplines. Within the first year, our apps have each been downloaded over 10,000 times by users in over 30 countries. While the apps were originally focused on materials for mass and energy balance courses, we have worked toward a vertical integration of the chemical engineering curriculum and the software now includes relevant data and modules for thermodynamics, fluids, separations, controls and other courses. The apps are available as free downloads on Apple's App Store, effectively making real chemical engineering course materials available to anyone in the world with an iOS device. By lowering the accessibility barriers to course materials, students can review past or future courses at any stage of their education. This readily accessible format also helps promote chemical engineering and STEM education. Our presentation will discuss the use of The Chemical Engineering Apps both inside and outside the classroom and our vision for tablets and smartphones in chemical engineering and STEM eduction.

(581b) Process Simulation As a Means to An End, Not Just "At the End"

Rockstraw, D. A., New Mexico State University

Originally presented on: 11/6/2013 15:40:00 - 16:05:00

Chemical process simulation is a common tool of the commercial sector for performing rigorous design and analysis calculations. Use of modern simulation software can be adapted to all levels in the college curriculum, thereby permitting students sufficient time to develop a deep understanding of the details of its use, including such advanced features as pseudocomponent definitions; accessing variables; integration of user-defined modules; process dynamics and control; and economic evaluation & optimization. Numerous resources are now freely available that permit faculty to learn the features of the software that align with the content of the courses they teach, including textbooks that directly integrate the use of simulation software, and extensive, easy-to-follow teaching resources published by the software companies. The adaptability of the software combined with the availability of these learning resources dictate that process simulation no longer be taught as a single final-semester lab employing a cookbook-style approach, rather process simulation should now be taught in parallel with the fundamentals content of the chemical engineering curriculum. Details of an example integrated curriculum will be discussed, and an assessment of the impact on student chemical process design projects will be presented.

(581c) What, If Anything, Is a Chemical Engineer ?

Jimenez Esteller, L., University Rovira i Virgili
Guillen-Gosalbez, G., Universitat Politecnica de Catalunya
Originally presented on: 11/6/2013 16:05:00 - 16:30:00

The educational model designed and implemented at the School of Chemical Engineering (ETSEQ) of University Rovira i Virgili in Tarragona (Spain), to enable ChE students to acquire and integrate technical and scientific knowledge through the simultaneous and gradual development of competencies encompassing social and management skills has been evolving during more than a decade. This model is based on the large-scale deployment of a project-based cooperative learning approach throughout the ChE curriculum. This extensive deployment can only be effective if it is supported by (semi-external) experts in change management and the systematic development of student teams that, in turn, requires that individual students develop key social and management skills.

Several educational moduled had been developed and implemented at the ETSEQ (University Rovira i Virgili, Spain) to facilitate the transition of first-year ChE students into a comprehensive project-based learning
environment. All first-year students participate in the first-year integrated design project with fourth-year ChE students acting as project managers for first-year project teams. The modules are taught during the first two weeks of the term and is structured to take into account the factors needed for students to understand and commit themselves to change. The first-year students' evaluation shows that the module helps them: (1) to identify what they need to accomplish to gain future employment as chemical engineers; (2) to understand what an integrated design project consists of and what the benefits of teamwork are, and (3) to realise that the integrated project and the related teamwork are great opportunities to acquire competencies that are essential in today's workplace.

The expertise in team development and change management methodologies, gained from more than a decade of work, has complemented the practice of the ETSEQ with experiential learning methodologies. A set of external training interventions has been designed to support the development of competencies by students. In the new educational system student teams grow from leader-directed teams in the first semester of the first academic year to self-directed or empowered teams at the end of the bachelor. In this empowerment journey, fourth-year students play a key role as they act as facilitative leaders of first-year and second-year project teams, adjusting their leadership role according to the team development stage. The core of the competency-based educational model is related to the transformation of the individual students (versatility, entrepreneurship and innovation, systemic thinking, etc.), of the organization (facilitative leadership, teamwork and cooperation), and of the institution (organizational development and performance, and organizational leadership). Results
show that student attendance has increased, that drop out has decreased, that more professors act as facilitators in the classroom, and that active-oriented and student-centered educational methodologies are increasingly applied.

(581d) A Vertically Integrated, Multidisciplinary Student Design Team Project to Develop Integrated Biodiesel and Biochar Technology for Sub-Saharan Africa

Seay, D. J., University of Kentucky

Originally presented on: 11/6/2013 16:30:00 - 16:55:00

This contribution will describe the organization and development of a multidisciplinary, vertically integrated student design team project involving the design, construction and implementation of a novel apparatus to produce biodiesel and biochar from locally available resources in sub-Saharan Africa. The project was conducted by the University of Kentucky Appropriate Technology and Sustainability (UKATS) Design Team. UKATS is a multidisciplinary, vertically integrated design team at the University of Kentucky – Paducah Extended Campus focused on the application of appropriate technology to solve sustainability and renewable energy problems in underdeveloped regions of the world. The UKATS team is comprised of students from Chemical and Mechanical Engineering and includes both Senior Capstone Design as well as underclass students. Additionally, a group of local high school students also participated, giving the project an outreach component as well.

The first project taken on by the UKATS Team was implemented through the newly established Embedded Study Abroad Program model at the University of Kentucky. This program allows a travel component to be embedded into a traditional core class or elective. This is significantly different from the traditional study abroad model, which typically includes a semester of international travel. For this vertically integrated student design team project, a traditional independent study elective was combined with the embedded study abroad model to create a unique design experience for the students. This model has the benefit of allowing students at different experience levels to work together in a way that allows all participants to participate at a level that is appropriate to their experience.

This project was conceived through a partnership with an NGO called the African Center for Renewable Energy and Sustainable Technology (ACREST), an NGO located in Bangang, Cameroon (www.acrest.org). ACREST operates a small technology center where they work to design and adapt renewable energy and sustainable technologies for the use of rural villagers in Cameroon and throughout sub-Saharan Africa. Recently ACREST has entered into partnerships with U.S. universities, whose students work to develop technologies as part of internships and capstone design projects. The primary objective of the project with the University of Kentucky - Paducah Extended Campus is to develop technologies to allow rural villagers in Cameroon to manufacture biodiesel from locally available vegetable oils such as palm oil or jatropha oil, and biochar from field crop residues and waste biomass.

Seven UKATS Team members and their faculty advisor have recently traveled to Cameroon to implement this unique vertically integrated, multidisciplinary project. This contribution will describe the organization of the project and the results achieved by the students in the field. In additional, student perspectives from the field will be included in the presentation. This project was funded in part by a US EPA People, Prosperity and the Planet Design Competition grant (No. SU835303).

(581e) Encouraging Information Transfer With a Pre-Capstone Design Project

Enszer, J. A., University of Notre Dame
Bayles, T. M., University of Maryland Baltimore County
Originally presented on: 11/6/2013 16:55:00 - 17:20:00

A common issue in our curriculum is information transfer, the ability to apply knowledge from one context into another. While efforts are made in individual courses to explicitly consider the connections between theories in the chemical engineering core (thermodynamics, transport, kinetics, and controls), there is significant resistance from students to apply skills they learned in previous courses in the context of their current courses. In one effort to improve information transfer, we have developed a month-long project that spans three courses – heat and mass transfer, reaction kinetics, and controls and safety – that puts students in teams across the courses to analyze a recent incident in the chemical process industry and redesign the unit(s) that failed. Each course instructor requires a specialized statement, aligned with their course, that analyzes the case study and defends their design. In the spring of 2013, students studied the T2 Laboratories, Inc. incident and suggested improvements to the heat exchanger, control scheme, and safety devices that would prevent the conditions that lead to the explosion. We share here how the project is approached across the core courses, the logistics of assembling the teams, and the assignments specific to each course.

(581f) Project Management and Teamwork: A Vertically Integrated Team Design Project for All Students in Chemical Engineering

Evans, E. A., University of Akron
Elliott, J. R. Jr., University of Akron
Leipzig, N. D., The University of Akron
Originally presented on: 11/6/2013 17:20:00 - 17:45:00

Over the last 10 y ears we have taught a one credit Project Management and Teamwork course that is taken four times (once each year) by every student in the Chemical Engineering and Corrosion Engineering programs. The course work focuses on the completion of a laboratory based design project using effective teaming skills. The teams are vertically integrated (Freshmen – Seniors) and every team member is expected to contribute. The seniors are expected to become effective team leaders by the time they complete this four course sequence. The project statements are carefully constructed so that there are opportunities for each team member to contribute. In addition, the project statements focus on a frontier area in chemical engineering that is being researched at The University of Akron (hydrolysis of waste paper, water filtration, CO2 capture, etc.).

The teams are constructed carefully following a set of guidelines. The teams present the results of their effort in three different formats (informal meeting, laboratory demonstration, and a formal oral presentation). The students are also evaluated based on weekly writing submissions, meeting participation, meeting minutes and the use of a project work plan.

The course is managed by two-three faculty members. Technical expertise is provided by a researcher within the department. The course begins with four lectures that cover teamwork, leadership, technical background and potential economic or societal impact of the project. For the remainder of the semester the teams meet separately to discuss progress that is made and make decisions.

We will present lessons learned over the last ten years that can be applied to project based as well as lecture based courses.

(588a) The Hi-Pele: A Da Vinci Foundry

Sanders, J. R., Tennessee Technological University
Loggins, L., TTU
Biernacki, J. J., Tennessee Technological University
Arce, P. E., Tennessee Technological University
Originally presented on: 11/7/2013 8:30:00 - 8:50:00

The Hi-PeLE^{TM} Learning and Thinking System is very efficient to develop the new (STEM) professional as outlined in the National Academic of Engineering 2020 Model. The platform incorporates numerous well established (pedagogically) methodologies including team-based learning, collaborative approaches and critical thinking, all strategically located in the " *Knowledge Acquisition"* section of the approach where students learn new skills and knowledge by using learning and documentation cycles. Once students has mastered an appropriate level, they have the opportunity to use such skills in the "*Knowledge Transfer"* section where they must identify a societal need, identify the "root cause" of such a need, propose an innovative solution, develop the proof of concept, and build and test a prototype. They never are told or given such a need but, instead, they are coached into a process to achieve it. Several pedagogical tools for innovation-driven learning are being calibrated and applied to help students in the quest for the development of the prototype for innovative technology. In this presentation, the co-authors will discuss a model for curriculum implementation of the Hi-PeLE^{TM}, give examples of courses to illustrate the success of the students' achieving of a high level of Da Vinci-style behaviors and make observations for the implementation in potentially other colleges or K-12 environments.

(588b) Integrating Theory and Practice Into Undergraduate Chemical Engineering Courses: Active Learning in Teaching Chemical Reaction Engineering and Unit Operations

Franca, W. L. Sr., Universidade Federal do Ceara
Lemos, C. M., Universidade Federal do Ceara
Bastos Neto, M.,
Cartaxo, S.,
Goncalves, L. R. B., Universidade Federal do Ceara
Originally presented on: 11/7/2013 8:50:00 - 9:10:00

Project-Based Learning was select as a tool to achieve active learning in chemical reaction engineering and Unit operation courses at Universidade Federal do Ceará. Chemical reaction Engineering (TF324) and Unit Operations II (TF325) are 4-credit-hour courses that are taught to chemical engineering students at the fourth year. The students were assigned a group-project to analyze some aspects involved in reactor design, especially heat transfer. The students worked as groups of 10–12 students and they were asked to design and construct a bench scale system (reactor and heat exchanger) on the assigned topic. The model reaction was the growth of *Saccharomyces cerevisiae* to produce biomass. Heat releasing is a general property of microorganism growth, regardless the type of carbon source or if the process is aerobic or anaerobic. However, the greater the cell catabolism, the greater the number of glucose molecules consumed per unit time, the higher the released energy and hence, a greater thermal effect is registered. Some studies have shown that heat production in fermentation can be useful in several ways, both in academic research and in industrial applications. The general correlation between heat production and the various other parameters suggest that heat production may be used as a measurement for real time fermentation monitoring, oxygen consumption or CO_2 production rate. It is known that temperatures around 36 ° C cause a higher growth of the *Saccharomyces cerevisiae*. Based on that information, students were encouraged to set up a system for *Saccharomyces cerevisiae* production, which should consists of a heat exchanger and a batch bioreactor. The evaluated parameters for the system efficiency were substrate consumption and yeast production, as well as parameters relating to the thermal exchange. The main goal of this project was to expose engineering students (future engineers) to the design and analysis of chemical reactor with heat exchange, aiming to enhance active learning.

(588c) Experiences Applying Active Learning in the Material and Energy Balance Course After Attending the Chemical Engineering Summer School

Seay, J. R., University of Kentucky

Originally presented on: 11/7/2013 9:10:00 - 9:30:00

This contribution will focus on experiences applying the Active Learning techniques presented in the workshop from Dr. Rich Felder and the Notes with Gaps technique presented by Dr. Lisa Bullard to the sophomore level Material and Energy Balances course at the University of Kentucky - Paducah Extended Campus after attending the 2012 ASEE Summer School held at the University of Maine in Orono, Maine. The techniques presented at the Summer School workshops were especially useful, as this was the instructor's first opportunity of teach the Material and Energy Balances course. Specific examples of how the techniques presented in the ASEE workshops were incorporated into the course will be provided. The instructor's experience incorporating short quizzes, group problem solving and individual problem solving into the course will be presented. In addition, anecdotes of how the students liked the Notes with Gaps techniques will be provided. The instructor will again be teaching the Mass and Energy Balance course in Fall 2013, so experiences to date from the second offering of the course will also be presented.

(588d) ChE Summer School Revisited: Course-Specific Applications of Active Learning Techniques

Bullard, L. G., North Carolina State University
Silverstein, D. L., University of Kentucky
Visco, D. P. Jr., The University of Akron
Keith, J. M., Michigan Technological University
Originally presented on: 11/7/2013 9:30:00 - 10:00:00

Chemical engineering faculty members, whether they are new to the profession or experienced educators, are often asked to teach a core course that they have not taught before. Many instructors would like to take the time required to come up with some new ideas to revolutionize the course in ways which engage students and maximize learning. The realities of the responsibilities of a faculty position often pose a significant challenge to accomplishing this goal, especially for young faculty. To aid in achieving the goal of improved quality of instruction in chemical engineering core courses, this paper summarizes the authors' selection of the most effective, innovative approaches reported recently in the literature or discussed at previous conferences for chemical engineering courses. This material was initially presented at the 2007 and 2012 ASEE Summer Schools for Chemical Engineering Faculty. The challenges associated with particular courses and solutions successfully applied to address those challenges are also described. Courses covered during the summer school workshops include mass and energy balances, separations, introductory thermo, transport, solution thermodynamics, kinetics and reactor design, process control, and process/product design. This presentation will cover a subset of these topics.

(588e) Teaching Assistant Training At University of South Carolina

Padak, B., University of South Carolina

Originally presented on: 11/7/2013 10:00:00 - 10:20:00

At University of South Carolina we started to run a training program for our graduate students after attending a teaching assistant (TA) training workshop at the ASEE summer school presented by Dr. Lisa Hwang and Dr. Robyn Dunbar. The main purpose of the training was to provide motivation for the students. The students who are planning to pursue a career in academia are already self-motivated and try to gain as much experience as they can from being a TA and improve their teaching skills. However, most of the students who plan to work in industry after graduation are usually not interested in teaching and do not see the benefit of being a TA for their career. We use the training to emphasize that this experience is part of their professional development and will provide them with other skills such as communication and leadership. In their future job they will likely work in a team environment and need to provide feedback to their teammates or employees. We try to teach them how critical it is to give feedback to the students and tell them what they did wrong and give suggestions on how to do it correctly as opposed to just deducting points while grading. We show them examples of previously graded homework/exams and do a grading exercise where they get to grade a sample exam paper. The key point here is to make sure they are consistent while grading and we discuss how to create a grading rubric before they start grading. Other topics discussed in the training include managing faculty/TA expectations and time management.

The first training session was conducted in the beginning of Fall 2012 for all of our current graduate students and we will continue to do it every Fall for our incoming students. We also started collecting evaluations from the students in class regarding the performance of their TA's and will use it to address the common issues in the future trainings, which will hopefully improve their teaching skills. More details on the training session and some of the feedback we received from the students will be discussed during the presentation.

(588f) Alternate Models to TA-Graded Homework

Ramsey, J. D., Oklahoma State University
Madihally, S. V., Oklahoma State University

Originally presented on: 11/7/2013 10:20:00 - 10:40:00

The graduate curriculum at Oklahoma State University requires graduates students to take two core courses each semester during their first year. The number of students enrolled in the courses can vary from as few as 10 to as many as 25 students. During semesters when enrollment is low, the number of students is too few to warrant a dedicated teaching assistant to grade homework but high enough to require significant time from the instructor. As a possible solution, we investigated two alternative models for grading homework, self-assessment and peer-assessment, and compared the results to the traditional approach of having a teaching assistant evaluate the homework. The same group of graduate students was used to investigate both models in two different core courses: i) Selected Diffusional Unit Operations and ii) Advanced Chemical Reaction Engineering. Using the self-assessment model, students' homework were collected, copied, and returned to the students to grade and report their scores. Using the peer-assessment model, students' homework were collected and redistributed to randomly selected classmates who graded the homework and reported the scores.

One obvious concern is grade inflation due to students who grade homework more generously than a teaching assistant. For one of the two courses where an alternative model was used, we can directly compare homework scores reported by a teaching assistant from previous semesters to homework scores reported from peer-assessment, and there does appear to be some minor overall grade inflation. This effect can be overcome, however, by adjusting the weighting of the homework in the overall course grade and taken alone should not be a reason to avoid one of the alternative methods. In fact, based on survey results at the completion of the courses, most of the students reported greater learning from the alternative models that required them to grade either their own or a peer's homework. Comparing the two alternative methods, students who graded their own homework expressed a higher level of satisfaction and confidence in the grading than students whose homework was graded by a peer.

(588g) Integrating Abstracts of Homework Problem Solutions Into a Material Balances Course

Dahm, K., Rowan University
Farrell, S., Rowan University

Originally presented on: 11/7/2013 10:40:00 - 11:00:00

This presentation will discuss the integration of writing abstracts for homework problems into a course on material balances. Students were assigned traditional weekly homework assignments which consisted of problems assigned from the course textbook. Students completed the assignments in teams of three and submitted one solution per team. However, in addition, each individual student was required to write an abstract for each problem. In the abstracts, students summarized the purpose of the problem, the system under consideration, the known and unknown information and the solution procedure. There were two purposes to assigning the abstracts. First, it was hypothesized that requiring students to write about their problem solutions in a reflective way could foster a more thorough understanding of the processes being modeled, and instill in students a conscious recognition of effective problem-solving strategies. Second, it was hypothesized that the abstracts would provide an effective tool for assessing individual contributions to the team assignments. The authors will present an assessment of the impact of the abstracts on the Fall 2012 course. Assessment will include a comparison to the Fall 2011 students, who completed comparable homework problems in teams of the same size, but who were not required to write abstracts. The abstracts will be used again in the Fall 2013 offering of the course and the presentation will discuss lessons learned and changes implemented for Fall 2013.

(622a) A Systematic Methodology for the Modeling of Intrinsic Chemical Reaction Kinetics: N-Hexane Hydroisomerization On a Bifunctional Zeolite

Toch, K., Ghent University

Thybaut, J. W., Ghent University

Marin, G. B., Ghent University

Originally presented on: 11/7/2013 8:30:00 - 8:52:00

Kinetic modeling of chemical reactions is, among others, an important item of chemical engineering education. The presented work focuses on a systematic methodology for modeling the non-isothermal intrinsic kinetics of chemical reactions allowing their later incorporation into more complex reactor models. While a dedicated experimental campaign should typically aim at the acquisition of intrinsic kinetics, transport phenomena at the industrial reactor scale can be accounted for via well-known correlations from literature [1].

The proposed methodology start with an analysis of the experimental data as a function of the operating conditions, see Figure 1. Insights in the actual reaction mechanisms are pursued and first considerations with respect to the structure of the corresponding rate equations can be made. As a second step in the construction of a kinetic model, a preliminary determination of the kinetic parameters is performed. This is necessary since the response surface determined by the objective function, i.e., the sum of squares of residuals of the responses, may contain multiple maxima and minima. This can be done via a literature survey or quantum chemical calculations. An alternative method to determine good initial parameter values for pre-exponential factors and activation energies/reaction enthalpies, is to perform an isothermal regression for each temperature and fit the results to the Arrhenius equation. Thirdly, the actual non-isothermal regression is performed. Lastly, the regression results are analyzed. This is based on a statistical analysis and verification if the kinetic model and parameters obtained are physically viable. The statistical analysis comprises the investigation of the confidence intervals, binary correlation coefficients, test for model adequacy, regression and parameter significance and residual analysis (performance plots, residual figures, parity diagrams and Q-Q plots).

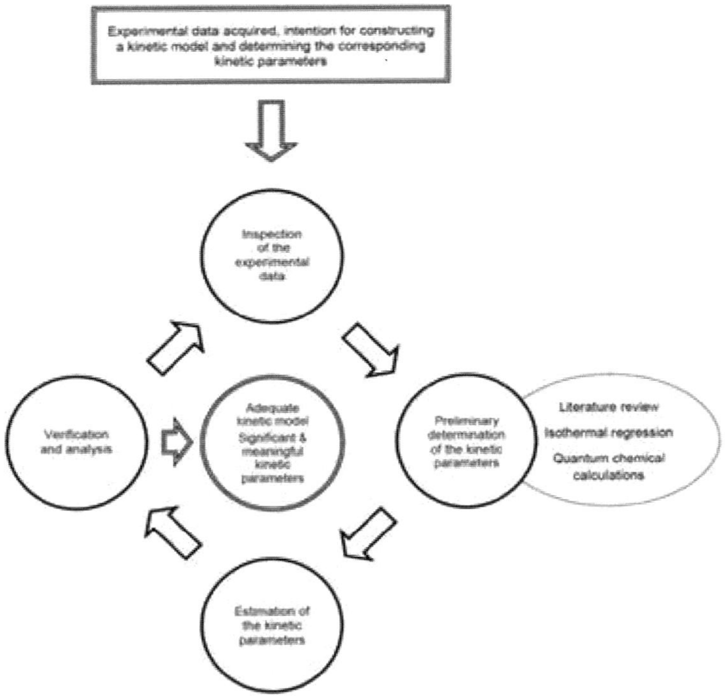

Figure 1: Proposed methodology for kinetic modeling

A simple, yet challenging, reaction is taken as example through this contribution, i.e., the hydroisomerization of *n*-hexane on a bifunctional catalyst MC-389, i.e., a with platinum impregnated USY derived catalyst. The metallic function of the catalyst allows to operate at milder conditions. This is due to the dehydrogenation of the physisorbed alkanes which occurs on the Pt sites and results in the formation of olefins, see Figure 2. It also provides resistance against coking.

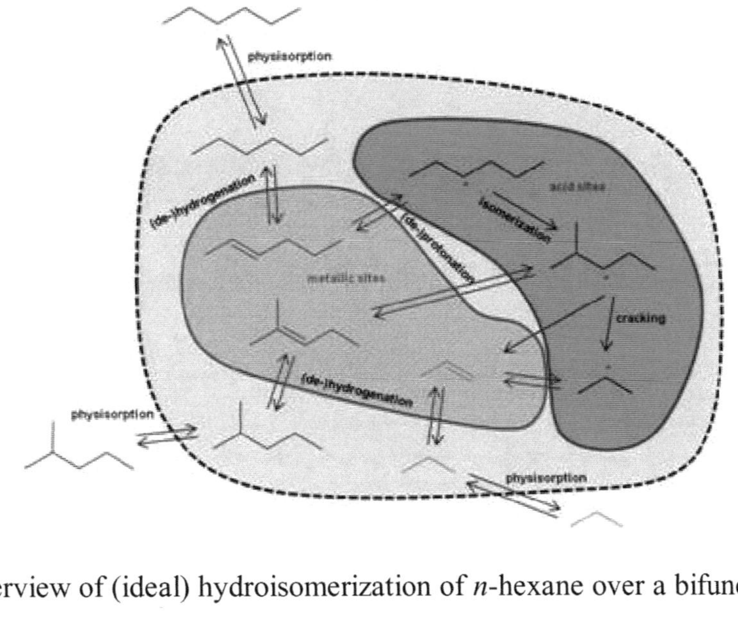

Figure 2: Schematic overview of (ideal) hydroisomerization of *n*-hexane over a bifunctional zeolite

These olefins are then relatively easily protonated on the acid sites, where the actual conversion to isomers and cracked products occurs. Only 3 reaction products are observed: 2-methyl-pentane, 3-methyl-pentane and propane. The experimental conditions ranged from 493 to 573 K, 1.0 to 2.0 MPa and a molar hydrogen to hydrocarbon ratio of 50 to 100 mol mol^{-1} at a space-time of 191.0 kg_{cat} s mol_{C6}^{-1}. These reaction conditions resulted in so-called ideal hydrocracking [2-4]. Ideal hydrocracking occurs if the acid catalyzed conversion reactions are the rate determining steps, which leads to very specific kinetic behavior. Figure 3 shows the experimental and simulated conversion of *n*-hexane by hydroisomerization on MC-389 catalyst as function of temperature at different hydrogen to *n*-hexane molar feed ratios. With increasing hydrogen to *n*-hexane ratio, the conversion is decreasing which indicates that (de-)hydrogenation is in quasi-equilibrium.

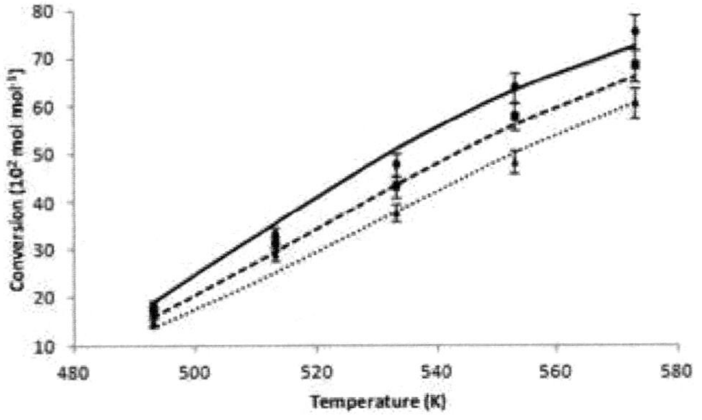

Figure 3: Conversion of n-hexane by hydroisomerization on MC-389 catalyst as function of temperature at different hydrogen to *n*-hexane molar feed ratios. Lines are calculated by the kinetic model. Circles, full line: $F^0_{H2} / F^0_{C6} = 50$ mol mol^{-1}; squares, dashed line: $F^0_{H2} / F^0_{C6} = 75$ mol mol^{-1}; triangles, dotted line: $F^0_{H2} / F^0_{C6} = 100$ mol mol^{-1}.

The reaction network was constructed under the validated assumptions of ideal hydrocracking kinetics and the corresponding rate equations were derived:

$$r_{4a} = \frac{k^{comp}_{pcp(1)} \dfrac{p_{C_6}}{p_{H_2}}}{1 + K_{phys}\left(p_{C_6} + p_{C_2MeC_5} + p_{C_3MeC_5}\right)}$$

Eq.1

in which p_{C6} is the partial pressure of *n*-hexane, K_{phys} is the equilibrium coefficient for physisorption, k_i^{comp} is the composite rate coefficient for pcp-branching or beta-scission. The rate coefficients and equilibrium coefficient are described by a reparametrized Arrhenius or Van ?t Hoff equation in order to decrease the binary correlation between the pre-exponential factor and activation energy:

$$k = k_{Tm} \cdot \exp\left(\frac{E_a}{R}\left(\frac{1}{T_m} - \frac{1}{T}\right)\right)$$

$$K = K_{Tm} \cdot \exp\left(\frac{\Delta H}{R}\left(\frac{1}{T} - \frac{1}{T_m}\right)\right)$$

In total, 8 kinetic parameters are determined, i.e., the rate and equilibrium coefficients at the mean temperature, i.e., k_{Tm} and K_{Tm}, and the corresponding activation energy and reaction enthalpy. Good initial guesses were acquired via isothermal regression, see Figure 4 and Table 1.

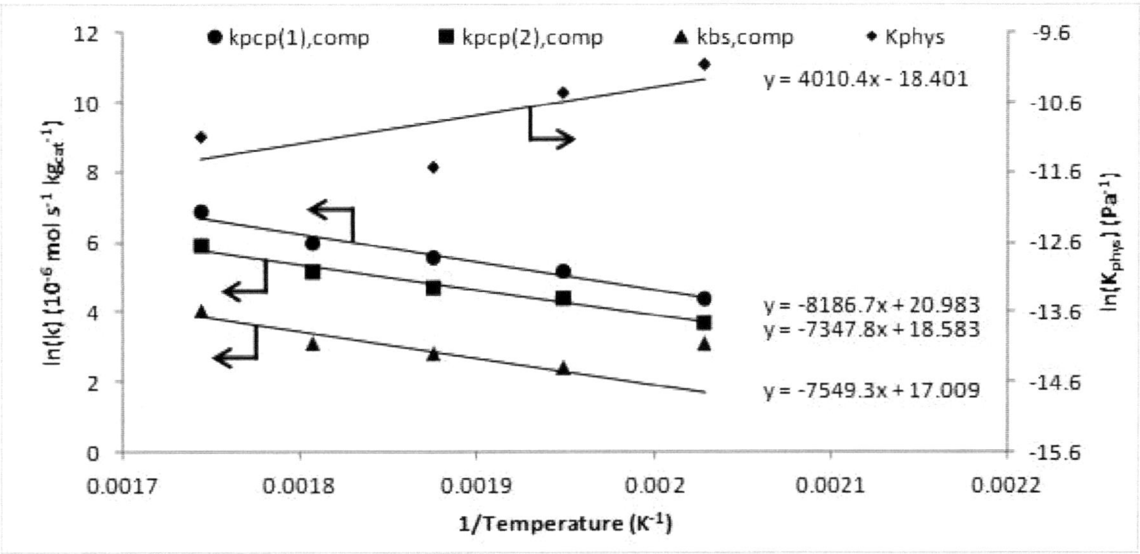

Figure 4: Arrhenius plots, ln(k) and ln(K) as function of the reciprocal of temperature for which an initial guesses for the pre-exponential factors and activation energies/reaction enthalpy are obtained for non-isothermal regression.

Table 1: Determined values of the kinetic/equilibrium coefficient at average temperature and activation energy and reaction enthalpy by the isothermal regression and the Arrhenius plots, see Figure 4.

	k_{Tm} or K_{Tm} (10^{-6} mol s^{-1} kg_{cat}^{-1} or Pa^{-1})	E_a or ΔH (kJ mol^{-1})
$k_{pcp(1),comp}$	264.9	68.0
$k_{pcp(2),comp}$	116.5	61.1
$k_{bs,comp}$	16.5	62.8
K_{phys}	1.2 10^{-5}	-79.5

Non-isothermal regression, i.e., regression taking into account of all experimental data at once, led to the significant determination of all 8 kinetic parameter, see Table 1.

Table 2: Parameter estimates, corresponding confidence interval and t values of the kinetic/equilibrium coefficients at average temperature and activation energies and reaction enthalpy determined by non-isothermal regression.

k_{Tm} or K_{Tm}	E_a or ΔH

	(10^{-6} mol s^{-1} kg_{cat}^{-1} or Pa^{-1})	(kJ mol^{-1})
$k_{4a,obs}$	258.6 ± 38.9	65.5 ± 9.2
$k_{4b,obs}$	115.7 ± 17.8	57.1 ± 8.8
$k_{5,obs}$	15.4 ± 5.8	65.8 ± 24.2
K_1	(1.1 ± 0.9) 10^{-5}	-83.4 ± 34.4

The model has a higher *F* value for the global significance of the regression, i.e., 758.4, than the corresponding tabulated *F* value, i.e., 3.9. All parameters are estimated significant as indicated by the relatively narrow confidence intervals. The parameter estimates also have a clear physical meaning. All binary correlation coefficient were lower than 0.95 which indicates that all parameters are uncorrelated with each other. A residual analysis showed that the model is capable of predicting the experimental observations, as shown the performance figure (Figure 5). Also the residuals seem to be normally distributed with an expected value equal to 0 and without having any systematic deviations as function of the independent variables as shown by the residual figure (Figure 6) and Q-Q plot (Figure 7).

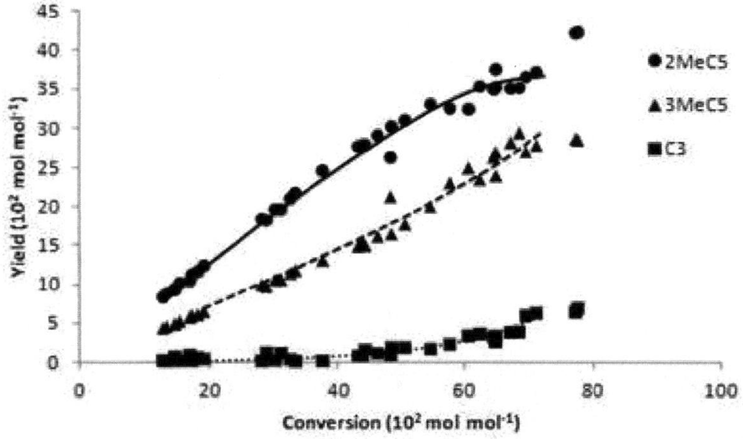

Figure 5: Product yield of *n*-hexane hydroisomerization on MC-389 catalyst as function conversion. Lines calculated by the kinetic model. Circles, full line: 2-methyl-pentane; squares, dashed line: 3-methyl-pentane; triangles, dotted line: propane.

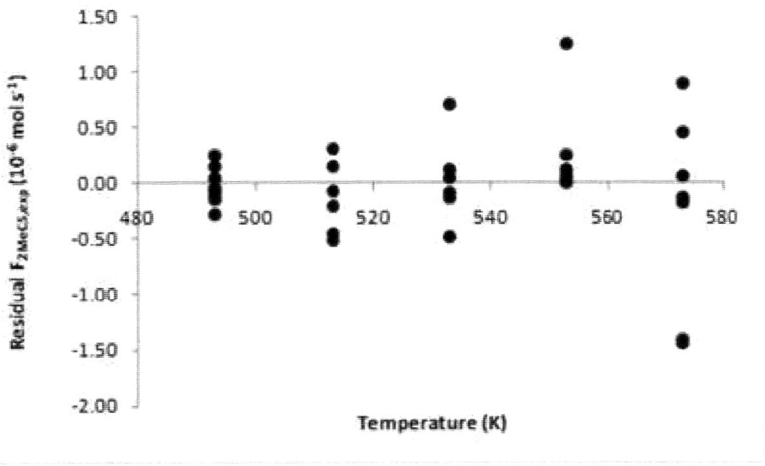

Figure 6: Residual figure for the molar outlet flow rate of 2-methyl pentane temperature determined by the kinetic model.

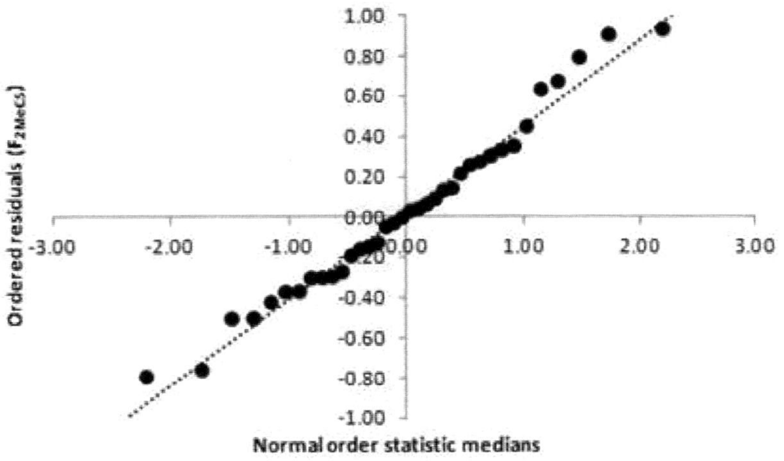

Figure 7: Q-Q plot of the residuals of the molar outlet flow rate of 2-methyl pentane determined by the kinetic model.

In summary, a systematic methodology for non-isothermal modeling of intrinsic chemical kinetics has been proposed. Inspection of the experimental data leads to an increased insight in the reaction mechanism and corresponding rate equations. Good initial parameter values are essential to obtain the optimal parameter values during non isothermal regression and can be acquired via isothermal regression. After regression, statistical tools can be used to evaluate the significance of the regression and performance of the model. This methodology has been successfully applied to *n*-hexane hydroisomerization on a bifunctional zeolite.

[1] G.F. Froment, K.B. Bischoff, Chemical reactor analysis and design (1990)

[2] T.F. Degnan, C.R. Kennedy, AIChE Journal. 39 (1993) 607-614.

[3] G.E. Gianetto, G.R. Perot, M.R. Guisnet, Ind. Eng. Chem. Res. 25 (1986) 481-490.

[4] J.W. Thybaut, C.S.L. Narasimhan, J.F. Denayer, G.V. Baron, P.A. Jacobs, J.A. Martens, G.B. Marin, Ind. Eng. Chem. Res. 44 (2005) 5159-5169.

(622b) Engineering in the Global and Societal Context: Four Credits, Three Weeks, Two Faculty, One Course

Vigeant, M., Bucknell University
Jablonski, E. L., Bucknell University
Vogel, B. M., Bucknell University
Raymond, T., Bucknell University
Originally presented on: 11/7/2013 8:52:00 - 9:14:00

Students who study abroad have higher graduation rates and higher GPAs than students who do not study abroad, and also report on the importance of this "transformative" experience to their lives. Unfortunately, semester-long study abroad can be challenging to schedule within the restrictive engineering curriculum. Therefore, a short-term intensive study abroad option was created specifically for engineering students. ENGR 290: Engineering in the Global and Societal Context was established in 2004 as a cross-disciplinary engineering elective to expose students to the impacts that environment, history, and culture have on engineering practice outside of the US.

ENGR 290 has been taught in Great Britain, Switzerland/Germany/France, Sweden/Norway, Brazil, China, Argentina, and Costa Rica. Students travel throughout the target country for three weeks in late May / early June, lead by 2-3 Engineering Faculty. Course goals include developing an understanding of the interrelationships between engineering practice, culture and regulation as well as developing a greater understanding of higher education within the host country. This is accomplished through visits to industrial sites, significant infrastructure, cultural and historical sites, and universities. Students are assessed on their engagement with the course, daily journal entries, and a significant term paper, among other assignments.

While not limited to chemical engineering students, the course has been taught by four of the chemical engineering faculty. In this talk, we will share specific benefits of this course for chemical engineering students and faculty. We will also discuss course outcomes and the extent to which the course has been successful in achieving them.

(622c) Teaching Classical Gas-Liquid Mass Transfer Models Using Multi-Physics Software

Uner, N. B.,
Uner, D., Middle East Technical University
Originally presented on: 11/7/2013 9:14:00 - 9:36:00

In this presentation, the sensitivities of the classical gas-liquid mass transfer models, i.e., the film theory, the penetration theory and the film-penetration theory, are compared with respect to the results of CFD simulations in a microfluidic domain. COMSOL® platform was used for CFD simulations. Two examples will be presented. The first example will demonstrate the inability of penetration theory to explain the phenomena at microscopic level. Due to large Fourier numbers in microfluidic domain, penetration theory fails. The second example will demonstrate how the two film theory depends on hydrodynamics. The film coefficients measured in double stirred tank reactors hardly reflect the reality. The real cases exhibit strong variations in concentrations both with respect to time and position in multiphase contactors. Furthermore, the film interface is not flat. Therefore, the accuracy of the predictions is still based on the empirical information averaged over the whole contactor geometry. Comparisons with the numerical simulations solving the species conservation equations (and in some cases also the equation of motion) using COMSOL® platform clearly reveals the limitations of the existing theories. This presentation intends to provide two examples for (under)graduate chemical engineering curriculum for teaching mass transfer using multi-physics software, hinting to the need for better theories of mass transfer.

(622d) Design Appraisal and Design Projects - What Would You Like to Design?

Ventura-Medina, E., The University of Manchester
Webb, C.,
Originally presented on: 11/7/2013 9:36:00 - 9:58:00

The final year design project represents, in most Chemical Engineering programmes, the point of convergence of all prior knowledge and skills acquired during earlier years and the culmination of the degree. As such Design Project is an essential part of any chemical engineering curriculum. Typically during the final year Design project, or Capstone design project, students work in teams on the design of a process or product over one semester mixing elements of group and individual work. Although there are a differences from university to university, the most common approach taken is generally based on providing the students with a brief that contains the key elements or constrains associated to the design problem. Plant capacity, location, product specification, raw material specifications, process or production route are some of the elements that might be typically contained in the general brief. Based in this brief and without having any input on the product or process selection students proceed over the semester to arduously work on the design of a process or product completing an incredible amount of work over a short period of time.

In some universities the initial part of the design project is to carry out a feasibility study of the process but in most cases this is not considered. Similarly, the most common way of delivering the Design project course is using the same project for the entire cohort and in cases introducing variations (e.g. location, plant capacity, product specifications) from team to team.

One of the requirements for IChemE accreditation in the UK is the need to demonstrate Advance Design and Design practice within the curriculum considering complex and open-ended tasks. Also, the development of critical thinking and problem solving are key skills considered for accreditation.

In 2010 a review of the Design project at Manchester called for some action to be taken in terms of the type of Design problems used, the focus of the project and the support in place prior to the Design Project itself. Based on this review few changes were made through the introduction of more variety of open-ended design problems and a different approach to supervision and support. Consequently, a new course, Synthesis and Design, was introduced in semester 1 to support design teaching and supervisors were paired up so that each pair will support 4 groups of students working in the same product. A selection of different products are then used for the project. However, maybe the two more radical changes were the introduction of a new project as part of the Synthesis and Design course, the Design Appraisal Project, and the lack of any design brief for either the Design Appraisal project or the Design Project. Students start the Design Appraisal project as a white paper.

Since its introduction the Design Appraisal Project has been delivered as part of the Synthesis and Design course, the semester before the Design Project, and constitutes the precursor of the Design Project. The Design Appraisal Project is an open-ended project that focuses on assessing the feasibility of the production of a given product. Students are asked to produce a Design Project Brief as the outcome of the project. This brief will allow them to carry out a Design Project on the same topic the following semester.

In 2010 students were given the task of assessing the feasibility of production of a given product among six different products pre-selected by the staff. From 2011 onwards, students have been asked to propose products that were interesting to them. Each team presents a Product Business Case Proposal to be considered by the Design teaching staff for further development. A selection of these proposals is then used to allocate products to teams.

This paper discusses the changes and approach introduced since 2010 to Design in the third year curriculum at Manchester. Positive outcomes, advantages, logistics and potential issues regarding the implementation of this approach will be presented from the perspe

(622e) Impact of the Science Without Borders Exchange Program On the Learning Experience of Students From the Chemical Engineering Course of the University of Campinas in Brazil

Rodrigues, M. T. M.,
de la Torre, L. G.,
Moraes, M., University of Campinas
Originally presented on: 11/7/2013 9:58:00 - 10:20:00

Science without Borders is a Brazilian Federal Program with the main goal of promoting consolidation and expansion of science, technology and innovation in Brazil through international exchange and mobility of students, professors and researchers. Two modalities of exchange activities are established: outgoing (for post doctorate candidates , undergraduate and graduate students) and incoming (for senior visitors and young talents). Until the year 2015, a total of around 100,000 grants will be given to fulfill this objective, from which close to 30% will be granted to undergraduate students of different areas going to several institutions around the world.

The School of Chemical Engineering at the University of Campinas has already sent a significant number of undergraduate students and the impact of the implementation of this program in the course and also on the evolution of learning styles and contents of students who had the experience abroad will be discussed. An immediate and obvious result observed in students taking their last year in Brazil is the more integrated approach to chemical engineering questions, what is undoubtedly extremely positive and desirable. Aspects such as curricular differences of several countries, forms of studying in the country and outside it, differences in attitude and specific contributions to student knowledge will be addressed together with the changes noticed on the side of the instructors.

(622f) Integrating Industrial Practice in the Chemical Engineering Curricula

Ventura-Medina, E., The University of Manchester
Hapgood, K. P., Monash University
Originally presented on: 11/7/2013 10:20:00 - 10:42:00

In recent years the department of Chemical Engineering at Monash has been increasingly incorporating contributions to the teaching delivery from industrial experts with the aim to enrich the curriculum content, provide up to date relevant expertise in engineering practice and improve the student experience.

Monash has successfully embedded chemical engineering practice within the curricula by having industrial experts contributing to significant components in the teaching delivery and in particular in years 3 and 4 where students have contact with industrial experts every semester. These contributions are via continuous delivery of material for 3-12 weeks of the semester and are additional to ad-hoc guest lectures and the industrial expertise of some academic staff. Units such as Process Control, Sustainable Processing I, Process Design, Engineers in Society, Design Project, Sustainable Processing II covering aspects such as control valves, sensors and control systems, sustainability, Life Cycle Analysis, carbon emissions and legislation, equipment design, HAZOP, major safety incidents, safety legislation, environmental management and legislation and waste management.

The paper will describe and discuss the approach taken to engage the industry overcome logistic issues, the synergies that have been built from these interactions and the perceived benefits and outcomes of this approach from the perspective of both students and staff.

(622g) Teaching Residence Time Distribution Concept and Applications through On-Line Active Learning

Barna, R., Ecole des Mines d'Albi-Carmaux
Sauceau, M., Processium
Veuillez, N.,
Berthiaux, H., UniversitÃ© de Toulouse, centre RAPSODEE, ecole des Mines dâ€™Albi-Carmaux
Originally presented on: 11/7/2013 10:42:00 - 11:04:00

Residence Time Distribution (RTD) theory and its application in the field of continuous reactor is one of the most concept-integrating and intellectually exciting course to give for a Chemical Engineering teacher. However, it is undoubtedly one of the hardest chemical engineering subject to teach, and more than often drives to a general feeling of disappointment from the lecturer. Its appropriation by the students requires a complete understanding of material balances for both reacting and non-reacting systems, as well as for steady-state and transitory flow systems. Advanced statistics, integral calculation, but also Laplace and Fourier transforms indeed represent a great deal of mathematical tools and concepts to handle properly, even for chemical engineering students approaching the end of their curriculum. The overall understanding by a student of the coupling between chemical reaction kinetics and the flow modeling through inert tracer perturbation requires time, maturation, and more than ever, dedication. This is certainly the reason why conventional ways of teaching, through lectures and classroom exercises, are usually failing in this field, as we have experienced over the past years.

In this work, we will present an active learning scheme that we have been using for two years and for two distinct populations, namely apprentices and students. It takes the form of a role play between groups of alumni and various teachers (or one teacher complying with various roles). Each group is part of an engineering office receiving an order from a customer who needs to comply with a specific conversion objective for its stirred industrial continuous reactor. Of course flow information is lacking, as well as the link between stirrer's speed and flow structure. Each group will therefore need to build and validate a flow model by defining and ordering RTD runs to an experimental station (in fact, a teacher) possessing the same type of reactors at a much smaller scale. Data back-treatment as well as model breakdown, allow to address customer's order through a formalized report.

All this is supported by an internet interface that allows exchange of information between groups, as well as the solicitation of an external expert, a role played by another teacher. The importance of the number of hits to the site is an indicator of the dedication and interest of the students. Basic knowledge tests performed at the end of the sequence, allow to validate the level attained by each alumnus. Both teachers and students are expressing their very positive experience of this way of teaching and learn.

(668a) Using Social Media in Chemical Engineering Courses

Zollars, R. L., Washington State University
Hundhausen, C., Washington State University
Originally presented on: 11/7/2013 12:30:00 - 12:51:00

Students use social media for many reasons including discussions amongst others within groups in which they may participate. This can be utilized to promote group interactions within the chemical engineering curriculum. We have used an asynchronous, social networking-type software developed by Chris Hundhausen (OSBLE – On-line Studio Based Learning Environment) in two different applications in our chemical engineering program.

The first application was in answer to the difficulties in assessing the ABET outcomes *3f – 3j* (an understanding of professional and ethical responsibility, an ability to communicate effectively, the broad education necessary to understand the impact of engineering solutions in a global, economic, environmental, and societal context, a recognition of the need for, and an ability to engage in, life-long learning, and a knowledge of contemporary issues). A procedure (the Engineering Professional Skills Assessment – EPSA) had been developed at Washington State University in which students would discuss a current event and have that discussion recorded by personnel from the Center of Teaching, Learning and Technology (CTLT). Transcripts of this discussion then were assessed for the level of ability in ABET *3f – 3j* by faculty using an established rubric. In 2010, however, budget reductions on the campus made it impossible for CTLT to continue to provide this service. Since the EPSA had received awards from both ASEE and ABET as a direct measurement of the ABET *3f – 3j* outcomes we desired to continue to use this process. OSBLE has been used for the past three years to fill the role formerly played by CTLT. Students now can participate in an on-line, asynchronous discussion about a current event just as they had in the past when personnel from CTLT recorded and transcribed the discussions. Now, however, this discussion may last over a much longer time (usually 1 – 2 weeks) since the students do not have to actually meet face-to-face and no transcription is required since OSBLE records all of the communication between students. Further details of how this has been implemented will be provided.

The second application is the use of a studio-based learning component in the material and energy balance class at Washington State University. In this class students are required to submit certain homework problems in an electronic format. The entire class then is divided into small groups (3 – 4 students) which then are assigned to review another student's problem solution. OSBLE again is used to facilitate this discussion. The asynchronous nature of OSBLE allows students to participate in this discussion on their own schedule, thus freeing the approach from the time constraints of a typical class schedule. OSBLE also anonymizes all of the student-to-student communication so that students do not know who submitted the original solution nor do they know which other students are in their discussion group. After a period of about 1 week the students are asked to come to a consensus as to the advice they (as a group) would give to the student who submitted the original solution on ways to improve that solution. This necessity to go beyond merely applying principles from the material and energy balance class to having to evaluate other's work thus promotes higher level learning skills (the "evaluation" level in Bloom's Taxonomy) than can typically be reached in this type of class.

The use of OSBLE has not only allowed us to use something students use all of the time (asynchronous, social networking) to provide a richer learning environment for students but also provides the opportunity to collect the student's communications to assess their level of understanding.

(668b) The Role Of Smartphones and Tablets In Numerical Problem Solving

Elly, M., Ben-Gurion University of the Negev

Shacham, M., Ben Gurion University of the Negev

Cutlip, M. B., University of Connecticut

Originally presented on: 11/7/2013 12:51:00 - 13:12:00

Numerical problem solving has been introduced into the Chemical Engineering education and practice in the early nineteen sixties (Shacham et al. 1996). At that time Fortran or other source code programming languages were used to implement the numerical solution algorithms on "mainframe" computers. Starting in the mid nineteen eighties the emphasis shifted to the use of general purpose mathematical software packages, such as MAPLE, MATHCAD, MATLAB, Mathematica and Polymath for numerical problem solving using the PC. PC based mathematical software packages are routinely used, nowadays, for numerical problem solving in engineering education and practice (Cutlip et al., 1998, Shacham and Cutlip, 1999) as they are easy to use and provide the most time efficient route for obtaining accurate solutions.

The dominant position of the PC in the computing field has been challenged lately by the mobile devices such as smartphones and tablets. Furthermore, it is predicted that by 2017 the dominant operating system for all computing devices will be Google's Android (Wingfield, 2013). Relying on these trends and predictions we decided to check the possibility of the use of Android based smartphones and tablets for numerical problem solving in chemical engineering. For this aim, we have developed the PolyMathLite application, a slightly simplified version of the Polymath software package (Polymath is a trademark of Polymath Software http://www.polymath-software.com). PolyMathLite solves systems of linear and nonlinear (NLE) algebraic equations, systems of ordinary differential equations (ODE) and carries out polynomial, multiple linear and nonlinear regression of data.

PolyMathLite was evaluated on both Android smartphones and tablets using the set of test problems listed in Table 1. Ten out of the eighteen problems are described in detail by Cutlip et al. (1998). These ten problems were solved by a group of educators on PC using the mathematical software packages Maple, Mathematica, MATLAB, MathCAD, Polymath and Excel. A comparison of the performance of the various packages in solving the set of 10 problems was reported by Shacham and Cutlip (1999). Six additional problems are described in the textbook of Cutlip and Shacham (2008). Sample problem 17 is from an NLE test problem library (Shacham et al., 2002) while problem 14 is described in the reference: Shacham et al., 2009.

The test problems used represent a great variety with respect to the complexity of the mathematical model, the level of difficulty of the numerical solution and the associated ChE course. The various problems are applicable to required courses, such as: introduction to Ch. E., thermodynamics, fluid dynamics, heat transfer, mass transfer, separation processes, reaction engineering, process dynamics and control; and to the elective courses: mathematical methods, biotechnology and optimization. The simplest problems consist of a single NLE (Problems 2 and 4), while the most complex consist of 12 NLEs (Problem 18) and 9 ODEs. Some of the problems pose special challenges to the solution algorithms. In Prob. 11 a differential-algebraic system (DAE) need to be solved, in Prob. 12 a partial differential equation (PDE) is solved using the method of lines, and multiple linear regression of the data in Prob. 14 with a model that includes a free parameter yields an ill conditioned regression model while removing the free parameter provides a statistically valid model. Problems 10 and 13 require special stiff integration algorithms to find the correct solution. Problems 17 and 18 represent special challenges to the NLE solver algorithms as the values of some of the variables (mole fractions) are very close to zero at the solution, however these variables are not allowed to obtain negative values during the solution process as their logarithm or square root needs to be computed.

In spite of the special challenges in terms of the problem size and algorithmic requirements, the solutions obtained using PolyMathLight on smartphones and tablets matched the solution obtained by Polymath 6.1 up to seven significant digits for all the problems included in the test problem set. The solutions were obtained in a very short time, even for the large scale, complex problems. On the negative side, the problem input via the touch screen has proven to be very cumbersome and time consuming.

In the extended abstract and the presentation the testing of PolyMathLite will be described in more detail and the potential effects of the availability of numerical problem solving tools for smartphones and tablets on the CHE education and practice will be discussed.

References

1. Cutlip, M., Hwalek, J.J., Nuttall, H.E., Shacham, M., Brule, J., Widman, J., Han, T., Finlayson, B., Rosen, E. M. and Taylor, R., ?A Collection of Ten Numerical Problems in Chemical Engineering Solved by Various Mathematical Software Packages.? *Comput. Appl. Eng. Educ.*, 6(3), 169-180(1998)

2. Cutlip, M. B. and Shacham, M. Problem Solving In Chemical and Biochemical Engineering with Polymath, Excel and MATLAB. Prentice-Hall, Upper Saddle River, New-Jersey, 2008.

3. Shacham, M., M. B. Cutlip, and N. Brauner, "General Purpose Software for Equation Solving and Modeling of Data", pp. 73-84 in Carnahan, B. (Ed), ?Computers in Chemical Engineering Education?, CACHE Corporation, Austin, TX, 1996.

4. Shacham, M. and M.B. Cutlip, ?A Comparison of Six Numerical Software Packages for Educational Use in the Chemical Engineering Curriculum?, Computers in Education Journal, IX(3), 9-15 (1999)

5. Shacham, M., N. Brauner and M. B. Cutlip, ?A Web-based Library for Testing Performance of Numerical Software for Solving Nonlinear Algebraic Equations?, Computers chem. Engng. 26(4-5), 547-554(2002)

6. Shacham, M., Cutlip, M. B. and Elly, M., "Live Problem Solving via Computer in the Classroom to Avoid "Death by PowerPoint"", Computer Applications in Engineering Education Vol 17, No. 3, 285-294(2009)

7. Wingfield, N., "PC Sales Still in a Slump, Despite New Offerings", The New York Times, April 10, 2013.

Table 1. List of Sample/Test Problems for PolyMathLite

No.	COURSE	PROBLEM TITLE	MATHEMATICAL MODEL
1	Introduction to Ch. E.	Steady State Material Balances on a Separation Train	Simultaneous Linear Equations[1]

2	Introduction to Ch. E. and Thermodynamics	Molar Volume and Compressibility Factor from Van Der Waals Equation	Single NLE[1]
3	Thermodynamics and Separation Processes	Three Phase Equilibrium - Bubble Point	NLE System[2]
4	Fluid Dynamics	Terminal Velocity of Falling Particles	Single NLE[1]
5	Thermodynamics and Reaction Engineering	Reaction Equilibrium for Multiple Gas Phase Reactions	NLE System[1]
6	Mathematical Methods	Vapor Pressure Data Representation by a Polynomial and Equations	Polynomial Fitting, Linear and Nonlinear Regression[1]
7	Heat Transfer	Unsteady State Heat Exchange in a Series of Agitated Tanks	ODE's with known initial conditions[1].
8	Mass Transfer & Reaction Engineering	Diffusion with Chemical Reaction in a One Dimensional Slab	ODE's with split boundary conditions[1].
9	Reaction Engineering	Reversible, Exothermic, Gas Phase Reaction in a Catalytic Reactor	ODE System[1]
10	Process Dynamics and Control	Dynamics of a Heated Tank with PI Temperature Control	Stiff ODE System[1]
11	Separation Processes	Binary Batch Distillation	Differential-Algebraic System[1]
12	Heat Transfer	Unsteady state Heat Conduction in a Slab	Partial Differential equations[2]
13	Reaction Engineering	Rate Data Analysis for a Catalytic Reforming Reaction	Nonlinear Regression[2]
14	Mathematical Methods	Regression of Heat of Hardening Data of Cement	Multiple Linear Regression[3]

15	Biotechnology	Stoichiometric Calculations for Biological Reactors	Simultaneous Linear Equations[2]
16	Biotechnology, Reaction Engineering and Mathematical Methods	Biological Reaction in a Batch Reactor	Stiff ODE System[2]
17	Reaction Engineering and Mathematical Methods	Combustion of Propane in Air	NLE system where some variables must be positive[4]
18	Reaction Engineering and Optimization	Complex Chemical Equilibrium	NLE system where some variables must be positive[2]

[1]Reference: Cutlip et al., 1998

[2]Reference: Cutlip and Shacham, 2008

[3]Reference: Shacham et al., 2009

[4]Reference: Shacham et al., 2002

(668c) Using Process Simulators in Chemical Engineering Lectures: Case Study, Ethyl Lactate Production

Vitery, T., Universidad Nacional de Colombia
Jaimes, D., Universidad Nacional de Colombia
Monroy, C., Universidad Nacional de Colombia
Originally presented on: 11/7/2013 13:12:00 - 13:33:00

The need for new products and a new culture of caring for the environment, saving and optimizing energy, the rationalized use of fuel, etc, makes necessary to Chemical Engineering look for feasible and economic alternatives to solve these new challenges. Process simulation is a process of physical or chemical transformation by a mathematical model involving the mass and energy balances coupled to phase equilibrium, transport equations and chemical kinetics. This is an efficient and effective tool for analysis, synthesis and optimization of the process.

The development of process simulators worldwide has unfolded rapidly. Mainly due to three aspects: a) Computers with faster processors, graphical interfaces that facilitate the handling of graphics, storage of large amounts of data. b) Structured programming languages that facilitate more efficiently the solution of systems of equations and the same simulation. c) Determination of properties of mixtures and components with less deviation of the data obtained experimentally.

From real data (used to obtain a mathematical model and/or validate the results of the model or simulation) can gain information that allows analysis, synthesis and optimization. Furthermore, there is not required a lot of experiments that increase the cost and destabilize the operation of any process plant, because if the pattern used is appropriate (after validation), the reproducibility of results is excellent.

Therefore, the Department of Chemical and Environmental Engineering of the National University of Colombia has been working since 1992 with the simulator HYSIM (later renamed HYSYS), since 2001 with ASPEN ENGINEERING SUITE and created the Process Simulation course in 2003. This course aims to integrate, reinforce and consolidate the knowledge of the discipline courses, present the advantages for parameter estimation, fit and model validation, design, simulation, analysis and optimization of process simulators aspenONE ® V .7.3, SuperPro Designer ® and BR & E ProMax ®, and the interaction of these with Excel and Matlab.

This article aims to present a study conducted by a group of students that make up the subjects (Process Engineering, Plant and Equipment Design) and Modeling and Simulation of Chemical Processes, which select a compound of interest, evaluating different synthesis routes thereof and select one. Simulate the selected route in ASPEN PLUS and ASPEN HYSYS and propose an utility equation to selection and subsequent equipment sizing and optimization. The case study is ethyl lactate production from lactic acid and ethanol.

Ethyl lactate, also known as green solvent is an organic ester, biodegradable, used as a food additive, flavoring and cellulose, cellulose acetate, cellulose ethers, and various resins solvent. It is a particularly attractive solvent in the coating industry because of its high solvency, high boiling point, low vapor pressure and low surface tension. Likewise, ethyl lactate is a desirable coating for wood, polystyrene and metals, and also acts as an effective separator and paint remover. Also, ethyl lactate holds promise as an effective and nontoxic replacement for petroleum-based solvents.

One of the production routes of ethyl lactate (L_1E) is the esterification from aqueous lactic acid solution (88 wt %) in the presence of Amberlyst 15 cation-exchange resin as a catalyst. This heterogeneous catalyst has the advantage of be friendly with the environment, is not corrosive, easy to separate, and easy handling and storage in comparison with other homogenous catalysts like sulfuric acid.

Reactions are carried out in a chemical reactor or in a reactive separation column, in the first are necessary more unit operations for ethyl lactate separation and recovery. The use of reactive distillation allows an efficient production of organic acids esters, because the reaction and separation process integration in a unique equipment. Furthermore, the higher conversion gained is comparable with the equilibrium conversion in a batch simple reactor. These synthesis pathways have undesired reactions because of the bifunctional nature lactic acid, which undergoes intermolecular esterification in aqueous solutions above about 30 wt % to form linear dimer (L_2) and higher oligomer acids (mainly L_3, L_4). The extent of homoesterification increases with increasing acid concentration, thus complicating the use of lactic acid as a reactive substrate. When mixed with an alcohol, lactic acid and its oligomers undergo esterification.

The extent of oligomerization is inversely related to the water content of the solution. In reactive distillation or any other lactic acid esterification scheme, oligomer formation and esterification adversely affect ethyl lactate yield and pose a considerable challenge in predicting process behavior. For accurate design, therefore, it is critical to include and characterize oligomer reactions in the process model. Ethyl lactate (L_1E), typically the desired product, can be recovered from the mixture, but its yield is reduced from the theoretical maximum by the presence of the oligomeric compounds.

The monomer (L_1) esterification and oligomer (L_2, L_3) formation and esterification, are completely described by the following set of reactions:

$$L_1 + EtOH \rightleftharpoons L_1E + H_2O$$

$$L_2 + EtOH \rightleftharpoons L_2E + H_2O$$

$$L_3 + EtOH \rightleftharpoons L_3E + H_2O$$

$$L_2 + H_2O \rightleftharpoons 2L_1$$

$$L_3 + H_2O \rightleftharpoons L_1 + L_2$$

In lactic acid esterification, where a series of relatively slow reactions occurs, a reliable knowledge of the kinetics is, thus, essential to simulate the process effectively. The kinetics of esterifying lactic acid and its oligomers with ethanol over Amberlyst 15 cation-exchange resin have been determined via batch reaction studies.

The kinetic model is useful in both batch and continuous process designs for lactic acid esterification. It uses nth-order, reversible rate expressions for esterification and oligomerization reactions. Because the reversible reactions, concentrated (88 wt %) lactic acid is the preferred feedstock because it contains relatively little water, also because less alcohol is required and is not necessary feed preheating.

Components such as diethyl lactate (L_2E) and triethyl lactate (L_3E) not found in ASPEN HYSYS and ASPEN PLUS databases, estimation of their properties (in both simulators) was performed using its structural formula and physical properties found in the literature. Also was performed the estimation of the thermodynamic model parameters and its validation using equilibrium data reported in the literature.

In this way a comparison was made between the process that use a single reactor another that use a reactive distillation column, from the operational, of safety, and economic point of view, presents the advantages and disadvantages of each. We performed the design, modeling, simulation, equipment sizing, optimization and energy integration of the process. To optimize the process, we performed a sensitivity analysis between the main independent variables of the process such as temperature, excess molar ratio, feed tray, Holdup time, among others, to see the behavior of the conversion, selectivity of lactate acetate and raised economic function for the two production routes.

(668d) Use of Arduino Microcontrollers in Chemical Engineering Curricula

Butterfield, A., University of Utah

Originally presented on: 11/7/2013 13:33:00 - 13:54:00

While the proper collection and analysis of real-world data using sensors, microcontrollers, and computer software are vital chemical engineering skills, most textbook problems given to our students assume data are obtained properly and will give no indication of the sort of error inevitable in physical measurements. Hands-on use of actuators and sensors in chemical engineering curricula is typically reserved for junior or senior students in unit operations or process control laboratories. Such a delay in incorporation of real-world data is in part due to the expense and complications found in giving students access to microcontrollers and sensors. To overcome such hurdles, we have introduced Arduino microcontrollers into our freshman design laboratory and have begun incorporating them into the remainder of our curriculum.

Arduino Uno boards are open-sourced, inexpensive ($23), and flexible microcontrollers. They have six analog inputs, and fourteen digital inputs/outputs, six of which may be pulse width modulated outputs and used as analog outputs with minimal additional circuitry. Each board may operate alone, using only its onboard compiler. Alternatively, they may be controlled through a USB cord using the Matlab programming language, which is used throughout our curriculum.

In order to train our students on the use of Arduino boards, we have created a body of publicly available teaching materials and instructional screencasts. As a result, we have been able to use these microcontrollers as a key component of our freshman design laboratory. Student teams in this course use Matlab and Arduino boards in a series of design competitions involving a wide range of core chemical engineering concepts. By the end of the first year in our program each student is capable of using these boards with Matlab on department computers or on their own laptop to take a wide variety of physical measurements. This new student skill has opened up a range of novel active and collaborative projects for many other core chemical engineering courses, and we have begun to bring hands-on design projects based on the Arduino boards into upper division courses.

We have also amassed a large collection of inexpensive sensors that may work in conjunction with Arduino boards and we have assessed them for pedagogical use. Such sensors may be used in a wide range of chemical engineering courses, and may measure a wide variety of properties, such as: pressure, temperature, humidity, ethanol and carbon monoxide concentration, light intensity, and so on. Teaching modules involving the characterization of mass and heat transfer, reaction kinetics, fluid mechanics, thermodynamics, and microbial growth have been developed using the Arduino platform.

The results from this work have been wide reaching. Student survey responses regarding projects involving these microcontrollers have been overwhelmingly positive. Faculty members are very receptive to bringing such hands-on projects into our core courses, once the foundational instruction on Arduino boards has been established in our freshmen class. Lastly, the Arduino boards have become a vital component of our department's outreach efforts, finding use in summer camps and K-12 visits.

(668e) Biomolecular Engineering - Case Studies and Web-Access Exams

Curtis, B. S., Berkeley
Curtis, W. R., The Pennsylvania State University
Originally presented on: 11/7/2013 13:54:00 - 14:15:00

SUMMARY: In 2004, an educational experiment was undertaken with 20 honors students at the Pennsylvania State University to teach Biomolecular Engineering in and 'open web' format. Students would have access to the Internet for a significant part of the course, including during exams. This teaching approach allowed in-course access to the burgeoning bioinformatics databases (GenBank, BRENDA, KEGG, &etc) while simultaneously emphasizing synthesis: complex problem solving that requires students to think critically about and apply fundamental principles from multiple chemical engineering disciplines (transport, thermodynamics, kinetics). This course has subsequently evolved into a required junior-level course. This presentation will discuss the challenges and potential advantages of this educational format, with emphasis on its applicability to larger class sizes (75+ students). Strategies for managing the format's unique technological (e.g. rapidly-evolving online resources) and administrative (e.g. academic integrity) challenges are considered. The discussion will include the perspective of a former student and later teaching assistant of this course, who has since taught an analogous Biomolecular Engineering course as a graduate student at UC Berkeley.

BACKGROUND: In the past 25 years, biological aspects of chemical engineering have become sufficiently 'mainstream' that many Departments have chosen to adopt some permutation of this field as part of their official departmental name. Although Penn State did not adopt a name change, a curriculum track in biochemical engineering was implemented in the early 1990's. As a result, at the time of the presentation format experiment in 2004, our department had nearly a decade of experience in teaching a curriculum track that included biochemistry, bioprocess engineering, bioreactor design, bioseparations and numerous professional and technical electives. This curriculum track accommodated around 25 students, but the electives often had enrollments of 50-75 students. Although discussions of integrating bio into all other courses periodically resurfaces, we have retained this separate curriculum. As we have increased the biological component of standard undergraduate curriculum to include additional biochemistry and molecular/cell biology, this resulted in flexibility in developing a course that is more advanced. This has lead to the dilemma of a Junior level course that touches on the breadth of chemical engineering fundamentals in a biological context. A case study format as discussed below was found to accommodate this situation, while also accommodating the lack of retention that some students experience from lower level courses - particularly biological content.

CASE STUDY FORMAT: The traditional format for Biochemical engineering follows 'chapters' of biochemistry, metabolism, enzyme and growth kinetics with typical divergence into various speciality areas of metabolic engineering, bioinformatics (and now synthetic & systems biology). An alternative approach is to focus on a series of case studies in which the fundamentals are introduced 'on demand' in the context of the production of a product. The case studies of high fructose corn syrup (HFCS), Insulin, Erythropoietin (EPO), Beer & Biofuels currently provide for the coverage of the content. This case study approach also provides for vertical themes (e,g, carbohydrates, organism production platforms, data-fitting and statistics) that advance with each additional case study. Recognizing the power of the internet to provide unprecedented access to information, the real 'experiment' was to change the fundamental format of a course to teach a significant part of of the course where students are on computers - including quizzes and midterm exams. Access to computers provides a decreased reliance on content details, and greater emphasis on concept. This also provides for a much needed synthesis of principles that has plagued the reductionist approach to chemical engineering curriculum where we have dissected our fundamentals into separate courses - where real world processes include considerations of competing considerations of intrinsic kinetics, mass transfer, heat transfer etc.

CONCLUSIONS & RECOMMENDATIONS: A divergence from 'chapters' received very divergent reactions as a mode of instruction. Some students thrive on the format, and others become extremely anxious about 'what they need to know'. It can be argued that this provides for an accommodation of a diversity of learning styles.

Special attention must be given to the issue of academic integrity, as having access to the internet during an exam opens up nearly unlimited opportunity to try to gain an advantage; very explicit rules on what can, and cannot be done … and extra precaution to be vigilant about maintaining confidence that being honest will not be a disadvantage. The ability to use the internet as both an instructional tool as well as a component of the examination assessment process is a paradigm shift away from traditional teaching of Biochemical Engineering. This places the emphasis on principles and thinking … which is more consistent with the nature of chemical engineering education. This experience has potential to be implemented for the myriad of specialty topics of the Junior/Senior year curriculum experience where one can get quickly mired down in the 'language' and content details of a special topic, instead of refocusing on the important aspects of an iteration of ChE fundamentals in a new context. A clear advantage of this approach to teaching is that it teaches students to 'walk-the-talk' of life-long learning. The clear downside of this approach is the effort required to keep up with the rapid pace of changes of web-based information, and the need to create novel problems each semester.

(668f) Not Your Average Flipped Classroom: An Online, Multi-Media "Textbook" for Introduction to Chemical Engineering

Hwang, L. Y., Stanford University
Dunn, B., Stanford University
Khosla, C., Stanford University
Originally presented on: 11/7/2013 14:15:00 - 14:36:00

Students have been learning from traditional textbooks that are static in content and organization, and instructors constantly face the challenge of identifying a textbook that aligns with their own teaching goals. Our goal has been to create an online, multi-media “textbook” for our *Introduction to Chemical Engineering* course that can be easily modified to fit the needs of different instructors and students and made freely available to the public. We have aimed to make our evolved “textbook” (1) easily adaptable by using a modular organization with associated short video screencasts, solved example problems, and practice problems; (2) able to integrate many media types including text, screencasts, video demonstrations, and eventually applets; (3) easily updated so that instructors can modernize the content when needed or simply use their own favorite examples to illustrate concepts; and (4) free of any copyrighted material so that we can make it freely available to others to use in their own classrooms. In our first iteration using this online resource, we have found that we now have the flexibility to use in-class meetings in a way similar to that of typical “flipped” classrooms, which is to discuss interesting current examples that apply concepts covered in the assigned module of the online textbook, discuss more practice problems, and integrate more interactive student learning activities.

(668g) Online Data Resources in Chemical Engineering Education: Impact of the Uncertainty Concept

Magee, J. W., National Institute of Standards and Technology
Kim, S. H., Korea University
Kang, J. W., Korea University
Kroenlein, K., National Institute of Standards and Technology
Diky, V., National Institute of Standards and Technology
Muzny, C., National Institute of Standards and Technology
Kazakov, A., National Institute of Standards and Technology
Chirico, R. D., National Institute of Standards and Technology
Frenkel, M., National Institute of Standards and Technology
Originally presented on: 11/7/2013 14:36:00 - 14:57:00

In traditional chemical engineering education practice, students have relied on either handbooks or online resources for a thermophysical property value that is relevant to an assigned problem. A property value consists of three parts, namely: the numerical value; its units of measurement; and, its uncertainty. Frequently, it's this third part that is inadequately evaluated or even unknown. That can spell trouble, since uncertainty propagates through a problem's solution to the result. We review the concept of uncertainty for thermophysical properties, its implementation in globally-adopted communication standards for properties data and the evolution of online data resources. To facilitate the translation of developments to engineering education, we employ NIST Web Thermo Tables to furnish properties data with their associated expanded uncertainties. We present pertinent examples for the engineering classroom [thermodynamics; fluid mechanics; heat and mass transfer] and an example application to research. These examples illustrate the practical combination of the concept of uncertainty with online data resources. Reference: S.H. Kim, J.W. Kang, K. Kroenlein, J.W. Magee, V. Diky, C.D. Muzny, A.F. Kazakov, R.D. Chirico and M. Frenkel, *Chemical Engineering Education*, 47(1),48-57(2013).

(676a) Just-In-Time Teaching in a Material and Energy Balances Course

Liberatore, M. W., Colorado School of Mines

Originally presented on: 11/7/2013 12:30:00 - 12:46:00

An attrition rate of 25 to 35% for the material and energy balances course is believed to be "average" based on previous years at the Colorado School of Mines, conversations with other faculty who teach the same course, and published conjecture by Felder. The motivation for this work derives from insufficient student learning related to a lack of practice (with problem solving). Recently, many, likely a majority of, students copied the solutions manual leading to a false sense of understanding the material and problem solving required in the Material and Energy Balances course. One student even printed off the solutions manual and turned it in UNCHANGED, except for putting their name on it. While the fundamentals of engineering disciplines are the same as in previous years and decades, the students think differently. Technology has dictated how these "digitally native" students think and pedagogical techniques need to change to challenge these students in new ways. Just-in-time teaching feedback from online homework from Sapling Learning and instructor-written online quizzes immediately identify conceptual understanding and problem solving skills of the class. The positive results with a recent online homework experiment (published in Chemical Engineering Education in 2011) showed a clear need for more faculty using online homework technology. Recent results (to be published in Chemical Engineering Education in 2013) will be the focus in this talk.

(676b) Professional Development Memos in Introduction to Engineering Courses

Duke, S. R., Auburn University

Originally presented on: 11/7/2013 12:46:00 - 13:02:00

First year students in Introduction to Engineering courses attend professional organization meetings and provide a memorandum that describes the meeting content and how the organization and meeting content affects their development and career direction. The memos have been part of the chemical and electrical sections of the Intro course at Auburn University for more than 7 years (more than 2000 students). Students must complete two memos as part of the course grade; additional memos may be completed for extra credit. One memo is due prior to mid-term: it is used to provide early intervention for students with writing skill deficiencies. The memo assignment is used as a platform to introduce professional and technical communication guides, notions of lifelong learning, and department expectations and policies. Students seek and select meetings or lectures sponsored by approved technical societies and student organizations. Anecdotal results include a number of student organization leaders who attribute the memo requirement with getting them involved early in engineering organizations and competition teams. Assessment results indicate that the memo experience assists in retention of engineering students and in attaining writing outcomes.

(676c) Successes and Challenges of Introducing Peer-Led-Team-Learning Early in the Chemical Engineering Curriculum

Willing, G. A., University of Louisville
Originally presented on: 11/7/2013 13:02:00 - 13:18:00

The Partnership for Retention In Math, Engineering, and Science (PRIMES) is an NSF STEP program being implemented across 9 different departments in 3 different schools (Engineering, Education, and Arts and Sciences) at the University of Louisville. The cornerstone of this program is the development of Peer-Led-Team-Learning (PLTL) communities in several of the foundation courses for each of the participating departments. In Chemical Engineering, these courses include Introduction to Chemical Engineering, Introduction to Materials Science and Material and Energy Balances. The exact implementation of PLTL varies from course to course, but most focus on the Supplemental Instruction (SI) model. The success of PLTL in the different courses is largely determined by how the PLTL has been implemented and the nature of the course. This talk will focus on the successes and challenges in implementing PLTL in the various courses. Additionally, initial data regarding the impact of PLTL on both the student and their peer-mentors will be discussed.

(676d) Influence of Human Social Networks On Engineering Education

Pease, L. F. III, University of Utah

Originally presented on: 11/7/2013 13:18:00 - 13:34:00

Here we elucidate the influence of special actors within class social networks on student academic performance and commitment to chemical engineering. Every chemical engineering department desires to enhance student performance and student commitment, expending significant resources in this pursuit. However, very little is understood about how extracurricular (non-virtual) networks affect engineering student performance and certainty of their major. Here we use social network analysis to evaluate the influence of three types of special actors within extracirrcular student networks, the top students (>93.5%), connectors (>7 links to others in the class), and persuaders (identified by >2 peers as convincing), on student performance and certainty of major in our introductory chemical engineering course. We find direct connectivity to these special actors to have a significant and positive influence on student performance as measured by final grades, particularly for women, working students (>7 h/wk), and entering freshmen. Direct connections to the least connected (≤1 link) or students with the lowest grades has a dramatically negative influence on the grades of transfer students. Students who work or have long commutes also report statistically significant commitment to the major. We also highlight the influence of randomly assigned study buddies on student connectivity and specific strategies to enhance student performance based on these findings.

(676e) Degrees of Freedom - Not Just for Sophomores

Biernacki, J., Tennessee Technological University

Originally presented on: 11/7/2013 13:34:00 - 13:50:00

Typically, chemical engineering students are introduced to the concept of degrees of freedom along with mass balances where the concept is used to help formulate balances and related equations and again in thermodynamics in the context of phase equilibrium. Unfortunately, the concept is not used beyond these contexts, the formalisms are not directly derivable one from the other and, as a consequence students do not ascertain that the *concept* is more generalizable than these specific applications would lead them to believe. Since, the degrees of freedom concept is generally not practiced outside of these isolated contexts and, in fact, in subsequent courses, students that learned these formalisms will revert to "hunt and peck" strategies for solving material balances and for doing thermodynamic phase or reaction equilibrium calculations; rather than apply a degrees of freedom analysis as a starting point. This lack of skill and understanding limits the students' ability to accurately formulate problems; an area that can be greatly improved by implementing a pervasive approach to the utilization of a robust and generalized degrees of freedom formalism throughout the curriculum. Therefore, an extended degrees of freedom formalism is proposed that can be applied to steady and dynamic processes. The new formalism is an extension of existing practices that in general do not collapse to the Gibbs Phase Rule and do not work for dynamic situations.

(676f) Facilitating Learning in Thermodynamics and Computations Using Technology

Lira, C. T., Michigan State University
Elliott, J. R., The University of Akron
Originally presented on: 11/7/2013 13:50:00 - 14:06:00

Enhanced technology is enabling greater individualized asynchronous delivery of information and learning. While uncertainties exist over the formal adoption of MOOCs for college credit, some of the tools can be integrated into traditional courses. Screencasts provide delivery of information, but by themselves do not provide an assessment of student skills. Over the past two years, screencasts have become available for many topics in thermodynamics, both from learncheme.com and chethermo.net. Such material provides opportunities for more personal interaction in the classroom, or group exercises.

For assessment of learning, individualized assessment is preferred. A disturbing trend in higher education is the proliferation of online homework solutions, which compromises the assessment of learning and the benefit of students working through the problems. Both physics and chemistry departments have been addressing scaling of delivery and assessment for almost a decade, making it logical to jump on their learning curve. Recently, some publishers and commercial ventures have been developing randomized homework for students, but primarily for the large service courses, such as chemistry, physics, and statics. It is unlikely that significant investment will occur for smaller programs such as chemical engineering. Michigan State has started to develop homework/quizzing materials on the open source lon-capa platform (http://www.lon-capa.org/) now available at 160 institutions world-wide. Unlike commercial packages, instructors can develop and share problems across institutions and change textbooks without loss of flexibility in use of the tools. Simple scripting in perl enables randomization, and running problems with a computer algebra system (maxima) is supported as well as real-time plotting using gnuplot. Any institution can adopt lon-capa by providing a server. Experience in use of lon-capa at MSU was extremely positive for multi-step problems as well as computer programming tutorials. MSU Students were requesting *more* lon-capa problems! Students liked: 1) the immediate feedback when a step was incorrect; 2) multiple trials to do self-correcting or consult with classmates (about the procedure since the values were individualized!); 3) the flexibility to move at their own pace; 4) use of technology to help them learn.

(676g) Team-Based Learning: Developing Workplace Skills Alongside Engineering Fundamentals

Lamm, M. H., Iowa State University
Originally presented on: 11/7/2013 14:06:00 - 14:22:00

Background: An instructional strategy called team-based learning (TBL) [1] was implemented in an introductory chemical engineering course on material and energy balances (MEB). TBL is an alternative to conventional classroom lecture instruction. In TBL students are placed in high-performance learning teams and work together, during class time, to apply course concepts and solve discipline-relevant problems called application exercises. The goal of introducing TBL in the MEB course was to enhance the quality of student learning by fostering teamwork skills and life-long learning strategies along with the acquisition and mastery of fundamental engineering concepts.

Description: TBL was used to teach a 15-week MEB course to second-year chemical engineering students. Outside of class, individual students prepared for each classroom session by completing a reading assignment, one or more assigned example problems, and one homework problem. Students submitted the homework problem to receive individual feedback on their work and a grade for the assignment. During class, the student learning teams solved application exercises and received immediate feedback from the instructor. The team scores on the application exercises were used to determine the course grades of individual students. The effectiveness of the TBL instructional strategy was evaluated using student performance on course exams and informal student feedback. The exam scores from students in the TBL course were compared to the exam scores of similar groups of students taking the MEB course taught with a conventional classroom lecture approach. The student comments on course evaluations were qualitatively analyzed by coding and identifying themes.

Evaluation: Students in the TBL course performed well on exams, when compared to students in previous MEB course offerings who received conventional classroom lecture instruction. Informal observations showed a high-level of student engagement in the TBL classroom. Student comments from formative and summative course evaluations indicated that students were also engaged with learning the course concepts outside of the classroom and that the students were forming strategies for life-long learning.

Conclusion: TBL was an effective instructional strategy for the MEB course. Students in the TBL course performed as well as students who had taken an MEB course taught with a conventional classroom lecture format on exams that measured the ability of students to solve problems by applying engineering concepts. The TBL format required individual accountability and provided experience in teamwork. Both of these behaviors are expected from students in upper-level engineering coursework, and are consistent with expectations in professional engineering practice. Due to a recent rise in student enrollment in the engineering college at the instructor's institution, the number of students taking the TBL course was 50% larger than previous MEB course offerings. Given that the quality of student learning, as measured by exam scores, was maintained despite the larger than usual student course enrollment suggests that teaching with the TBL instructional strategy is an effective way to deliver high quality instruction with limited faculty time. Further investigation of TBL in engineering education is underway.

References

1. L. K. Michaelsen, A. B. Knight, and L. D. Fink (Eds), Team-Based Learning: *A Transformative Use of Small Groups in College Teaching*, Stylus Publishing, Sterling (VA) (2004).

(676h) Impact of Learning Style Preferences On Student Performance and Perception in Chemical Engineering

Miskioglu, E., The Ohio State University
Wood, D. W., the Ohio State University
Originally presented on: 11/7/2013 14:22:00 - 14:38:00

Studies on learning styles are largely focused on the use of hypermedia, rather than traditional classroom lectures, as a teaching technique. The validity of learning styles indicators has also been a subject of debate in education. Felder and Silverman established the Felder-Silverman learning styles, which led to the Felder-Solomon Index of Learning Styles (ILS), in the context of engineering education specifically. The Felder-Silverman model identifies five dimensions of learning styles: active/reflective, sensing/intuitive, visual/auditory (verbal), sequential/global, and inductive/deductive. Inductive/deductive was later removed and the model is currently accepted as having four dimensions. The ILS is a 44-question survey designed to identify learning preferences in these four dimensions.

Learning styles are emphasized as being preferences for how an individual learns and indicators of how studying should be best approached, rather than indicators of success. This has not, however, prevented a few researchers from exploring whether or not student performance can be dependent on matching question type to learning style. A study by Cook, et al., with medical students failed to establish a connection between learning style and performance. The authors offer a number of reasons why no correlation may have been established, including small sample size and the inherently high inclination of the sample population to succeed. Also of consideration is the fact that this study was conducted with students in an advanced point in their educations, and that preferences can be influenced (adapted) by education. For these reasons, we are interested in the correlations between learning style preferences, assigned problem or task perception, and student performance in the context of an introductory chemical engineering material balances course. Students in this class come from a variety of educational backgrounds (differing high schools and an array of general education electives) and their learning style preferences are not yet influenced by a shared chemical engineering curriculum.

Assignments have a natural degree of bias towards specific learning style components, and we are looking to see if student preferences correlate with their performance and perceptions when these biases are taken into consideration (*e.g.* whether students who have a visual preference over verbal perform better on and prefer homework problems with information presented visually rather than verbally). The sample size of this study is approximately 100 students per semester. Data collected include student grades by specific task/problem and self-efficacy survey responses. Surveys are administered with weekly homework assignments, as well as exam problems, with questions & statements evaluating student responses to problems based on a Likert scale. Surveys include items such as: I understood how to solve problem 5 immediately; I am confident in my answer for problem 5; I am uncertain I solved problem 5 correctly; I am comfortable solving problems similar to problem 5. Results will provide evidence regarding whether learning styles can affect student performance, as well as identify naturally occurring biases in chemical engineering problems that are translatable to other fields. It is commonly accepted that every field attracts a certain "type" of individual, but a correlation between learning styles and success that demonstrates similarities among students who do not succeed in introductory chemical engineering would suggest that our curriculum repels students otherwise interested in the field. A better understanding of student background and learning preferences is invaluable in developing improved curriculum, especially in the ever-changing modern classroom.

This study is designed to have minimal impact on the course. The only thing we have added is the survey. By evaluating the traditionally run lecture course as it is, we are able to establish a baseline for understanding the role of learning styles in our curriculum.

(676i) Results From a First-Year Chemical Engineering Design Laboratory

Butterfield, A., University of Utah

Originally presented on: 11/7/2013 14:38:00 - 14:54:00

Students within chemical engineering departments do not, on average, encounter a laboratory course specifically geared towards chemical engineering until their junior or senior years. Up to that point, laboratory experiences are found within chemistry and physics departments or general engineering labs, and chemical engineering courses often keep to a traditional lecture style. However, this structure may leave valuable pedagogical tools underutilized. Inclusion of hands-on interactive and collaborative teaching methods, such as those typically found in lab courses, have been found to increase student conceptual learning, student satisfaction with their department, and student retention, particularly in underrepresented groups.

We have developed a freshman chemical engineering design course and laboratory, *Chemical Engineering Design & Innovation*. This course is intended for students in the second semester of our curriculum, to be taken after a more traditional lecture course covering an introduction to chemical engineering. Students in this laboratory collaborate in teams, and, through guided inquiry, achieve a variety of design goals related to many chemical engineering core concepts.

Teaching modules developed for this course include: 1. an introduction to and assembly of basic sensors; 2. design of a spectrophotometer and determination of reaction kinetic parameters; 3. creation of a process to generate an alginate drug delivery product and determination of its mass transfer characteristics; 4. photobioreactor design and determination of microbial growth kinetics. 5. biodiesel creation, characterization, and process scale-up. Students also complete a collaborative project to assist our seniors in their capstone laboratory course in order to create social connections, transfer important knowledge about courses and internships across cohorts, and develop job application skills needed to obtain internships.

This freshman course is also meant to act as a launching platform for effective, evidence-based pedagogy throughout the remainder of our curriculum. Each module acts as an anchor point to which students may affix important chemical engineering concepts, and through which difficult theory may be revisited later in their academic career. As such, some of the teaching modules have been brought into upper-level courses to create active-learning projects and address relevant theory with newly-gained sophistication.

Results from student surveys indicate that the goals of this course have been successfully achieved. Students have used their hands-on experience to develop insights into important chemical engineering phenomena. They have developed prototyping and data-analysis skills that have greatly increased the employability of our freshmen in faculty research programs and local industry. Students also report well above average satisfaction with the course and its open-ended design projects, and they have effectively developed valuable social connections with both peers and graduating seniors.

(733a) A West Point Pinot Noir

Bozic, R. G., United States Military Academy

Originally presented on: 11/7/2013 15:15:00 - 15:33:00

The first West Point Chemical Engineering Cadet Winemaking experience was planned, developed, and executed in order to increase interest and educate chemical engineering majors, prospective students, and parents about chemical engineering. Inspired by the beer brewing operations in the Kicking Mule Brewing Company, the brewing section of the cadet AIChE club, and the book by George M. Taber, *Judgment of Paris* about the 1976 Paris wine tasting, members of Kicking Mule embarked on a wine making venture in January 2013. After 4 successful years of brewing beer for the cadet senior club on West Point and based on research, discussion with venders, and lessons learned from chemical engineering classes, the Kicking Mule Brewing Company purchased wine making equipment and supplies and made a ~13 vol% alcohol Pinot Noir wine. The process was scaled to approximately twenty kilograms (i.e. in this case ~44lbs) of de-stemmed Pinot Noir grapes from Lodi, CA grapes and 5 grams of Bourgovin RC 212 (Saccaromyces cereviciae) yeast in a 5 gallon primary reaction vessel based on available supplies and cost. After 11 days of the oxic fermentation, the must was pressed using a "3 bucket" press system developed in part by one of the cadets, and the wine transferred to the secondary anoxic vessel. After a total of 72 days, the wine was siphoned, sulfites were added, and 10 ea bottles of red wine produced. A shorter than typical aging period was used in order to ensure student involvement in examination all parts of the process in a single semester. Special emphasis was placed on chemical engineering and chemistry of each step this wine making process. Cadets also learned about the economics of the process. At a 6% interest rate over 10 periods, in this case months to match the academic year, a $22.62 annuity, was calculated using the capital-recovery factor with all supplies and equipment for this project costing $166.45. This project was a highlight in a multi-year successful venture into brewing for the AIChE club and has also contributed to a responsible atmosphere surrounding alcoholic beverages at the Academy for all those involved.

(733b) Writing Journal Reflections in Engineering Courses: Effect On Critical Thinking Skills

Piergiovanni, P. R., Lafayette College

Originally presented on: 11/7/2013 15:33:00 - 15:51:00

Many humanity and social science courses require students to keep journals, where they reflect on the class materials and discussion. The journals from a recent first year seminar course clearly showed how the students worked out their ideas, and began to think critically about the ideas presented in class. The final papers written by students who wrote in their journals daily showed increases in four areas of critical thinking (question and reasoning; recognizing assumptions; presenting and analyzing data, and drawing conclusions).

As a result, last year daily online journal entries became a required component of ChE 321 (Applied Fluid Flow and Heat Transfer). The prompts were designed to encourage students to reflect on what they had learned in class and relate it to ideas or objects they observe on campus. The student responses from two semesters were analyzed for the elements of critical thinking. Did requiring students to summarize what they learned in words – instead of equations – increase their understanding and problem solving ability? Was it possible to see misconceptions early, before the students began solving problems for the homework? Did any journal prompt trigger an "aha!" moment, when the student grasped the essence of the subject?

This presentation will present the rubric used to analyze the journal entries, document how the journal reflections encouraged the students to think deeply and process what they learned in class, and how they contributed to the four areas of critical thinking as well as exam scores.

(733c) Heat Transfer for Students: An Iphone and Ipad App for Chemical Engineering Enthusiasts

Keith, J. M., Michigan Technological University
Nelson, G.,
Thompson, A.,
Sprabery, R.,
Gazzini, J.,
Originally presented on: 11/7/2013 15:51:00 - 16:09:00

The educational literature indicates students learn best by doing. This was the case a generation or two ago when this research was being conducted, but is also true now as students get their information through Facebook and twitter. Since students tend to want to use their smartphones in class, why not direct their use to help them learn? In this paper, we discuss the design and implementation of a heat conduction app which allows students (or grandmothers) to visualize the solution of unsteady-state heat conduction. Most students are afraid of the mathematical solution to this problem, and either get stuck in the details of the solution and never really understanding the result. The app, titled "Heat Transfer for Students," allows them to easily see the temperature profile in the object as a function of time for heating or cooling, as well as the impact of the system parameters such as the thermal conductivity, heat capacity, density, length, and heat transfer coefficient on the result.

(733d) Impact of Learning Styles Preferences On Team Dynamics in Chemical Engineering Unit Operations

Miskioglu, E., The Ohio State University
Wood, D. W., the Ohio State University
Originally presented on: 11/7/2013 16:09:00 - 16:27:00

Engineering is a highly team-oriented discipline. Teamwork is emphasized in high-level engineering courses to help further prepare students for the career their degree entails. When left to themselves, students will often form groups that are homogenous with respect to several indicators, including personality and learning style. While homogeneous groups may occasionally be preferred for specific tasks, heterogeneous groups have been shown to have better performance over a more broad range of tasks. Qualitative data from interviews by Martin and Paredes, 2004, suggests that students also grasp the value of a heterogeneous group and that, despite some difficulties because of clashing styles, are able to recognize the gains of such diversity. Quantitative data on the matter regarding teamwork and learning styles consists of student surveys conducted on a numerical scale corresponding to agreement (*e.g.* Likert scale). We are interested in the relationship between learning styles and team dynamics in advanced chemical engineering students, specifically, those taking the team-based senior-level unit operations course.

The course begins with a lecture during the spring semester in which students are given a learning styles questionnaire (Felder-Soloman Index of Learning Styles) to determine their preferences. These preferences will be used to identify potential correlations between team make-up (heterogeneous or homogeneous with respect to learning style) with team performance and student perceptions during the team-based laboratory portion in the Maymester. Course enrollment was approximately 180 students for 2013, and is consistently 150 to 200 students each year. Teams are assigned at the end of the spring term lecture course, with normalization for GPA. The Maymester laboratory course runs 4 weeks, with teams responsible for one experiment per week. With their laboratory report, they turn in an individual survey evaluating team dynamics. Survey items include: All group members participated in executing lab protocol; I would have preferred to do the protocol by myself; group members contributed equally to writing the lab report; the final product (lab report) benefitted from input from all group members; I would have preferred to write the lab report myself. By administering this survey with each experiment, we are able to control for variations by lab protocol, as well as identify any changes in team dynamics that may occur through the term as students become more accustomed to working together. We are looking to correlate team composition with respect to learning styles preferences (classified as heterogeneous or homogeneous by dimension) with team performance and student perceptions of the team's efficacy. This study is designed to have minimal impact on the course. The only thing we have added is the survey. By evaluating the traditionally run laboratory course as it is, we are able to establish a baseline for understanding the role of learning styles and influences of team dynamics in our curriculum.

Students at this advanced point in their education have shared much of their discipline-specific training. An understanding of the variability in performance based on learning style and team composition can help elucidate the complex nature of team dynamics. Armed with such information, instructors can make decisions regarding how to create teams in a manner most beneficial to the students, and students will have a better understanding of teams as they prepare to finish their degrees.

(733e) Building the Community Through Service Learning: Assessing the Effect of Service-Learning Group Projects On Intrinsic Motivation, Depth of Understanding, and Leadership Development During a Junior-Level Core Chemical Engineering Class

Bundy, B. C., Brigham Young University
Originally presented on: 11/7/2013 16:27:00 - 16:45:00

Group projects have been shown in many cases to improve student learning by giving students the opportunity to apply course topics to real experiments. Here we seek to expand the influence of a group project beyond the participating undergraduate students to the community at large without sacrificing the learning experience. Specifically, we have integrated a service-learning group project into a Junior-level Heat and Mass Transfer Course which has directly influenced over 1,500 K-12 students. An IRB-approved survey was performed which asked students to rate how the service-learning project impacted their intrinsic motivation, depth of understanding and leadership development. Herein we present the results of this survey and our experience on how projects can be optimized for both undergraduate student learning and service to the community.

(733f) Understanding What Makes Mathematical Modeling Hard for Capstone Design Students

Cole, J., Northwestern University

Originally presented on: 11/7/2013 16:45:00 - 17:03:00

Mathematical modeling is a practical skill from which engineering students can benefit. This is especially true in capstone design, where open-ended, analytically complex problems are often solved by innovatively applying engineering fundamentals from previous coursework. Ideally students would break down the components or systems within a proposed design solution into distinct elements. The fundamental relationships between the elements could be examined, manipulated, and put into mathematical representation. The resulting mathematical models would then be used to test assumptions about the components or find the limitations of the system.

However, in our experience and in our research we have found that students are not adept at applying mathematical engineering fundamentals to open-ended design problems. More often students take a "trial and error" approach, preferring to match up their open-ended design problem to several somewhat similar problems encountered in previous coursework until they get an answer that seems to provide a reasonable solution. Little effort is made by the students to understand the fundamental behaviors governing the elements in their system or what a reasonable output looks like. Instructors often see this when students use process modeling software, such as Aspen. Students may enter process conditions and constraints that may not make sense but lead to a converged solution.

This project investigated students' abilities to generate mathematical models that they can use in the development of innovative design solutions to open-ended problems. In particular, we studied how students' approached creating, solving, and interpreting mathematical models in a chemical engineering capstone design course. Two modeling problems were analyzed in this study. The first was a simple optimization problem and the second was a simulation of a distillation column in Aspen HYSYS. The instruction and the modeling problems were broken down into steps, from defining the model elements to generating the mathematical equations and interpreting model outputs. Student performance was measured on each of these steps, in order to determine what aspects of mathematical modeling were the most difficult.

Our research has shown that students struggle with creating, manipulating, and critiquing mathematical models to assist in the design of a product or process. The results of this work will be presented in this paper and will be used to revise instruction in order to improve students' abilities in mathematical modeling in the context of design.

(733g) Course Packages for Thermodynamics and Material and Energy Balances

Falconer, J. L., University of Colorado
deGrazia, J., University of Colorado
Medlin, J. W., University of Colorado
Nicodemus, G. D., University of Colorado
Originally presented on: 11/7/2013 17:03:00 - 17:21:00

One of the primary obstacles to the use of best practices in engineering education is overcoming implementation barriers for faculty. To combat this, we have prepared structured course packages for chemical engineering thermodynamics and material and energy balances courses using Microsoft OneNote. These contain easy-to-use and effective active learning resources, thereby facilitating the adoption of educational best practices and improving student learning. A course package includes a set of teaching materials that is organized according to learning goals, is easy to navigate, and provides all the necessary resources for active learning and effective assessment. The course packages include links to screencasts (short videos with narration) that can be viewed before class as part of a flipped classroom as well as a complete set of class notes that will utilize active learning (ConcepTests with peer instruction, in-class group exercises, etc.). In addition, the course packages incorporate measureable learning objectives for each class period, a database of solved example problems, homework assignments, and exams. Instructions for the use of the course materials as well as how to adapt them for other courses is an integral part of the package. An essential element of these course packages is explanations of the underlying theories for these teaching approaches and links to additional references. These course packages are created in Microsoft OneNote to provide users a central location for all their course materials and to make them easy to use. In addition, the materials are mapped to popular textbooks, thus aiding novice instructors to most effectively use these textbooks to promote student learning. A generic OneNote template was also prepared that can readily be used for other courses. More information about the course packages and our inventory of conceptests and screencasts can be found at www.learncheme.com. The site contains more than 1,300 ConcepTests and 900 screencasts. The screencasts are also available on iTunesU and YouTube/learncheme.

(733h) Quantitative Analysis Of Motivation, Learning Strategies and Conceptual Understanding

Matusovich, H. M., Virginia Tech
McCord, R., Virginia Tech
Originally presented on: 11/7/2013 17:21:00 - 17:39:00

Quantitative Analysis on Motivation, Learning Strategies and Conceptual Understanding

Introduction

Recent studies have shown that graduates of engineering programs are entering the workforce while still holding onto robust misconceptions in areas such as statics, electricity and magnetism, and thermodynamics (Streveler, Litzinger, Miller, & Steif,2008). While significant research has focused on the assessment of conceptual understanding and how to identify and repair misconceptions, little research has focused on the intentional ways that students engage in learning or the effects of this engagement on conceptual change. This quantitative analysis portion is part of a larger overall project, situated in a framework of intentional conceptual change that looks at the effect of student motivation and learning strategy choice on conceptual change in thermodynamics. Our three-phased study will identify the motivational factors and learning strategies students use that affect conceptual change, explore those factors more in depth through real-time data collection in the thermodynamics classroom, and then finally test interventions developed based on the factors identified.

This analysis focuses on the identification of relevant motivation and learning strategies factors through survey data collection and follow-up interviews. We developed a survey instrument to measure motivational factors and learning strategies that affect conceptual change drawing on existing scales such as the Motivated Strategies of Learning Questionnaire (Pintrich, Smith, Garcia, & McKeachie,1991), the Self-Regulated Learning Inventory (Lindner, Harris, & Gordon,1996), and the Thermal and Transport Concept Inventory (Streveler et al.,2011).

Literature Review

A critical challenge facing engineering education developing engineers with the technical expertise needed to lead innovation. In order to meet this challenge, we must first understand how students learn in order to create learning environments that support the development of deep, conceptual understanding. Research has shown that students enter into undergraduate programs with misconceptions about how the world works. We also know that students are graduating from undergraduate institutions still holding on to robust misconceptions in areas such as physics, electricity and thermodynamics (Streveler et al.,2008). To better understand how to correct misconceptions, we need to first understand how students engage in the process of conceptual understanding.

"*Hot cognition,*" which merges traditional cognitive approaches with learner's *intentionality* – i.e. why and how they choose to engage in learning – provides a framework that can be used to better understand student engagement in conceptual understanding (Boyle, Pintrich, & Marx,1993). This framework is different from "cold" approaches in that it considers not only knowledge structures, but adds intentionality, recognizing that learners make choices about their engagement in learning and their selection of learning strategies. We know little about students' motivation and their corresponding learning approaches and the connection to conceptual understanding in engineering classrooms (Streveler et al.,2008).

Linnenbrink and Pintrich (2003) proposed a framework to explain how *hot* conceptual change occurs. The core precept of their model is that motivation for conceptual change determines how the learner will approach the task of learning. This model starts with a primary categorization of students' motivation under *Achievement Goals* then considers *Other Motivational Beliefs,* such as self-efficacy and interest, and the associated *Learning Strategies* as contributing to *Intentional Conceptual Change.* This research study uses the framework posed by Linnenbrink and Pintrich to inform the development of instruments to measure motivation, learning strategies and conceptual change.

Methods

To better understand how students engage in conceptual learning, this first phase of our study answers the research question: What are the intentionality factors (i.e., student motivation and learning strategy choice) and relationships among these factors that lead to successful conceptual change? (RQ 1)

To answer this question, a survey instrument was developed using existing instruments to measure motivation constructs, learning strategies and conceptual understanding. The motivation constructs considered for this survey instrument included extrinsic and intrinsic motivation, interest, attainment value, cost, identification with academics, self-efficacy and instrumentality. The learning strategies section of the survey consisted of elements from the Motivated Strategies of Learning Questionnaire (Pintrich et al.,1991), the Self-Regulated Learning Inventory (Lindner et al.,1996) and several questions designed by the researchers. The conceptual understanding portion of the survey consisted of questions from the Thermal and Transport Concept Inventory (Streveler et al.,2011). A detailed discussion of the survey development can be found here (McCord & Matusovich,2013). The survey was distributed to universities and colleges around the US by contacting thermodynamics professors and asking them to send the link to their students. In all, eleven universities from around the United States have participated in the survey.

Analysis

A total of 81 student participants have completed the entire survey. While no significant correlations have been found linking specific motivational or learning strategies profiles to conceptual understanding, several significant correlations have been found between different motivational profiles and different choices for learning strategies that students engage in. These correlations are shown in Table 1, which show the Pearson correlations between the motivational factors and the learning strategies. For example, positive correlations have been found between the motivation constructs of interest and intrinsic motivation and engagement in learning strategies such as self-consequences, self-evaluation and monitoring. We have also found correlations between the learning strategy of seeking help from a teacher and motivational constructs such as interest, attainment value, self-efficacy and endogenous instrumentation. A preview of the correlation analysis conducted is shown in Table 1.

Table 1: Correlations between Motivation Constructs and Learning Strategies

Learning Strategies

		Time and Study Environment	Keeping Records	Self-Consequences	Self-Evaluation	Monitoring	Setting Goals	Memorization	Seeking Help from Teachers	Seeking Help from Peers	Rehearsal	Transformation
	Extrinsic	0.209	-0.046	.259*	0.117	.301**	0.189	0.131	0.179	-0.2	0.149	-0.062
	Intrinsic	0.115	0.035	.278**	.337**	.295**	.270*	0.137	0.149	-0.005	.338**	0.196
	Interest	0.204	-.226*	.396**	.354**	.264*	0.185	0.091	.307**	0.187	.423**	0.205
	Attainment Value	.566**	-0.178	.389**	.242*	0.208	.370**	0.152	.294**	0.216	.503**	0.112
Motivation	**Cost**	0.101	-0.206	.231*	.438**	.407**	0.133	0.202	0.204	-0.213	0.005	0.114
	Identification with Academics	.369**	-0.182	.465**	.326**	.372**	.377**	0.139	.469**	0.03	.449**	0.143
	Self-Efficacy	.219*	-0.129	.387**	.491**	.569**	0.126	0.138	.301**	-0.135	.298**	0.032

Endogenous Instrumentation	.250*	-0.085	.328**	.280**	.380**	0.148	0.144	.362**	0.078	.376**	0.035
Exogenous Instrumentation	0.094	0.083	0.184	.238*	0.179	0.128	0.038	0.149	0.012	0.177	-0.026

**. Correlation is significant at the 0.01 level (2-tailed).

*. Correlation is significant at the 0.05 level (2-tailed)

Discussion and Conclusions

We found significant connections between a student's motivational profile and the learning strategies that they choose to engage in when learning thermodynamics. This is an important finding because it tells us that students motivational beliefs are directly tied to the actions that they engage in to learn content. Understanding these connections may provide valuable information on how develop pedagogical interventions based on students' motivational profiles.

In the first round of data collection, we did not find significant ties between motivation, learning strategies and conceptual understanding. While this result goes against our initial hypothesis that there would be a connection, there may be methodological explanations as to why no connection was found. One possibility is that the survey questions were directed towards asking students about their experiences in their thermodynamics class. While we wanted to focus to be on the thermodynamics class, it may be difficult for students to relate their responses to thermodynamics as a whole. For example, questions asked students about their interest in thermodynamics yet concept inventory questions about specific content such as entropy or internal energy. It may be more appropriate to ask students about their interest in specific topics covered in thermodynamics. This may be more easily done in the context of interviews than through survey data collection.

References:

Boyle, R. A., Pintrich, P. R., & Marx, R. W. (1993). Beyond Cold Conceptual Change: The Role of Motivational Beliefs and Classroom Contextual Factors in the Process of Conceptual Change. *Review of educational research, 63*(2), 167-199.

Lindner, R. W., Harris, B. R., & Gordon, W. I. (1996). The Design and Development of the "Self-Regulated Learning Inventory": A Status Report: Western Illinois University.

Linnenbrink, E. A., & Pintrich, P. R. (2003). Achievement Goals and Intentional Conceptual Change. In G. M. Sinatra & P. R. Pintrich (Eds.), *Intentional Conceptual Change* (pp. 347-374). Mahwah, NJ: Lawrence Erlbaum Associates, Inc.

McCord, R., & Matusovich, H. M. (2013). *Developing an Instrument to Measure Motivation, Learning Strategies and Conceptual Change*. Paper presented at the 120th ASEE Annual Conference & Exposition, Atlanta, GA.

Pintrich, P. R., Smith, D. A. F., Garcia, T., & McKeachie, W. J. (1991). *A Manual for the use of the motivated strategies for learning questionnaire (MSLQ)*. Ann Arbor, Mich.; [Washington, DC]: University of Michigan ; U.S. Dept. of Education, Office of Educational Research and Improvement, Educational Resources Information Center.

Streveler, R. A., Litzinger, T. A., Miller, R. L., & Steif, P. S. (2008). Learning Conceptual Knowledge in the Engineering Sciences: Overview and Future Research Directions. [Article]. *Journal of Engineering Education, 97*(3), 279-294.

Streveler, R. A., Miller, R. L., Santiago-Roman, A. I., Nelson, M. A., Geist, M. R., & Olds, B. M. (2011). Rigorous Methodology for Concept Inventory Development: Using the 'Assessment Triangle' to Develop and Test the Thermal and Transport Science Concept Inventory (TTCI). *The International journal of engineering education, 27*(5), 968-984.

AUTHOR INDEX

AUTHOR INDEX

AUTHOR INDEX

AUTHOR INDEX

AUTHOR INDEX